River Basin Management - Under a Changing Climate

Edited by Ram L. Ray,
Dionysia G. Panagoulia
and Nimal Shantha Abeysingha

Published in London, United Kingdom

River Basin Management - Under a Changing Climate
http://dx.doi.org/10.5772/intechopen.100758
Edited by Ram L. Ray, Dionysia G. Panagoulia and Nimal Shantha Abeysingha

Contributors
Benny Syahputra, Berkah Fajar, Sudarno, Muhammad Xaaceph Khan, Abida Butt, Youichi Yasuda, Nozomi Fuchino, Cheikh Faye, Nader Heydari, Ehsan Atazadeh, Shannon K. Brewer, Robert Mollenhauer, Desiree Moore, Dusty Swedberg, Maeghen Wedgeworth, Rebeca Gonzalez-Villela, Alfonso Banderas Tarabay, Marco Mijangos Carro, Daniel O. Omokpariola, Eucharia N. Nweke, Victor U. Okechukwu, Theresa C. Umeh, Nwanneamaka R. Oze, Utkarsh Kumar, Rashmi, Dhirendra Kumar Singh, Suresh Chandra Panday, Manoj Parihar, Jaideep Kumar Bisht, Lakshmi Kant, Quinex W. Wintalamu Chiluwe, Bimo A. Nkhata, Dev Tewari, Saly Sambou, Rene Ndimag Diouf, Joseph Sarr, Lawrence E. Edward Stevens, Raymond R. Johnson, Christopher Estes, Ram L. Ray, Nimal Shantha Abeysingha

First published in London, United Kingdom, 2023 by IntechOpen
IntechOpen is the global imprint of INTECHOPEN LIMITED, registered in England and Wales, registration number: 11086078, 5 Princes Gate Court, London, SW7 2QJ, United Kingdom

British Library Cataloguing-in-Publication Data
A catalogue record for this book is available from the British Library

Additional hard and PDF copies can be obtained from orders@intechopen.com

River Basin Management - Under a Changing Climate
Edited by Ram L. Ray, Dionysia G. Panagoulia and Nimal Shantha Abeysingha
p. cm.
Print ISBN 978-1-80355-558-4
Online ISBN 978-1-80355-559-1
eBook (PDF) ISBN 978-1-80355-560-7

For EU product safety concerns:
IN TECH d.o.o., Prolaz Marije Krucifikse Kozulić 3, 51000 Rijeka, Croatia,
info@intechopen.com or visit our website at intechopen.com.

Meet the editors

Ram L. Ray is an associate professor at Prairie View A&M University, Prairie View, Texas. He holds a Ph.D. in Civil Engineering, an MS in Soil Science, and a BS in Civil Engineering. He has more than 20 years of experience as a faculty and research scientist at Prairie View A&M University and as an engineer in the federal government. Dr. Ray's research interests include climate-smart agriculture, carbon sequestration, climate change, hydrology-land surface-atmosphere interactions, soil water dynamics, watershed hydrology, and modeling. He has authored two books, several book chapters, and articles in peer-reviewed journals. He is an editor and editorial board member of several journals, including *Agronomy*, *Agricultural & Environmental Letters*, *Remote Sensing in Earth System Sciences*, *Remote Sensing*, *Sustainability*, and *Civil Engineering and Constructions*.

Dr. Dionysia G. Panagoulia is an associate professor at the National Technical University of Athens, Greece, with expertise in hydrology, hydroclimatology, and water resources. She is the author of more than 140 published research works and lately has extended her research work to water economics theory and complex time/dark matter approaches. She has cooperated in joint research with the University of Stuttgart, Germany; CNRS Laboratory of Oceanology and Geosciences Wimereux, France; Wageningen University, Netherlands; and McGill University, Canada. She was/is a member of twelve scientific societies, a reviewer and guest editor for thirty-two international journals, and a recipient of the 2018 Outstanding Reviewer Award from Water MDPI.

Dr. Nimal Shantha Abeysingha is a professor in the Department of Agricultural Engineering and Soil Science, at the Rajarata University of Sri Lanka. He has a Ph.D. in Water Science and Technology. He has published more than forty-five research works in refereed journals, two books, and three book chapters. Dr. Abeysingha is an editorial board member of several journals including the *Journal of Environmental & Earth Sciences*, the *Sri Lankan Journal of Agriculture and Ecosystem*, the *Indian Journal of Soil Conservation*, the *Journal of Agriculture and Applied Biotechnology*, and has reviewed a large number of manuscripts in more than 25 reputed journals. His research interests include climate change assessments in agriculture, hydrological modeling and simulation, and soil and water conservation in an agroecosystem.

Contents

Preface

While water management under changing climate is approached uniformly under well-documented theoretical approaches in developed countries, the same does not happen in developing countries because the application circumstances differ sharply. This book examines this issue via a multi-country, multi-aspect approach.

Climate change impacts natural resources, the most important of which is water. To protect water resources, it is critical to plan, conserve, and manage the river basin, which is a primary water source. This book discusses several aspects of water management in terms of their impact under particular circumstances with an emphasis on diversity extending to goal, country locality, and point of view.

Several authors belonging to research and public institutions from different parts of the world have collaborated in the technical discussion of this topic, reporting their experience and presenting advances in critical research for water resources planning, conservation, and management on selected river basins globally. The book contains five sections and fourteen chapters covering important research aspects of river basin management under a changing climate using field experiments, modeling, and analysis at various scales.

Section 1, "Introduction", includes an introductory chapter that presents an overview of water resources planning, management, conservation, and monitoring of any river basin that must address complex science and issues using available resources, tools, and techniques under a changing climate.

Section 2, "Water Resources Planning and Conservation", includes three chapters focusing on approaches to improve water productivity and livelihood resilience, the efficacy of artificially assembled boulder installations in improving migration routes for aquatic animals, and a hierarchical approach to fish conservation in semiarid landscapes.

Section 3, "Water Resources Management", includes four chapters on trend analysis of streamflow and rainfall, characteristics and process interactions in natural fluvial riparian ecosystems, the impact of hydraulic structures on water resources management, and assessing water availability for the environmental flow in selected river basins.

Section 4, "Water Quality Monitoring and Management", includes three chapters on monitoring conditions of rivers and streams using biological indices, pollution evaluation of industrial effluents from consolidated breweries, and assessment of water quality using physicochemical parameters and aquatic insect diversity.

Section 5, "Community's Role in River Basin Management", includes three chapters that discuss community participation in river basin management. They also discuss managing water demands and investigating water control, impacts, and sub-regional cooperation around a transboundary hydrological system.

This book contributes to the understanding of water resource communities and the potential approaches for planning, managing, conserving, and monitoring water resources in river basins under climate change.

The editors wish to thank all authors for their valuable contributions. We also wish to thank the staff at IntechOpen, particularly Author Service Manager Ms. Ana Cink for her assistance in finalizing the work.

Ram L. Ray
Prairie View, A&M University,
Prairie View, USA

Dionysia G. Panagoulia
National Technical University of Athens,
Athens, Greece

Nimal Shantha Abeysingha
Rajarata University of Sri Lanka,
Anuradhapura, Sri Lanka

Section 1

Introduction

Chapter 1

Introductory Chapter: Water Resources Planning, Monitoring, Conservation, and Management

Ram L. Ray and Nimal Shantha Abeysingha

1. Introduction

River Basin Management (RBM) is an integral process to protect several wildlife species, sources of drinking water for animals, plants, and humans, sources of navigation channels, flood regulation, and others. On the other hand, RBM relies on effective public participation and management from all beneficiaries. RBM can be considered as the integration of strategic planning and management of quantity and quality of water resources through sustainable development [1, 2].

The rapid growth of population and water demand globally developed stress on the river basin to meet water demands, including municipal, agricultural, recreational, industrial, and other water demands. In addition, climate change impact is imposing threats/stresses on small to large river basins globally [3]. With projected global climate change and water demands, there are potential risks to the river basin, including loss of native biodiversity, ecosystems, and humans from increased flood and drought disasters [4]. Global climate change and global warming might cause frequent droughts, shifts in precipitation, lower water levels in water bodies, and consequently, less water to dilute pollutants.

Therefore, an integrated approach that includes planning, monitoring, and management using in situ and satellite measurements and modeling should be implemented to reduce the stresses on the river basin. This integrated approach requires significant contribution and participation from stakeholders, such as policymakers, watershed/water resources managers, researchers, forest managers, industries, farmers, growers, and several other natural resource users who are directly or indirectly responsible for additional stresses to the river basins.

The primary goal of this book is to address some of the critical issues of river basins through effective planning and management under the changing climate. This book includes the following four sections and 13 book chapters.

2. Water resources planning and conservation

Currently, we are concerned about the potential water scarcity in the face of increasing, primarily population-driven, water demands and its impact on energy and food production and food security [5]. In addition, the combined pressures of population growth and climate change significantly increased water demand [6]. Therefore, water resource planning and conservation are critically important globally. The effective planning and conservation of water resources provide water resource security, which is important for agricultural, municipal, and industrial water demands.

Water resource planning and conservation are strongly connected because both processes complement each other. For example, the best and most effective planning is needed to conserve water resources. The best management practices in any watershed, whether agricultural, urban, or forested watersheds, help conserve water resources. For example, if farmers practice climate-smart farming for an agricultural watershed, they can conserve water resources and increase food production.

This section includes three highly diverse chapters focusing directly and indirectly on water resources planning and conservation. For example, while one chapter focused on approaches to improve water productivity, the other discussed the multiscale perspective for the conservation planning of riverine fishes. Further, this section focused on multiscale environmental relationships, which are important for water resource conservation and river basin planning. This section also discussed potential approaches to improve migration routes for aquatic animals, which are critical for biodiversity and the ecosystem.

3. Water resources management

Freshwater, which includes water in glaciers, lakes, reservoirs, ponds, rivers, streams, wetlands, and groundwater, is a limited natural resource. Therefore, many countries globally are experiencing water scarcity due to increased water demand due to the increasing population and living standards [7]. Some of the potential challenges associated with water resources management are low availability per capita, uneven temporal and spatial distributions, inconsistency in spatial distributions and productivity, and fragile water ecology and environment [7]. Traditional approaches for water resource management were to provide adequate water for municipal use without paying enough attention to its sustainable development and management [8].

Water resource management includes political, economic, cultural, social, technical, legislative, and organizational ingredients in one river basin or a total water cycle [8]. River basin management is indeed a complex process that requires several components to be incorporated, such as precipitation, evaporation, evapotranspiration, infiltration, and other inputs and withdrawals from the river basin system (**Figure 1**). Therefore, new strategies, advanced tools, techniques, monitoring, and evaluation system are critical and must include most aspects for effective water resource management in a river basin.

The challenge is that integrated water resources management (IWRM) should address complex water issues to maximize economic and social welfare equitably without compromising the sustainability of vital ecosystems [9, 10]. In addition, climate change impacts increased water demand and the challenges for integrated water resources management [11]. Despite several challenges, at a conceptual level, IWRM has gradually become an accepted framework for good water governance [12].

This section includes four chapters focusing on water resources management. For example, while one chapter focused on trend analysis of streamflow and precipitation in a river basin to support water resources management, the other discussed the characteristics and process interactions in natural fluvial riparian ecosystems. Further, this section investigated the impact of hydraulic infrastructures on the water resources management of the river basin. This section also conducted a comparative

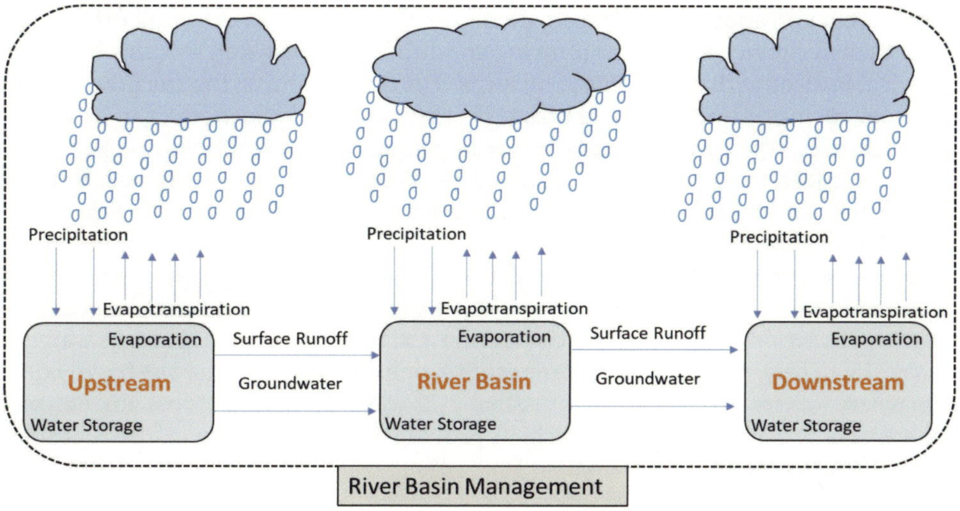

Figure 1.
Driving parameters essential to quantify for effective river basin management.

analysis of the precipitation variation in relation to the water availability in the rivers for the previous period and subsequent periods to determine the change in the availability of water in the ecosystem.

4. Water quality monitoring and management

Water quality monitoring is a basic tool for managing freshwater resources. Monitoring water quality in a river and its status gives clues for the health of the river and also the health of the river basin. River basin approach has been introduced to monitor and manage water quality in all most all countries in the world [13]. Therefore, this book introduces the section on water quality monitoring and management under the river basin.

There are many ways to monitor water quality. The conventional method of assessing water quality is evaluating the physical, chemical, esthetic, and biological properties of water. However, biological measurements of the abundance of animal life and aquatic plant and the use of bioanalytics, use of remote sensing, and IoT are becoming popular. Biomonitoring is considered more efficient and effective than traditional methods. It is widely used worldwide to monitor river pollution as bioindicators are sensitive enough to detect environmental change [14]. Biomonitoring is of two types active and passive. Active biomonitoring uses organisms under controlled conditions into the site to be monitored, whereas passive biomonitoring uses organisms, organism associations, and parts of organisms that are a natural component of the ecosystem and appear there spontaneously [15]. Aquatic insects can be used as bioindicators in aquatic ecosystems. Each aquatic insect has a different tolerance value to environmental conditions in which only a few species can survive in polluted ecosystems. One chapter discusses the use of aquatic insects to assess water quality along with some physical and chemical parameters. In addition to aquatic insects, algal communities are robust indicators of the physical, chemical, and biological changes of water induced by environmental flows which alter nutrient concentration, salinity, and alkalinity. One chapter in the book offers a comprehensive review of the monitoring conditions of rivers and streams using biological indices with an emphasis on algae. However, monitoring the point source pollution, such as effluent disposal from the industry to a waterbody, can be done efficiently using the conventional evaluation technique for the chemical, physical, and biological status of the water. One another chapter in this section deals with the pollution evaluation of industrial effluents from consolidated breweries using conventional approaches.

Improving access to clean water for drinking, bathing, and irrigation as per the standard is a top priority in all countries. The nature-based solution that leverages ecosystem functions is gaining more attention and is widely used to manage water quality [16]. The watershed management approach considers the nature-based concepts where forests, wetlands, and grasslands, as well as soils and crops, are managed properly. These well-managed watersheds provide high-value green infrastructure for enhancing source water protection.

5. Community's role in river basin management

It is now well recognized that the use of engineering measures with regulation-based management strategies has limited capacity to manage river basin and their water resources. Participatory approaches that engage stakeholders and the public in river basin management are promising and sustainable. Different basin users have conflicting water quantity, quality, flood risk, and ecological health demands. These demands can be managed while introducing best management practices to different land uses of the basin through a participatory approach. In addition, river basin management is data-intensive as it requires a picture of the entire socioeconomic and ecological health of the basin where stakeholders have these fragmented data in different scales [17]. Therefore, the role of the stakeholders is well understood.

Once the river basin or watershed of the river basin management plan is set, stakeholders are the entities that implement the agreed management activities. Therefore, getting their consent is highly encouraged even though sometimes agreed decision is not that quality [18]. Community participation has the power to make decisions autonomously in order to be able to solve the needs and interests of life and improve the standard of their living. Thus, one chapter of this section discusses in detail the community participation in river basin management.

Particularly, when managing the water demand and conserving the water in the basin, the stakeholder's role is highly acknowledged, and giving incentives is one way of getting users involved in conserving water in a river basin or a catchment. One chapter of this section discusses how water users use the property rights theory to conserve water. The results from the analyses indicated that property rights would be significant in curtailing water demands in a catchment by acting as incentives in water resource utilization, specifically by motivating water users to conserve water.

The catchment area becomes transboundary when it extends between two or more countries. The cooperation of the stakeholders is important for better managing the water of such river basins. Transboundary cooperation incorporating robust water diplomacy pathways for sustained water management is required rather than technical water management [19]. One of the chapters of this section discusses the impacts and sub-regional cooperation around a transboundary hydrological system.

DOI: http://dx.doi.org/10.5772/intechopen.109176

6. Concluding remarks

Water resource planning, management, conservation, and monitoring of any river basin must address complex science and issues using available resources, tools, and techniques under a changing climate. While much of the complexities of river basin management are human-induced, climate change has increased the complexities and challenges. Each river basin/water resources system is unique with respect to its management issues, challenges, and climatic and environmental conditions. Therefore, river basin planning, management, and conservation approaches must adapt to these situations. River basin planning, management, and conservation strategy should be based on those situations, especially when we have to plan, manage, and conserve water resources of the river basin under changing climate.

This book has focused on using an integrated approach, which includes modeling, trend analysis, the role of infrastructure, and community participation in river basin management under changing climate. The participation and contribution of stakeholders in river basin management are critical to building sustainable water resource management in a river basin. This chapter has summarized the 13 chapters of the book in four sections for water resources planning, conservation, monitoring, and management, including modeling.

Author details

Ram L. Ray[1*] and Nimal Shantha Abeysingha[2]

1 College of Agriculture and Human Sciences, Prairie View A&M University, Prairie View, USA

2 Department of Agricultural Engineering and Soil Science, Rajarata University of Sri Lanka, Anuradhapura, Sri Lanka

*Address all correspondence to: raray@pvamu.edu

References

[1] Evers M. Integrative river basin management: Challenges and methodologies within the German planning system. Environmental Earth Sciences. 2016;**75**(14):1085

[2] Bandaragoda DJ, Babel MS. Institutional development for IWRM: An international perspective. International Journal of River Basin Management. 2010;**8**(3-4):215-224

[3] Yamamoto K, Sayama T, Apip. Impact of climate change on flood inundation in a tropical river basin in Indonesia. Progress in Earth and Planetary Science. 2021;**8**(1):5

[4] Palmer MA et al. Climate change and the world's river basins: Anticipating management options. Frontiers in Ecology and the Environment. 2008;**6**(2):81-89

[5] Cosgrove WJ, Loucks DP. Water management: Current and future challenges and research directions. Water Resources Research. 2015;**51**(6):4823-4839

[6] Bagatin R et al. Conservation and improvements in water resource management: A global challenge. Journal of Cleaner Production. 2014;**77**:1-9

[7] Wang X-j et al. The new concept of water resources management in China: Ensuring water security in changing environment. Environment, Development and Sustainability. 2018;**20**(2):897-909

[8] Ren J-L et al. An informetric profile of water resources management literatures. Water Resources Management. 2013;**27**(13):4679-4696

[9] Fulazzaky MA. Challenges of integrated water resources management in Indonesia. Water. 2014;**6**(7):2000-2020

[10] Ako AA, Eyong GET, Nkeng GE. Water resources management and integrated water resources management (IWRM) in Cameroon. Water Resources Management. 2010;**24**(5):871-888

[11] Al-Jawad JY et al. A comprehensive optimum integrated water resources management approach for multidisciplinary water resources management problems. Journal of Environmental Management. 2019;**239**:211-224

[12] Boutkan E, Stikker A. Enhanced water resource base for sustainable integrated water resource management. Natural Resources Forum. 2004;**28**(2):150-154

[13] Katarzyna S, Rafal U, Katarzyna K. Monitoring and modelling of water quality. In: Kostas V, Dimitra V, editors. Water Quality. IntechOpen: Rijeka; 2012 Ch. 8

[14] Saad Abdelkarim M. Biomonitoring and bioassessment of running water quality in developing countries: A case study from Egypt. The Egyptian Journal of Aquatic Research. 2020;**46**(4):313-324

[15] Ronci L et al. Comparison between active and passive biomonitoring strategies for the assessment of genotoxicity and metal bioaccumulation in *Echinogammarus veneris* (*Crustacea: Amphipoda*). Italian Journal of Zoology. 2016;**83**(2):162-172

[16] Hamel P, Tan L. Blue–Green infrastructure for flood and water quality

management in Southeast Asia: Evidence and knowledge gaps. Environmental Management. 2022;**69**(4):699-718

[17] Carr G. Stakeholder and public participation in river basin management—An introduction. WIREs Water. 2015;**2**(4):393-405

[18] Coglianese C. The limits of consensus: The environmental protection system in transition: Toward a more desirable future. Environment: Science and Policy for Sustainable Development. 1999;**41**(3):28-33

[19] Pohl B, Swain A. Leveraging diplomacy for resolving transboundary water problems. In: Madani K, Shafiqul S, editors. Water Diplomacy in Action: Contingent Approaches to Managing Complex Water Problems. New York, US: Anthem Press; 2017. pp. 19-34

Section 2

Water Resources Planning and Conservation

Chapter 2

Approaches to Improve Water Productivity and Livelihood Resilience in the Karkheh River Basin: A Case Study from Iran

Nader Heydari

Abstract

The CGIAR Challenge Program on Water and Food (CPWF) was conducted in the Karkheh River Basin (KRB) as one of the nine benchmark basins and a representative basin in the arid and semi-arid regions of the world in its phase one of implementation. The CPWF Program in the KRB began in 2003 and ended in 2008. Four focal projects, namely: 1) Water Productivity Improvement (WP), 2) Improvement of Livelihood Resilience of Local Communities in the Upper Catchments (LR), 3) Karkheh Basin Focal Project (BFP), and 4) a small grant project for Stakeholder Participation (SG) were implemented in the KRB. The international focal institutes in charge of these projects were ICARDA, IWMI, and the CENESTA Center (an NGO), respectively. These institutes implemented the mentioned projects in collaboration with the Iran country's national institutes/centers (NARES). In this chapter, the approaches of CPWF in implementing the program in KRB are explained and discussed. Moreover, some major achievements and lessons learned from implementing the mentioned focal projects are provided. Based on the results, it can be stated that the roadmap for improving water productivity from the plant to the basin scales, by considering the policy and institutional issues, has been drawn in the CPWF program in KRB. Moreover, for better management of upstream watersheds in KRB, integrated watershed management principles have been well formulated and developed, taking into account the livelihood resilience improvement issues of local communities. In general, it can be stated that the development of multidisciplinary research of national research institutes and collaboration with related international institutes is necessary to improve water productivity and integrated management of catchments and to solve water crisis in the KRB and Iran as a whole. The lessons learned from implementing the CPWF program in KRB could be used as a suitable model to improve the quality of future similar studies in Iran and Central Asia.

Keywords: CPWF, Karkheh River basin, stakeholder, water, agriculture

1. Introduction

The Karkheh River Basin (KRB) is one of the benchmark basins in the first phase of the CGIAR[1] challenge program on water and food (CPWF). The basin is located in the west south part of Iran, and an arid to semi-arid region (**Figure 1**). Most of the upstream area of the KRB is located in the Kermanshah, Lorestan, and to some extent in Hamedan provinces (Figur 1). Rainfed agriculture is the major agricultural system and farmers' agricultural production and incomes are quite low [1, 2].

The downstream region of the basin is mainly located in the provinces of Khuzestan and Ilam, where irrigated agriculture is the predominant farming system (**Figure 1**). The last part and outlet of the basin is "Hur al-Azim" wetland on the border of Iran and Iraq.

In general, KRB was not in good condition regarding water use, agricultural production, water productivity (WP), and the livelihood of the beneficiaries. These conditions provided the ideal situation for the CPWF to select KRB as a benchmark basin, indicating a representative basin from arid to semi-arid regions and directly impact improving people's livelihoods and enhancing basin WP for preserving it for the environment.

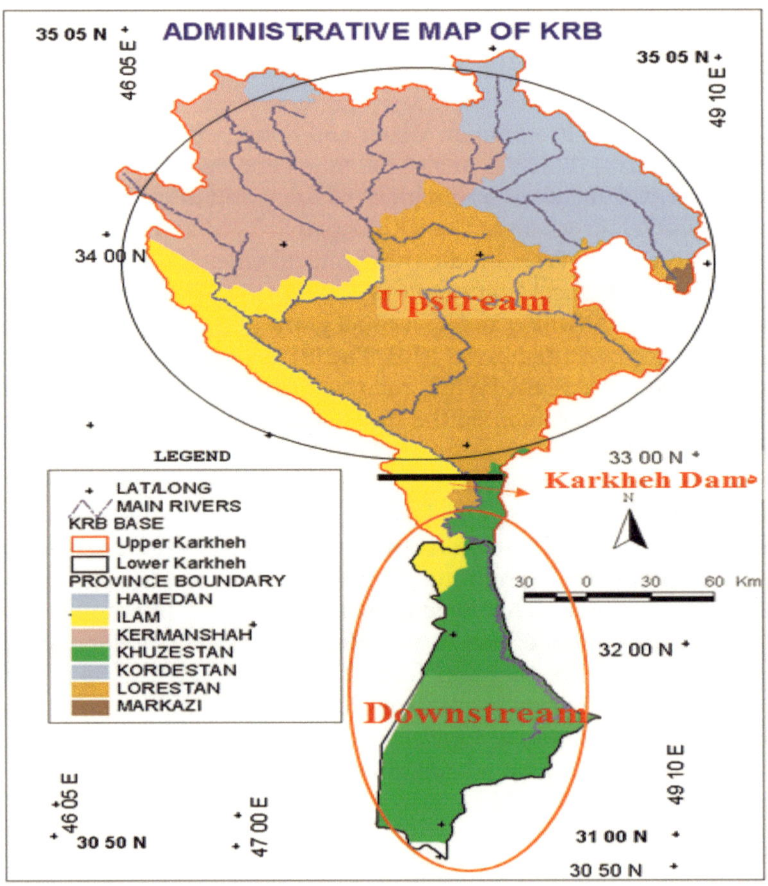

Figure 1.
Administrative map of Karkheh River basin and location of upstream and downstream areas of the basin.

[1] Consultative Group for International Agricultural Research

KRB was one of the nine benchmark basins of the CPWF Phase 1. It was implemented by CGIAR worldwide in the year 2004 and ended in 2008.

Three major research projects were launched in this basin mainly by the International Center for Agricultural Research in Dry Areas (ICARDA) and the International Water Management Institute (IWMI), and with the cooperation and participation of national institutions affiliated with the Agricultural-Research-Education-Extension Organization (AREEO) of the Ministry of Jihad Agriculture of Iran.

The four major international projects approved for the KRB were: 1) Water productivity improvement (WP), 2) Strengthening livelihood resilience in upper catchments of dry areas by integrated natural resources management (LR), 3) Basin Focal Project of KRB (BFP), and 4) a small grant project namely as SG.

In terms of the dimensions and scope of the CPWF program in KRB, it can be acknowledged that the program covered all issues related to improving WP and socio-economic issues related to the use of natural resources to improve the livelihoods of local communities in the basin. Regarding geographical distribution and main focus of the pivot projects, the WP project is located mostly in the irrigated areas in the Khuzestan (basin downstream), Kermanshah, and Lorestan (basin upstream) provinces and focusing on WP improvement issues including supplementary irrigation in basin upstream [3–5] (**Figure 1**). The LR project covered the entire area of upstream basin (**Figure 1**) [6, 7]. BFP project was also a large-scale pivot project that mostly dealt with large-scale WP and poverty issues for the entire basin area [2, 8].

Each of these focal projects had major and minor sub-projects and various core activities in areas related to the main project theme. **Table 1** summarizes the key sub-

Title of the main project	Leading Institute	Sub-projects and or activity	Major Report/ Publication
Improving On-farm Agricultural Water Productivity in the Karkheh River Basin (WP)	ICARDA	Determining and evaluating farm WP in the irrigated lands in the south of KRB (non-saline and saline lands), Determining and evaluating the farm WP in the rain-fed lands upstream of KRB and the effect of supplementary irrigation (SI) of rain-fed fields (mainly wheat and barley) on crop WP, Development issues of SI irrigation in upstream areas of the basin and its impacts on the quantity and quality of water flow in the downstream basin, Socio-Economic assessments of WP in field level Policy-institution issues related to improving WP in the KRB	[4]; [5]; [3]
Strengthening Livelihood Resilience in Upper Catchments of Dry Areas by Integrated Natural Resources Management (LR)	ICARDA	Review and development of the principles of integrated watersheds management for KRB, Erosion and sedimentation issues upstream of KRB, Drought studies in the basin, Rangeland and forest management in the basin, Issues of land use change and preparation of agro-ecological zoning map of the basin, Natural resources management issues and agricultural production policies in the basin, Water resources issues in the basin Development and participatory transfer of production technologies and improvement of WP in the basin, Study of gender issues and water management.	[6]; [7]

Table 1.
Specifications of the approved CPWF projects in KRB and their key sub-projects and or activities.

projects or activities considered in the two comprehensive pivot projects, i.e., WP and LR projects.

In general, the purposes of these two participatory focal projects were to identify the basic issues, assess the status of agricultural WP, identify its sources of inefficiencies, provide the necessary technical and managerial solutions to improve poor agricultural and rural communities and increase income and improve their living conditions through comprehensive and appropriate management of natural resources. Ultimately, training and capacity buildings of all stakeholders, researchers, and experts of national agricultural research and extension services (NARES), and related local organizations were other goals of the projects [9].

It can be claimed that the CPWF program in the KRB is the first comprehensive international project in Iran that addresses water management and agricultural WP issues with participatory approach from field to basin scales and with the cooperation of international institutes, executive organizations, universities, and research institutes of the country.

2. Objectives and approaches of CPWF for river basin management

The CPWF's main goal in KRB was to produce more food with less water for the Iran's growing population over the next 20years, and to achieve this ultimate goal by reducing malnutrition, alleviating rural poverty, improving the health of local communities and protecting the environment. CPWF also sought to provide the research, extension, and capacity building needed to significantly increase basin WP and livelihood resilience of local communities while protecting the basin's natural resources and environment, especially in drought conditions and climate change (CC) in the region.

Therefore, the key features of the CPWF were: 1) long-term goal to increase WP for food and livelihood of communities in a way that also preserves the environment and is socially acceptable; 2) medium-term goal to maintain the current status of water allocation for the agricultural sector at the end of the program, as per the current level of 2000 (beginning of the program), while increasing food and agricultural products. Achieving international environmental goals to reduce malnutrition and alleviate poverty by the end of 2015, especially in rural areas and small urban areas in low-income catchments with physical or economic scarcity or in the aquatic systems with a particular focus on high-poverty groups was also a medium-term goal and 3) short-term goal to include food security, alleviating poverty, improving the health level, reducing pollution, and environmental security [10].

2.1 CPWF research themes

The CPWF program organized and presented the following five research themes to achieve its goals. **Table 2** provides the specification of the research themes along with the coverage rate of the KRB focal projects.

2.2 Approaches and activities for implementation of CPWF projects in the Karkheh River basin

The approaches and activities for implementation of CPWF projects in Karkheh River Basin (KRB) included three main steps, 1) Holding of a Kick-off or Launching

Themes	Objective	Research sub-themes	Leading Institute	Rate of Coverage of Projects with the Research Themes (%)			
				*WP**	*LR*	*BFP*	*SG*
T1: Crop-Water Productivity improvement	To promote and increase food and livelihood security through the produce "more crop per drop" approach	• Scope of work at the plant scale: effects and future direction of plant breeding activities, • Crop and field scope: New opportunities for integrated natural resource management, • Outlook and scope of agro-ecological system: integrating water and land management, • Policy and institutional issues facilitating the transfer and adoption of improvement measures.	International Rice Research Institute (IRRI)	60	20	—	—
T2: Multipurpose use of upstream catchments (Water and people in catchments)	To improve water management in upstream catchments	• Water, poverty, and risks hazards in upstream catchments, • Potentials for improving water management, • Empowering communities and people to use improved land and water resources management.	International Center for Tropical Agriculture Research (CIAT)	20	65	—	—
T3: Aquatic ecosystems and fisheries	Conservation of aquatic ecosystems and fish and aquaculture for greater security of livelihood of people and biodiversity in catchments	• Issues of policy, institution, and governance, • Valuing ecosystem services and products and the costs of destroying resources and the environment, • Environmental water needs, • Improving water productivity.	World Aquaculture Center (ICLARM)	0	0	—	—
T4: Integrated basin water management systems	Management of catchments in a comprehensive and integrated manner	• Communications and impacts and scales of analysis, • Integrated decision support tools, • Good governance.	International Water Management Institute (IWMI)	10	5	—	—
T5: The Global and national food and water system	Evaluate water resources and food production in the water-food system on a national and global scale	• Globalization, trade, macroeconomics, and sectorial policies, • Investment and financial issues for agricultural water development and water supply, • Common cross-border (trans-boundary) water policies and related institutions, • Changes in the global water cycle.	International Food Policy Research Institute (IFPRI)	10	10	—	—

*: *Improving On-farm Agricultural Water Productivity in the Karkheh River Basin (WP); Strengthening Livelihood Resilience in Upper Catchments of Dry Areas by Integrated Natural Resources Management (LR); Basin Focal Project (BFP); Small Grant Project (SG).*

Table 2.
Specification of the research themes along with the coverage rate of the KRB focal projects.

Workshop of KRB, 2) Development of criteria for selection of pilot research sites in the basin, 3) Explanation and elaboration of the research projects to the project team and stakeholders by the project leaders and development of the project management structure.

2.2.1 Kick-off or launching workshop of KRB

The workshop, entitled CPWF program start-up workshop, was held on June 8–10, 2004 at the SPII[1] campus of Agricultural-Research-Education-Extension Organization (AREEO) with the participation of officials and all stakeholders. The workshop described the CPWF program and how the KRB was nominated and selected as one of the benchmark basins of the program.

In this workshop, various sessions and discussions were held in the form of brainstorming dialogs among the participants (experts and stakeholders). The two main areas of discussion were: a) Improving WP, food security, and livelihood of farmers in the aquatic ecosystem (irrigated agriculture) of KRB and mainly related to Theme1 of CPWF, and b) Issues of basin upstream, related to the Theme 2 of CPWF program, that is explained below.

2.2.1.1 Results of the CPWF theme one - workshop discussions

Regarding the first question of the workshop, i.e., high-priority research issues in the KRB, the summary of the Theme1 (improving water productivity at plant and farm scales) group as follows:

soil and water salinity, soil nutrition and fertility, irrigation efficiency, crop cultivars and plant species, thermal stress on crops, land drainage, cropping systems, land preparation, irrigation management, irrigation systems, drought stress, socio-economic, institutional, and environmental issues.

The above issues were divided into the following five groups:

1. Improvement of crop cultivars, including cultivars resistant to environmental stresses and having good crop yield and quality.

2. Water and irrigation management, including irrigation systems, water conservation and saving, and increasing irrigation efficiency.

3. Soil management issues, including soil nutrition, salinity, biological and physical–chemical aspects of soil, and land preparation.

4. Aspects of agro-technical management, including plant management and agro-systems; and 5- Political and institutional issues at the level of local communities.

Comments on the second question, i.e., "What are the best and most promising research methods that can cover at least three priorities and meet its research, are summarized as follows:

[1] Seed and Plant Improvement Research Institute (SPII) (in Karaj city near to Tehran, Iran capital)

a. Improving plant cultivars (Germplasm): common and molecular methods of breeding cultivars and using methods of participatory breeding,

b. Irrigation management: methods based on the local community, application of more efficient irrigation systems, special methods of farmer education (Farmer Field School), and water saving methods,

c. Soil management: the same methods mentioned for case (b),

d. Aspects of agro-technical management: including agricultural meteorological studies, plant modeling, plant diversity and intensive cultivation (intensification),

e. Policies and institutions at the local community level: including product insurance policies, the establishment of water users associations, methods of reducing energy consumption, motivation and desire strategies in the application of improved or new technologies to ensure compliance with them.

Table 3 summarizes comments regarding the key indicators questions in the priority areas.

For creating a credible basis lining, it was suggested that benchmark analysis of socio-economic and biophysical status of the basin at the beginning of the CPWF program and at the end of the program to be reviewed.

Also, a special working group for rainfed agriculture, mainly in the upstream areas of the basin, was formed. The specific results for this type of agriculture and in the form of Theme1 of CPWF program in KRB were as follows:

1. Three important issues in rain-fed agriculture of the KRB include: Low yield and yield instability, drought, inadequate existence of agricultural systems;

2. Promising research methods to address the above issues include: identifying areas in terms of Agro-ecological characterization, improving germplasm to withstand environmental and non-environmental stresses to the plant and thus improve plant WP, improving cropping systems and proper soil and crop management, application of supplementary irrigation, rainwater harvesting, application of participatory research methods, and crop modeling;

Row	Five years period	Fifteen years period
1	Adoption and application of varieties and new production technologies	Control of soil erosion
2	Improving farmers' incomes	Control of soil and water salinity
3	Reducing water consumption and improving agricultural WP	Creating confidence and ensuring economic benefits for farmers
4	Improving food quality	Increasing and creating security in the agricultural sector
5	Improving environmental quality	Improving the livelihoods of local communities
6	—	Improving and preserving the environment

Table 3.
Key indicators questions of the priority areas for the different horizons.

3. Key indicators in priority areas include cultivar adaptation, production technology adaptation rate, cropping system, soil erosion rate, soil quality, vegetation cover, and farmers' income.

2.2.1.2 Results of the CPWF-theme two workshop discussions

Summary for Theme two of CPWF program in KRB, i.e., the issues of the upstream catchments were as follows:

The problem of water availability for rainfed lands (especially for supplementary irrigation), the small size of agricultural lands, improper use of forest lands for the production of medicinal and industrial products and coal production, destruction of forests and their improper control, uncontrolled grazing in pastures, lack of integrated plan or management for watersheds, erosion of cultivated lands and destroyed pastures, and surface flows pollution due to mining activities, etc.

Also, the summary of three important and priority issues of the basin upstream was: 1) Agricultural issues, including reduction of productivity due to soil erosion, drought (especially in dryland areas), and land ownership issues; 2) Livestock issues, including the destruction of forests and pastures and productivity of livestock production; 3) Side effects including pollution in the downstream of the basin, floods and sediment, issues of water allocation for upstream and its impacts on downstream of the basin and the Karkheh dam in downstream.

Promising research methods were: 1) Integrated Natural Resources Management (INRM) method; 2) Integrated catchment management method focusing on interactions and communication between water consumption and landowners; 3) Participatory research and stakeholder participation; 4) Diversification in the sources of income of the inhabitants of the basin; 5) On-farm research including production of drought-tolerant crops, rainwater harvesting, and supplementary irrigation of crops in rainfed lands; 6- The effects of land ownership and issues of land use policies and regulations on land degradation; 7) Socio-economic issues including specifying livelihood issues, etc.; 8) Livestock issues and rangeland rehabilitation; 9) Cultural issues and information communications; 10) Water quality monitoring; and 11) Establishing a link between research on the scale of catchment and rain-fed agriculture.

Following the workshop, discussions on planning for future research and implementation activities in the KRB were as follows:

At first, a conceptual discussion entitled "Integrated River Basin Management (IRBM)" was introduced. Based on its model, the KRB area was divided into two parts of upstream of Karkheh Dam (Downstream) and downstream of the dam (Downstream) and its issues were investigated as follows:

1. In the upstream areas of Karkheh Dam the following studies were potentially proposed:

    a. Watershed studies, effects of natural and human processes on river hydrology as well as Karkheh Dam, runoff, flood and sediment, pollution, water allocation issues, surface and groundwater interactions, and identification and evaluation of water users (farmers) and their efficiency and productivity studies;

2. In the downstream areas of Karkheh Dam, potential studies proposed as follows:

   a. General issues include studies of soil and water management and their relationship, issues of land salinity improvement, long-term effects of development activities in the region (such as construction of new irrigation and drainage networks, etc.), effects of drought and water shortage on the downstream basin and especially on Hur al-Azim wetland, study of available water and available technologies for reclaiming saline and sodium lands in the region, Karkheh dam management issues, study of water pollution issues from various urban, industrial and agricultural sources.

   b. Specific issues of irrigated lands, including evaluation of irrigation efficiency and WP at the irrigation network and basin scales, water balance, supply and demand issues, determination of standards and indicators necessary for investment planning at the basin scale, irrigation methods and technologies, drainage, study of optimal cropping pattern, and the relationship between users and water distribution system in large-scale irrigation networks in the basin.

The workshop also provided supportive scientific lectures on integrated natural resource management (INRM) in arid areas and definitions and issues of participatory research in the basin.

In presenting integrated natural resource management in arid areas, the challenge of natural resource management and the concepts and principles of the integrated natural resource management framework was defined and explained, and several important tools were introduced. Accordingly, the IMRM framework is a method that: integrates research into different types of natural resources using stakeholder participation processes, and adaptive management and innovation.

It also aims to improve livelihoods, agro-systemic resilience, improve agricultural productivity and environmental services. Its solutions and effects of which operate at different scales and levels of local, ecological and global communities. Therefore, the principles of INRM are: integrating research and development, creating a system for adaptation and learning, creating a balance between hard and software sciences, focusing on choosing the right type of science and knowledge and applying it at the appropriate level, and changing the scientific and organizational culture.. INRM tools generally include three types of tools: process tools, cognitive tools, and problem-solving tools and investment on opportunities.

2.2.2 Development of criteria for selection of pilot research sites in the basin

2.2.2.1 Criteria for site selection in the upstream areas of the KRB

In the upstream areas of the basin, the main purpose of selecting the research site was mainly to implement the LR project. The main goal of the LR project upstream of the KRB was to improve livelihood resilience in arid watersheds through diversification and integrated natural resource management. According to the relevant conceptual model, resource degradation is not a reason for the change but a sign of changes in the socio-economic environment. Based on the results of field studies and considering diversity upstream of KRB, four research sites were initially selected based on the criteria and methods mentioned in the Mid-annual meeting [11] and Kick-off Workshop [12], as follows:

2.2.2.2 Criteria for site selection in downstream of the KRB

The criteria presented in the selection of research sites in downstream (south of KRB) were mainly derived from the results of field visits to the area by a group entitled "Agro-ecological plants and areas" in the early stages of the CPWF projects performed, which its results are presented in the followings.

The mentioned group categorized and presented their results of field visits to the entire basin in three categories: a) required data and information, b) quantitative determination of WP, and c) agricultural products in the basin [13, 14], which are presented only for case "a" in below[2].

2.2.2.2.1 Required data and information

A-1-Hydrology includes inflows and outflows (including seasonal and non-seasonal streams and rivers, water withdrawals for irrigation and runoff) (if possible daily data), rainfall and its time distribution and location of rain gauge stations, groundwater aquifer and its characteristics, and water quality.

A-2- Climates include: temperature, humidity and wind (ideally hourly, daily desirable but at least monthly), the amount of radiation (if data is available), evaporation data from the pan or reference evapotranspiration (ET).

A- 3- Water, soil, and land include soil texture and type, soil depth and root limiting layer depth (if any), soil water holding capacity, drainage characteristics, groundwater depth, general soil fertility level, soil salinity and sodium content, land slope, and gravel and coarse rocks level (if the area is pasture).

A-4- Crops and other agronomic issues include: type of plant species (under cultivation including trees), rangeland and forest, cultivation calendar (planting and harvesting time, etc.), crop yield (including the number of crop residues obtained if used as fodder), plowing and tillage operations, if the land is pasture, type of rehabilitation activities performed, estimation of vegetation cover and weed cover to crop cover ratio and date (time) of data collection, inputs include fertilizers and pest and disease control (insects, diseases, and weeds), plant coefficient (if locally determined or approved).

A-5- Irrigation operations include: the amount of applied water, its frequency and duration, irrigation water application system (and application efficiencies, storage and transfer, if available), leveling of farms with surface irrigation system, source (sources) of irrigation water, a ratio of water used for leaching salts (leaching fraction), reuse of drainage and farm effluent, quality of irrigation water,

A- 6- Livestock status includes: livestock species, time of production or purchase of livestock, or sale (live or killed), livestock density and their mobility, livestock grazing operations, and production products, including meat, milk, wool, eggs, poultry, or livestock labor, type of feed, quantity, source and seasonal amount.

---

[2] The category "a: is more related to the objectives of this chapter. The elaborations on categories of "b" and "c" have already been presented in their specialized reports and it may be refer to the relevant references.

A-7- Fishery and aquaculture include: species, water needs, products (including type and amount), its effects on water quality, the possibility of potential for artificial aquaculture and its expansion.

A- 8- Use of environmental water include: type, amount (in the presence of sound).

A-9- Economic-social and organizational-institutional issues including the level of land ownership and its type, and other information related to the upstream of the basin.

As discussed above, summarizing the criteria for selecting research sites downstream of the basin were [13, 14]: 1) Representation of soils in the region, 2) Appropriateness of water quality, 3) Having problematic soils, 4) Importance of agriculture in the region, 5) Low water and land productivity, 6) Access and existence of infrastructure, 7) Proximity to meteorological stations, existence of old and modern irrigation systems, 8) Existence of poverty in local communities, 9) Existence of diversity in agricultural systems, 10) Existence of associations and institutions (such as water users associations), 11) Existence of data and secondary information, 12) Cultivation of strategic irrigated agricultural products, 13) Existence of mechanized agriculture, 14) Location of the region within the borders of a province, 15) Access to water resources, 16) Existence of agricultural plots of different sizes (small, medium, large), 17) Existence of famous local communities, and 18- Proximity to a research center or station.

2.2.3 CPWF-KRB explanation and elaboration of the research projects

Table 4 summarizes CPWF-KRB main projects (WP and LR).

Project Title	Main Objectives	Sub-Objectives/ Components
Improving On-farm Agricultural Water Productivity in the Karkheh River Basin (WP)	Increase in food security and improve farmers' livelihoods, Increasing and sustaining WP and thus increasing farmers' income in dryland and irrigated farming. options to improve WP in rain-fed and irrigated areas of the basin, awareness and promotion of adaptation of new technologies, upcoming policies and appropriate organizational-institutional arrangements, capacity building in National Research Systems (NARES) and local community leaders, Evaluate WP and the organizational and institutional structure needed to improve it. Finding options for sustainable improvement of WP in irrigated and rain-fed agriculture of the basin, Familiarity and application of recommendations and technologies by farmers, Provide sustainable organizational arrangements and necessary policies, Training and capacity building among local communities, experts and researchers in the basin and even the country,	Improving rainwater productivity (RWP) through supplementary irrigation and agronomic practices in rain-fed lands upstream of KRB, Determining and improving agricultural WP in the irrigated lands (non-saline and saline lands) downstream of KRB, Socio-economic issues of WP improvement, Policy and institutional issues related to improving WP and water management in agriculture of basin, Study of impacts of upstream on downstream (Upper / Lower interactions).

Project Title	Main Objectives	Sub-Objectives/ Components
	Evaluate WP and propose related structures and policies for Karkheh, Euphrates and Amu Darya basin, Determining agricultural WP at farm and basin scales.	
Strengthening Livelihood Resilience in Upper Catchments of Dry Areas by Integrated Natural Resources Management (LR)	Strengthen the livelihood resilience of poor rural communities Improve the environmental integrity of the upstream watersheds Increase the adaptation capacity of the basin stakeholders in order to improve their livelihood and living conditions in arid and harsh areas in a sustainable way Develop an appropriate methodology that could link livelihood enhancement strategies to watershed management principles	Development of a framework for assessing vulnerability and livelihood resilience in upstream watersheds, Identification and evaluation of management principles of upstream watersheds in arid areas, Creating the capacity of local communities to strengthen livelihoods and manage of their watershed in a sustainable way, Effective strategic development to spread research results and experiences and achievements horizontally and upwards, Improve coordination, communication, and process skills. Climate change (CC) scenarios and plant adoption studies in KRB, Drought analysis for upstream areas of KRB Basin, Semi-detailed soil surveys Soil erosion studies inside and outside research sites, Land use changes studies, Participatory Technology Development activities in farmers' fields and rural areas upstream of the basin (called PTD), Water issues for integrated basin management, Develop a decision support model (DSS) to model integrated watershed management, Water resources of Honam watershed Gender and livelihood issues of catchment communities Experiences learned from agricultural management activities in KRB, Stakeholders and institutions in the basin, Livelihood analysis of local communities, Integrated basin water resources management and livelihoods resilience: Experiences of CPWF projects in benchmark basins and project impact assessments (IA), Conducting land use planning studies, Creating participation among communities, Creating employment to reduce the pressure on natural resources and soil erosion control, Rainwater water harvesting for optimal use of water and to improve the livelihood resilience of communities, Training of farmers (land users) and

Project Title	Main Objectives	Sub-Objectives/ Components
		develop their skills, Encourage women's participation, Soil erosion studies at LR project pilot sites in upstream of KRB, Impact Pathway of LR project, Agro-ecological zoning of KRB, Identify farmers' innovations, Development of principles of comprehensive watershed management and integrated management of water resources in the upstream areas of KRB.

Table 4.
A summary information on KRB main projects in the CPWF program.

3. Highlights of some innovations and thematic outputs of CPWF projects in KRB

3.1 Innovations and value-added studied subjects

- Biophysical similarity analysis was used to map similar areas to out scale the results of KRB projects to other basins in the WANA[3] region,
- Achieving the method of preparing the map of potential areas for the application of supplementary irrigation (SI) method in the catchments areas and taking into account the base flow of rivers,
- Systematic identification of innovations of local farmers,
- Identification and development of salinity-resistant cultivars of barley, wheat, and sorghum,
- Participatory Technology Development (PTD) for selected technologies effective in improving livelihood resilience, WP, and food production in eight local communities located in two selected pilot research sites in the major provinces of the basin (Honam in the Lorestan province and Merek in the Kermanshah province),
- Development and extension of Azetobacter biological fertilizer in the upstream areas of the catchment (mainly in Lorestan province),
- Development and extension of new chickpea varieties along with autumn cultivation in the upstream areas of the basin (mainly in Kermanshah province),
- Development and extension of salinity resistant cultivars, management methods, and water management technologies to improve WP in saline lands downstream of the basin.

[3] West Asia and North Africa

3.2 Thematic outputs from CPWF-KRB projects (for WP and LR main projects)

3.2.1 WP project

3.2.1.1 General

- Identifying and determining the characteristics and location of pilot research sites in the upstream and downstream areas of the basin (joint with the LR project),
- Characterization and development of agro-ecological map of the basin (in collaboration with the LR project),
- Determining and evaluating agricultural WP in the basin (this activity was planned to be done for the Tigris-Euphrates and Amu Darya basins in the region),
- Options for improving WP were developed in rain-fed and irrigated areas both for the farm and basin scales. These options mainly included supplementary irrigation, deficit irrigation, full irrigation, and salinity management.
- Development of zoning map of potential areas for application of supplementary irrigation at upstream and the effects of this measure at downstream of KRB.
- Organizational and policy institutional arrangements and regulations related to water management were documented and analyzed,
- Capacity building in the NARES, along with training human resources in the form of Ph.D. studies,
- Publications (project reports, scientific research and review articles, etc.) and dissemination of the project's findings.

3.2.1.2 Basin scale

- Assessment of water resources and drought in KRB at the basin scale,
- Assessment of areas prone to supplementary irrigation in upstream areas,
- Upstream developments and their impacts and interactions with the downstream areas,
- Effects of land use on sediment flow,
- Agro-ecological characterization and similarity analysis of KRB.

3.2.1.3 Field-scale

- Improving rainwater productivity using supplementary irrigation,
- Evaluation and improvement of WP in the non-saline southern areas of KRB,

- Evaluation and improvement of WP in the saline and waterlogged southern areas of KRB,
- Cropping options to improve WP under salinity of water and soil resources in the south of KRB,
- Socio-economic factors involved in WP in the southern lands of KRB.

3.2.2 LR project

3.2.2.1 General

- Integration of activities of different disciplines, organizations and institutions in the watershed through conducting participatory research in selected pilot research sites in the basin, participatory development of technologies, and watershed management principles;
- Involvement and active participation of provincial research and extension services in the project, and necessary collaborations with WP project (PN8) in common subjects,
- Development of a map of the African region based on climatic and edaphic similarities with the KRB,
- Analysis and preparation of runoff map in upstream areas of the basin,
- Drought analysis in the upstream areas of the basin,
- Effects of land use changes (1975–2002) on the amount of sediment in the upstream of the basin,
- Detailed studies of monitoring and analysis of runoff and water consumption in Honam and Merek watersheds,
- Development of recommendations for the restoration and management of rangelands and forests in the basin,
- Preparation of spatial GIS-based database for Honam and Merek watershed (in upstream), to support appropriate land analysis, erosion modeling, land use planning and development of decision-making models,
- Spatial GIS-based erosion modeling to investigate the effects of land use change on erosion in the Honam and Merek watersheds upstream of the KRB,
- Development of land suitability map for wheat cultivation for use in decision-making models.
- Livelihood analysis of eight local communities in Merck and Honam sub-catchments and their modeling to assess the effects of policy changes (such as inputs, subsidies and bank loans) on people's livelihoods,

- Participatory development of technologies to improve WP, food production, and livelihoods resilience in four local communities in two watersheds and preparation of the necessary bases for further research in the field of water management of the local communities. Some of the technologies tested were: autumn cultivation of chickpeas, potato management, mushroom production, rare medicinal plants, production of fast-growing trees (poplar), improvement of wheat and barley cultivars, application of bio fertilizers for wheat and barley, chemical fertilizers management, almond tree pest management, and wheat pest management,

- Conducting field research on women's issues and their participation and cooperation with projects,

- The use of improved varieties of chickpea and its autumn cultivation doubled the productivity of this crop,

- Application of Azetobacter and Azospirillium fertilizers increased the yield of wheat and barley and rain-fed barley (by 11–36%) and was welcomed by farmers,

- Watershed management principles were prepared and developed for upstream catchments in arid areas,

- Development of Integrated Watershed Management (IWM) Principles,

- Water resources of Merek and Honam catchments,

- Resources and rangeland management,

- Nutrition management,

- Erosion studies,

- Spatial analysis (GIS) based on a spatial decision support system.

3.2.2.2 Participatory development of technologies (PTD)

- Principles and concepts of participatory technology development (PTD),

- Barley seed breeding and improvement in a participatory manner,

- Improving the management of legumes and family of fodder products.

3.2.2.3 Socio-economic and policy-institution issues

- Women's participation in improving rural livelihood,

- Market access and its effects on low-income from farm residents,

- Rural livelihood zoning, effects of access to water resources and consideration of policy effects upstream of KRB,

- Principles of integrated water resources management and related institutional issues,

- Identify the livelihoods of rural communities, the effects of access to water resources, and policy measures in the upstream areas of the basin;

- The role and participation of women in improving the livelihood of rural communities,

- Market access and its effects on the income of small households in the basin,

- Policy-institutions issues in the basin,

- Principles of integrated water resources management and its institutional issues.

3.2.2.4 Development of integrated watershed management (IWM) principles

One of the project's important methodologies and outputs was the development of principles of integrated management of upstream watersheds in arid mountainous areas [15, 16]. The logic and hypothesis were that watershed management principles for tropical and temperate regions were well defined and documented. But mountainous areas with arid climates have their own characteristics in terms of climate, geomorphology, and the effects of water access on local communities' livelihood. Therefore, it is necessary to redefine and adapt these principles for the mentioned areas (such as upstream watersheds in the basins of Iran). The considered criteria and principles were developed under the following categories ([7, 15]): 1) issues and subjects of stakeholders and institutions, 2) issues and subjects of decision-making processes, 3) issues of stakeholder consultation and development of watershed management principles with the help of SWOT analysis, 4) issues of water scarcity and drought, 5) soil issues, 6) soil erosion/soil conservation issues, 7) issues of forest and rangeland ecosystems, 8) issues of areas with irrigated and rain-fed agriculture, 9) land use issues, 10) agronomic and crop selection issues, 11) on issues of participatory technology development, 12) policy matters, 13) livelihood issues, and 14) on gender issues.

4. Climate change impacts on land suitability in KRB

Considering the CPWF objectives, few studies were done on river basin climate change (CC) issues. The study done by Ghaffari et al. [17] on land suitability under current and CC scenarios in the KRB was the main study, and the results are summarized as follows;

Assessing the suitability of an area for crop production requires considerable effort in terms of information collection that presents both opportunities and limitations to decision-makers. Land suitability is assessed as part of a 'rational' cropping system, and optimizing a piece of land for a specified use should be based on its attributes. Furthermore, land may be considered either in its present condition or after specified improvements. Although criteria may vary, they are essentially based on climate, soil, topography, and water availability which are the most important categories of natural environmental information required for assessing land suitability.

The CC study in KRB describes a climate-soil-site model to assess CC impacts on land suitability for dryland winter wheat, focusing on the potential effects of temperature increase and rainfall variables on the land suitability in KRB. Assessments were made for the current climate condition and future climate scenarios by 2025 with GIS maps generated through a Simple Limitation Approach (SLA). Ghaffari et al. [17] used topography maps (10 m resolution), 25 years of climate data (1973–1998), physiological and phonological crop parameters, CC scenarios, soil management domain (SMD), and a simple limitation approach (SLA).

In **Table 5**, results of the projection of impacts of CC on land suitability classes for dryland winter wheat in upstream areas of KRB are provided.

It was concluded that by increasing temperature alone, it would be expected that highly and moderately suitable areas increases in the KRB. Increasing temperature and increasing precipitation increase highly and moderately suitable areas. Decreasing precipitation alone or increasing temperature will lower highly and moderately suitable areas. The main reason for this is water stress risk, not the direct effect of temperature [17].

5. CPWF approaches for continuing of the river basin management in phase 2

As mentioned earlier, the CPWF's aim was to identify a set of agricultural solutions (technical, socio-economic, and organizational-institutional) related to water management that can lead to greater resilience of local communities in catchments and protect the environment. It was expected that these strategies would lead to a significant increase in agricultural WP to help solve the "water crisis for food" problem in the world. Therefore, the selected research topics and priorities in phase 2 of the CPWF program (2009–2012) were: Compliance with the CPWF's research agenda in the areas of research topics or basin priorities; to be interdisciplinary and include cross-scale analysis and adaptation (social, ecological); help to improve the interrelationships of water, poverty, productivity, and ecosystems and within a global change context; address issues of rainfall management improvement, benefit sharing, multiple uses of water, drivers, and change processes.

In general, the research priorities for the implementation of Phase 2 of the CPWF program were as follow:

A. Upgrading of rain-fed agricultural systems

Research on this topic aims to increase food security and alleviate poverty by further increasing rainwater for food production in rain-fed agricultural production systems.

T:temperature/P: precipitation	No Change	0.0 °C +20%	0.0°C −20%	+1.5°C 0%	+1.5°C +20%	+1.5 °C −20%
Highly suitable (HS)	8.7	28	−91	6	53	−91
Moderately suitable (MS)	7.6	154	39	176	69	39
Marginally suitable (MG)	28	−50	15	−46	−25	18
Unsuitable (U)	55.7	0	1	−2	−5	0

Table 5
Percentage area of suitability classes for dryland winter wheat by CC scenarios in KRB upstream [17].

B. Benefits sharing

This research focuses on livelihood and income promotion, equity, and agricultural sustainability based on catchment/area boundaries.

C. Multipurpose use of water

This will focus on improving water storage capacity and distribution systems, such as irrigation networks and small reservoirs for food security and poverty alleviation.

D. Drivers and processes of change

This research theme is formed on the knowledge produced by the three previous research themes (A to C). Its approach is based on recognizing the opportunities and threats to people's livelihoods resulting from changes in how they use water to produce food (including livestock and aquatic ecosystems).

E. Other research fields (remained 20%)

Includes research into the potential effects of climate change (CC) on food production, hydrological issues and degradation of natural resources, community livelihoods, and the development of strategies to improve and enhance resilience in selected benchmark basins in phase 2 of the program.

The selection criteria for the benchmark basins in Phase 2 of the CPWF program were based on Phase 1 experiences and recommendations from the external evaluation of Phase 1 in 2007. The criteria were as follows:

- More focus on the effects of the implementation of the program in the next ten years at the basin scale and through the guidance and coordination of selected projects focused on several key issues and among a smaller number of benchmark basins,
- Focus on basins among low-income countries while coordinating with the needs and priorities of CPWF funding providers and donors,
- Overall, CPWF research should ensure that it leads to the creation and production of global achievements and international public goods (IPGS).

However, despite not selecting the KRB catchment in the second phase of the CPWF program for various reasons and according to the criteria presented above, the proposed research activities for future research in this basin were presented as follows [18]:

- Continue research on improving WP in the KRB,
- Paying attention to issues of adaptation to climate change and drought in the basin,
- Considering the priorities and strategies of water allocation in the basin,
- Environmental effects of water management and construction of dams, especially in the upstream areas of the basin,

- Research on fisheries and livestock production issues,
- Further activities in the field of spatial modeling and decision-making systems for planning and development,
- Continuation of activities and research related to PTD,
- Continuation of development activities and capacity building of existing organizations in local communities,
- Out scaling of technologies and other achievements of the projects,
- Evaluation of projects and analysis of their impacts from institutional and technological aspects,
- Establish a communication network with other CPWF benchmark basins, especially the Ganges, Mekong, Nile, and Andes basins, regarding WP improvement technologies and resource management methods.

6. Discussions

The CPWF program was a new approach by the CGIAR system to collaborative and multidisciplinary research that was implemented in two phases and in several selected benchmark basins worldwide.

The implementation of the CPWF program, with the contribution of national and international research institutes and centers, had a lot of capacity building for experts and faculty members at the national level (NARES). In addition to the scientific and research results, practicing the spirit of multidisciplinary and collaborative cooperation with each other, training and capacity building, identifying the KRB and Iran in international forums and events, recognizing international researchers and professors, and the possibility of working together were other benefits of the CPWF program in KRB.

A comparison between the approach and work organization used by the CPWF program in KRB with the other similar international joint projects, studied in Iran country, e.g., the United Nations Development Program (UNDP) and the Wetlands Conservation Project, with the support of the government of Japan and FAO-SIDA, in the Urmia Lake in the northwest of Iran, and on issues such as water accounting, improvement of WP, drought management, etc., indicate that:

- The CPWF program focused more on the research-development approach, while other projects, especially in recent years, have more of an implementation-extension nature (such as the implementation of the Urmia Lake Rehabilitation Project) than scientific field research,
- The CPWF program was implemented with foreign exchange resources and local in-kind resources. It differs from other projects in that the CPWF donors allocate their budget to the international institutions participating in the program and the cash flow is optimally allocated to the projects by confirming the proper progress of the work through them.

- Another advantage of the CPWF program was foreign parties' direct scientific and technical participation in its approved projects. In other words, international researchers directly participate in the scientific-executive and supervisory activities of the projects, and it was worked in the form of direct cooperation with the organization and the team of national collaborators. In other projects, however, international organizations usually play only the role of budget allocators and do not have much technical or advisory involvement in the technical and scientific issues of the project until the project is completed and the final work report is received and assessed.

- The CPWF program in the KRB was made up of and used to the advantage of many international stakeholders including international institutions, renowned universities in developed countries, various international associations and institutions, and well-known international professors carrying out project activities and having direct participation and contribution. While in other projects, this international participation and support are usually very low and to the extent of one or two major scientific organizations that mostly play the role of administrative and financial support rather than a scientific and project management contributor.

- The CPWF program was structured and its implementation was followed by a step-by-step review and scientific-technical reports and work progress during project implementation. Examples were the convening of steering committee meetings, technical committee meetings, publishing mid-term and timely articles and reports, and so on. The program was also very efficient and systematic in sharing its data and information and creating a database and network to inform stakeholders of the results of its stages. Networking is much emphasized in this program. For example, during the implementation and completion of projects, the program held several specialized workshops, conferences and international forums in different countries to inform the results. Such an integrated networking system, informing, monitoring and delivering timely outputs, is usually much less common in other similar international projects in Iran.

Significant outputs of CPWF in KRB could be mentioned as: Development of supplementary irrigation in rain-fed agricultural systems in upstream of KRB, provide necessary solutions and measures to further disseminate the results and apply new methods and cropping patterns of agricultural crops in the downstream areas of the basin, application of land drainage technologies and salinity management in saline lands downstream of KRB, transition from one dimensional and researcher-oriented agricultural research to demand-driven research, as well as the development of technologies related to providing more participation of farmers in research process, establishing a balance between rural development and food security goals in terms of protection of soil and water resources and sustainability of mountainous watersheds in arid and semi-arid regions, development and promotion of production and marketing of chickpeas in order to increase farm income and ultimately the income of watershed residents, and mainstreaming of participatory development of technologies in agriculture.

However, the KRB was not selected in phase 2 of the CPWF program. This was a failure and had many reasons that are out of this chapter's scope. The reasons for not selecting KRB for phase 2 are presented in detail by Heydari et al. [9].

Continuation of the work and dissemination of the results of the CPWF program in the KRB, or so-called out-scaling and up-scaling of the results to policy-makers and planners, was one of the things which should have been continued after the completion of the program. However, this task was followed to some extent but not fully, and stopped after a while so its issues should be investigated. This shortcoming occurs for most R&D projects in most developing countries, including Iran.

However, the project archives and databases (documented fully in [9]), as well as its many national and international research reports and published articles (e.g., [3, 5–7, 10]), along with the information presented in this chapter, could be used as important resources for future use.

The present collection can be a good roadmap to explain the process of water and food challenges in KRB and to continue the road in this basin and other basins in Iran country and even similar basins in the region such as the Amu Darya in Central Asia and the Tigris-Euphrates in the Middle East.

As a result, the following common issues and strategies can be used as key indicators for multidisciplinary and multi-scale studies in similar studies in related basins:

6.1 Basin scale issues

Water resources and drought, areas prone to supplementary irrigation, upstream impacts and interactions on the downstream basin, land use effects on sediment flow in the basin, agro- ecological characterization, and similarity analysis of the basin.

6.2 Assessment and evaluation of options for improving WP in the basin

Rainwater productivity improvement by using supplementary irrigation, evaluating and improving WP in irrigated lands (saline areas with salinity and drainage limitations), recognizing and presenting agricultural options to improve WP under salinity of water and soil resources, socio-economic factors affecting WP in the basin.

6.3 Improving the resilience and livelihood of local communities in the basin

Integrated watershed management (including: water resources of the pilot sites, rangeland resources and its management, soil fertility management, soil erosion studies, and spatial decision support system); participatory technology development (PTD) in the basin (including: mainstreaming of the principles and concepts of participatory technology development, seed modification of common and major crops under participatory method, and improving the management of common and major crops); socio-economic and policy-institution issues (including women's participation in improving rural livelihoods, market access and its effects on low-income farm residents, rural livelihood zoning, effects of access to water resources, and policy considerations in upstream developments; policy-making and organizational-institutional issues, and the principles of integrated water resources management and its related institutional aspects).

7. Conclusions

The CPWF program was a new approach to the CGIAR system to participatory and multidisciplinary research that was implemented in two phases and several selected basins in several global basins, including the KRB in Iran.

Methodology and the achieved results are a proper roadmap for water management in KRB and under increasing water scarcity and climate change conditions that can be used to continue the research and development in this basin and other similar basins in Iran. It can also be applied to other parts of the worlds, such as the Amu Darya basin in Central Asia and the Tigris-Euphrates basin in the Middle East.

Based on the results of the various projects mentioned above as well as the opinions of basin stakeholders and implementers, policies, plans, and various measures were extracted and suggested as outcome of CPWF in KRB.

The CPWF projects in KRB has been an example for biophysical similarlity analysis used for similar areas mapping to out-scale the results of projects to other similar basins in the Asia WANA region, potential area mapping for supplimentory irrigation, systematic identification of innovations of local farmers, participatory technology development, formulation of future research needs and activities, and etc.

Author details

Nader Heydari
Irrigation and Drainage Department, Iranian Agricultural Engineering Research Institute (AERI), Agricultural Research, Education and Extension Organization (AREEO), Karaj, Iran

*Address all correspondence to: nrheydari@yahoo.com

References

[1] Heydari N. Study of the approach used in the CGIAR challenge program on water and food in the management of catchments of Karkheh River. Iranian Journal of Watershed Engineering and Management. 2022. DOI: 10.22092/ijwmse.2022.356602.1928 (In Publish) (In Persian)

[2] Qureshi AS. Land, water and environmental Management in the Karkheh River Basin: Challenges and opportunities. In: Presentation in Karkheh Basin Focal Project Workshop on Land-Water and Environment Management, May 24, 2006. Karaj, Iran: International Water Management Institute (IWMI); 2006

[3] Farahani H, Oweis T, Saidat H, Abbasi F, Bruggeman A, Anthofer J, et al. Improving Water Productivity and Livelihood Resilience in Karkheh River Basin. Report No. 2. Aleppo, Syria: ICARDA; 2008 iv+169 pp

[4] Oweis T, Farahani H, Qadir M, Anthofer J, Saidat H, Abbasi F, et al. Improving on-farm agricultural water productivity in Karkheh River basin. Research report no. 1. In: A Compendium of Review Papers. Aleppo, Syria: ICARDA; 2008 iv+103 pp

[5] Oweis T, Siadat H, Abbasi F. Improving on-farm agricultural water productivity in the Karkheh River basin (KRB). In: PN8 Project Report, CGIAR Challenge Program on Water and Food (CPWF). Aleppo, Syria: ICARDA; 2009. 2009 78 p

[6] Bruggeman A, Turkelboom F, Porhemmat J, Ghafouri M, Moosavi SB, Mirghasemi SA, et al. Strengthening Livelihood Resilience in Upper Catchments of Dry Areas by Integrated Natural Resources Management. PN24 Project report, CGIAR Challenge Program on Water and Food (CPWF). Aleppo, Syria: ICARDA; 2010. 75 pp

[7] Ghafouri M, Siadat H, Oweis T. Integrated watershed management in the upper catchments of Karkheh River basin of Iran. In: Research Report No. 12. Aleppo, Syria: ICARDA; 2012 x + 89 pp

[8] Ahmad MD, Giordano M. Karkheh Basin Focal Project. CPWF Project Report no. 57. International Water Management Institute (IWMI). Colombo, Sri Lanka: IWMI; 2010. 43 p

[9] Heydari N, Abbasi F, Morsali E. Study and Documentation of the Methodology of Research Used for Improvement of Water Productivity from Field to Basin Scales in the CGIAR Int'l Project in Karkheh River Basin. Iran. Res. Rep. No. 60742, Agricultural Engineering Research Institute (AERI). Karaj, Iran: AERI institute; 2021. p. 390. (In Persian)

[10] Woolley J, Cook SE, Molden D, Harrington L. Water, food and development: The CGIAR challenge program on water and food. Water International. 2009;**34**(1):4-12

[11] Mid-Annual Meeting. Mid-annual meeting of WP and LR projects of Karkheh River basin. In: Agricultural-Research- Education- Extension Organization (AREEO), SPII Campus, Feb. 26–27. Karaj, Iran: AREEO; 2005 (In Persian)

[12] Workshop K-o. Kick-off Workshop of CPWF projects in Karkheh River basin. June 8-10. In: SPII Campus of Agricultural-Research-Education-Extension Organization (AREEO). Karaj, Alborz Province, Iran: AREEO; 2004 (In Persian)

[13] Field Visit Report. 2005a. Report of Field Visits from Southern Karkheh River Basin during Selection of Research Site Activities (Mainly for WP-PN8 Project): First Group Visit. Tehran, Iran: AREEO; March 2-5. (In Persian)

[14] Field Visit Report. 2005b. Report of Field Visits from Southern Karkheh River Basin during Selection of Research Site Activities (Mainly for WP-PN8 Project): Second Group Visit. Tehran, Iran: AREEO; March 12-14. (In Persian)

[15] Workshop IWM. Livelihood Resilience Integrated Watershed Management International Workshop. June 22–28. Aleppo, Syria: ICARDA; 2008

[16] Mid-annual meeting. Mid-annual meeting of Karkheh River basin projects. Feb. 4-5, 2008. In: Agricultural Engineering Research Institute (AERI). Karaj, Alborz: AREEO; 2008 (In Persian)

[17] Ghaffari A, De Pauw E, Mirghasemi SA. A land suitability study under current and climate change scenarios in the KRB, Iran. In: Presentation in the Int'l Workshop: Improving Water Productivity and Livelihood Resilience in Karkheh River Basin, Sep. 10–11, SPII Campus. Karaj, Iran: AREEO; 2007

[18] Workshop F. Final and Wrap Up Workshop of CPWF-Karkheh River Basin Projects: Final Workshop March 2–3 (in SPII Campus, Karaj), Warp Up Workshop March 4. Tehran, Iran: Agricultural- Research- Education-Extension Organization (AREEO); 2009 (In Persian)

Chapter 3

The Efficacy of Artificially Assembled Boulder Installations in Improving Migration Routes for Aquatic Animals

Youichi Yasuda and Nozomi Fuchino

Abstract

Naturally assembled rocks can be found in the natural rivers with gravel beds, and assembled boulders with large-scale rocks are stable after many floods. This chapter focused on assembled rocks from the point of view of the stability of the structure during many floods and the preservation of the aquatic habitat under normal conditions. In renovated rivers, the installation of hydraulic structures may result in the degradation of the river bed, local scouring behind the structure, bank erosion due to floods, and obstacles in the aquatic migration routes of aquatic animals. A balance between the prevention of disasters and the preservation of aquatic habitats should be maintained. Artificially assembled boulders should be recommended for river engineering technology. Unfortunately, there is no information on the hydraulics of artificially assembled boulders or on the imbrication of boulders in any part of the world. This chapter presents experimental research on the hydraulics of assembled boulders under normal and flooding conditions. A shallow water flow was formed near the water side under normal conditions. The stability of the structure of assembled boulders was confirmed under flooding conditions. These findings were confirmed by field inspection.

Keywords: artificially assembled boulders, stability of structure, shallow water flow, normal stage, flood stage

1. Introduction

In natural rivers, imbrication and assembled rocks can be observed. Imbrication is an overlapping arrangement of boulders which are different from the assembled rocks. In particular, in imbrication (**Figure 1**), collisions between boulders or driftwood during floods improve the ease of transport. On the other hand, naturally assembled rocks (**Figure 2**) can be stabilized against many floods, and they contain a stable habitat space for several kinds of aquatic animals under normal conditions. Moreover, the space may be useful as a refuge area during floods because of the flow velocity, including the turbulence is always low due to the formation of a seepage flow.

Figure 1.
Imbrication of boulders.

Figure 2.
Naturally assembled rocks.

There is a regular assembly method for forming the assembled rocks that makes sense in nature. Still, it has not been focused on as a technical method in civil engineering [1]. The mechanism of imbrication and its relationship with the habitat have been studied from the perspective of geomorphology and ecology. According to Hassan [2], the macroform is defined as transverse ribs of rubble in the direction of flow, and stone cells of rubble in a circle. The microform is defined as a stone cluster which is an accumulation of gravel of different sizes around a large piece of gravel; imbrication is a folding of gravel pieces along the direction of flow. Some researchers have further classified stone clusters into several forms (e.g. Strom et al. [3]). The formation of macroforms and stone clusters has been reported using field surveys to carefully record the arrangement of stones that have been moved and those that have not (Church et al. [4], Lamarre and Roy [5], Wittenberg and Newson [6]). The developmental process of clusters has been examined in laboratory experiments with simpler conditions (Papanicolaou et al. [7]). Strom et al. [3] studied the shape characteristics of a large number of samples from five different types of stone clusters in the field. Once imbrication and various types of stone clusters are formed, they are less likely to move with the flow than stone and gravel on their own. From the perspective of sediment hydraulics, the effects of microforms on flow resistance and sediment volume have been investigated by Hassan and Reid [8], as well as by Strom et al. [9]. From the perspective of river ecology, studies have been conducted focusing on the function of flow refuge against benthic disturbance (Biggs et al. [10]). Accordingly, imbrication

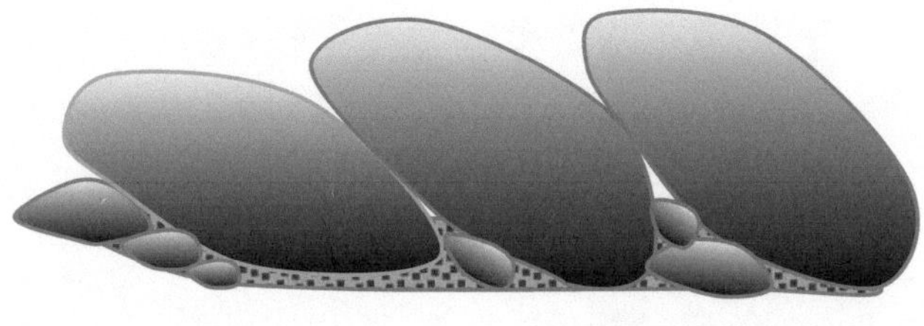

Figure 3.
Structure of assembled boulders.

and stone clusters can be formed artificially by arranging stones, and this process may be developed into a simple method for contributing to the improvement of the river bed environment in small and medium-sized rivers. However, these approaches were not applied for large floods, because the stability of the assembled boulders was not examined for a wide range of discharges. Recently, the authors focused on artificially assembled boulders, as shown in **Figure 3**, as naturally assembled rocks from the point of view of energy dissipation during floods and on the possibility of upstream migration under normal conditions [11, 12]. Based on field measurements, Rickenmann and Recking [13] as well as Hey [14], for example, investigated the flow resistance in rough river beds. Still, the stability of assembled rocks during floods and the formation of multiple flows under normal conditions were not found to be associated.

This chapter presents findings from practical and experimental research on the hydraulics passing through artificially assembled boulders. The focus was on consecutively assembled boulders installed at the drop structure. In order to clarify the flow conditions around the assembled boulders and the possibility of upstream migration, the hydraulics of consecutively assembled boulders were investigated experimentally based on three different downward slopes and discharges as the first stage of the research project on assembled boulders. A shallow water flow was formed under normal conditions as an upstream migration route if consecutively assembled boulders with transverse mild slopes were installed. Moreover, a surface jet flow passing over the assembled boulders was always formed, even during floods. Then, the structure of the consecutively assembled boulders was stable, even if the assembled boulders were installed without fixing the bottom part. The consecutively assembled boulders were practically installed in a check dam, introducing a flow in the assembled boulders, whose details could not be covered by the experiments, and stability in the assembled boulders during flooding. These results can help to provide practical applications in installing assembled boulders for fish passage, energy dissipation for low head structures, ground sills, protruding stones, and fishing reefs.

2. Experimental setup

The experiments were conducted in the Environmental Hydraulic Laboratory of Nihon University, College of Science and Technology, in order to investigate the hydraulics passing over consecutively assembled boulders. Three different downward slopes (i.e. 1/8.5, 1/12.5, and 1/25) were tested. In the case of 1/8.5 and 1/25 slopes, as shown in **Figure 4**, a half model was installed in a rectangular channel with a width of 0.4 m, length of 17 m, and height of 0.6 m. For the slope of 1/12.5, a symmetric model was installed in a rectangular channel with a width of 0.8 m, length of 15 m,

(a) (b)

Figure 4.
Half model of consecutively assembled boulders. (a) Half model with 1/25 slope. (b) Half model with 1/8.5 slope.

(a) (b)

Figure 5.
Symmetric model of consecutively assembled boulders (1/12.5 slope). (a) Stepped channel as a base. (b) Consecutively assembled boulders.

and height of 0.6 m. These models were constructed as 1/10 scale models, and it was assumed that the flow condition might be represented under a Froude similarity. The size of the boulders was set to around 0.6 m in the prototype. The stepped channel model was used as a base of consecutively assembled boulders (**Figure 5**). In order to change the water width in accordance with the discharge, a transverse stepped slope was installed at around 1/10. The assembled boulders were installed on each staircase without hardening the base, and the downstream end of the boulders was stabilized with L-shaped fittings (in the field, use stopper blocks). The experimental conditions are shown in **Tables 1–3** for slopes of 1/8.5, 1/12.5, and 1/25, respectively. Here, B is the channel width, h_c is the critical depth, Q is the discharge, and $W.L._d$ is the downstream water level based on the lowest bottom of the stepped channel at the downstream end of the consecutively assembled boulders. Moreover, the subscript.

Case	Q (m³/s)	Q_p (m³/s)	h_c (m)	h_{cp} (m)	$W.L._d$ (m)	$W.L._{dp}$ (m)
1	0.00348	1.10	0.020	0.20	0.075	0.75
2	0.00632	2.00	0.029	0.29	0.086	0.86
3	0.00949	3.00	0.039	0.39	0.097	0.97
4	0.0456	14.4	0.110	1.10	0.201	2.01

Note: B = 0.4 m; transverse slope = 1/11; h_c = critical flow depth ($=[(Q/B)^2/g]^{1/3}$).

Table 1.
Experimental conditions for 1/8.5 downward slope.

Case	Q (m^3/s)	Q_p (m^3/s)	h_c (m)	h_{cp} (m)	$W.L._d$ (m)	$W.L._{dp}$ (m)
1	0.0070	2.20	0.020	0.20	0.079	0.79
2	0.0111	3.52	0.027	0.27	0.089	0.89
3	0.0168	5.31	0.036	0.36	0.099	0.99
4	0.1211	38.30	0.133	1.33	0.232	2.32
5	0.1456	46.03	0.150	1.50	0.259	2.59

Note: B = 0.8 m; transverse slope = 1/11; h_c = critical flow depth ($=[(Q/B)^2/g]^{1/3}$).

Table 2.
Experimental conditions for 1/12.5 downward slope.

Case	Q (m^3/s)	Q_p (m^3/s)	h_c (m)	h_{cp} (m)	$W.L._d$ (m)	$W.L._{dp}$ (m)
1	0.00143	0.451	0.011	0.11	0.044	0.44
2	0.00595	1.88	0.028	0.28	0.079	0.79
3	0.0103	3.25	0.041	0.41	0.096	0.96
4	0.00312	0.986	0.018	0.18	0.066	0.66

Note: B = 0.4 m; transverse slope = 1/6.9, 1/10, and 1/9.0 from right side; $h_c = [(Q/B)^2/g]^{1/3}$.

Table 3.
Experimental conditions for 1/25 downward slope.

"p" indicates the prototype. The downstream water level was controlled with a sluice gate located at the channel end. The velocity was measured using a propeller current meter 0.03 m in diameter [15] and the sampling time was set to 20 seconds because of steady supercritical flow. The water and bottom levels were measured using a point gauge with a 0.1 mm reading.

3. Experimental results

The clarification of flow condition, water depth, and velocity on consecutively assembled boulders under normal conditions might help to clarify the upstream migration route of aquatic animals. Experimental investigation of flow condition, the water surface profiles, bed shape, and velocity field yields is helpful for hydraulic design on the consecutively assembled boulders. The important points are summarized as follows:

1. The main flow passing over the consecutively assembled boulders is formed along the water surface without jump formation with a surface roller and it is easy for aquatic animals to find the upstream migration route.

2. Multi-aquatic animals are able to migrate upstream through the consecutively assembled boulders in a wide range of discharges because the formations of shallow water flow near the water side and the gap flow of the assembled boulders produce a low velocity with low turbulence. In this case, the boulders that were utilized averaged about 0.6 m in size and installed at a transverse slope of about 1/10.

3. The velocity and depth of the migration route were independent of the downward slope in slopes ranging from 1/8.5 to 1/25, because the flow condition among the assembled boulders was locally independent.

4. Concentration of the flow at the lowest part of the assembled boulders, in addition to a priming effect, can be expected for the upstream migration route.

3.1 Flow condition under normal conditions

The flow conditions on the consecutively assembled boulders are shown in **Figures 6–8**. Under normal conditions, the main flow concentrated at the lowest part of the consecutively assembled boulders. On the other side, a shallow water flow on the assembled boulders was formed as a small cascade at each combination of assembled boulders. Moreover, pools with a low velocity were formed in the gaps between the boulders.

In the case of a 1/25 downward slope and a transverse slope around 1/8, even when the discharge in the prototype changes from 0.5 to 3.0 m^3/s, multi-aquatic animals (e.g. swimming fish, benthic fish, crustaceans, and shells) may migrate upstream in the shallow water flow region (**Figure 6**).

In the case of a 1/12.5 downward slope and a transverse slope around 1/11, the flow condition for the upstream migration route was formed up to the discharge in the prototype 5.3 m^3/s in the shallow water flow region (**Figure 7**). In this case,

(a) (b)

Figure 6.
Flow conditions for a 1/25 downward slope and a transverse slope around 1/8. (a) Flow condition for Case 4. (b) Flow condition for Case 2.

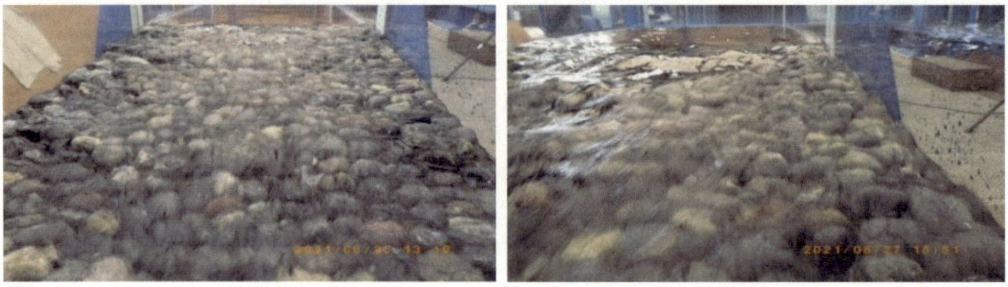

(a) (b)

Figure 7.
Flow conditions for a 1/12.5 downward slope and a 1/11 transverse slope. (a) Flow condition for Case 1. (b) Flow condition for Case 2.

Figure 8.
Flow conditions for a 1/8.5 downward slope and a 1/11 transverse slope. (a) Flow condition for Case 1. (b) Flow condition for Case 2.

since the cross-section of consecutively assembled boulders was symmetrical (8 m in width in the prototype), the upper discharge was 2.65 m^3/s for a half section 4 m in width.

In the case of a 1/8.5 downward slope and a transverse slope around 1/11, the flow condition for the upstream migration route was formed up to the discharge in the prototype 2.0 m^3/s in the shallow water flow region (**Figure 8**).

In the three different downward slopes in the flow direction, the main stream passing over the consecutively assembled boulders was always formed along the water surface, and enabling aquatic animals to find the upstream migration route easily.

3.2 Velocity fields under normal conditions

Figures 9–11 show the velocity profiles on the assembled boulders at cross-sections for three different downward slopes. Here, u_p is the mean velocity above the boulders in the flow direction, X_p is the streamwise direction coordinate from the upstream end of the consecutively assembled boulders, y_p is the transverse direction coordinate from the right side wall, and Y_p is the transverse direction coordinate from the center line of the symmetric cross-section. Moreover, these values are expressed according to the prototype scale. The velocity distribution depends on the gradient in the flow direction and the discharge, except for the upstream migration route of multi-aquatic animals. In the shallow water flow region, the flow velocity above the assembled boulders varied from 0 to 2.2 m/s on the prototype scale. The flow velocities between the assembled boulders might be further reduced, although they might be difficult to measure on a 1/10 scale model.

Figures 12–14 show the relationship between evaluated depth [12] and mean velocity for three different downward slopes in the flow direction. The data in these figures were recorded in a shallow water flow region. As shown in these figures, similar results were obtained, although there were variations. Accordingly, the flow condition in a shallow water flow region might be less sensitive to the downward slopes and variations in discharges under normal conditions. These results reveal that multiple flow with various water depths and velocities with low turbulence was formed in the shallow water flow region by installing assembled boulders on a transverse inclined slope. Furthermore, aquatic animals might migrate upstream amid various downward slopes and discharges because the shallow water flow is formed for three different downward slopes (1/8.5, 1/12.5, and 1/25).

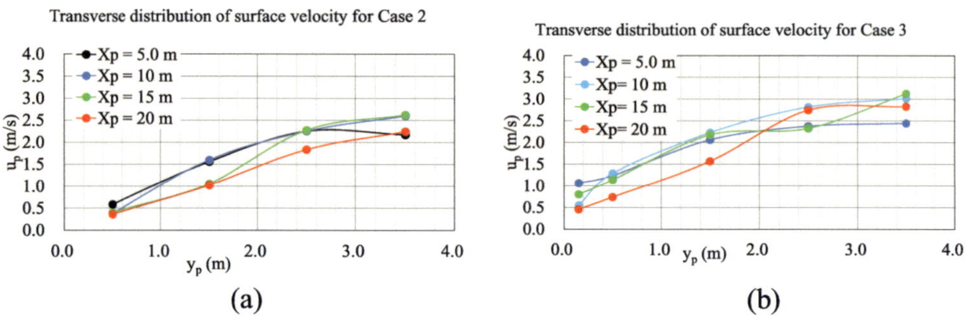

Figure 9.
Velocity profiles for a 1/25 downward slope and a transverse slope around 1/8. (a) Velocity profiles for Case 2. (b) Velocity profiles for Case 3.

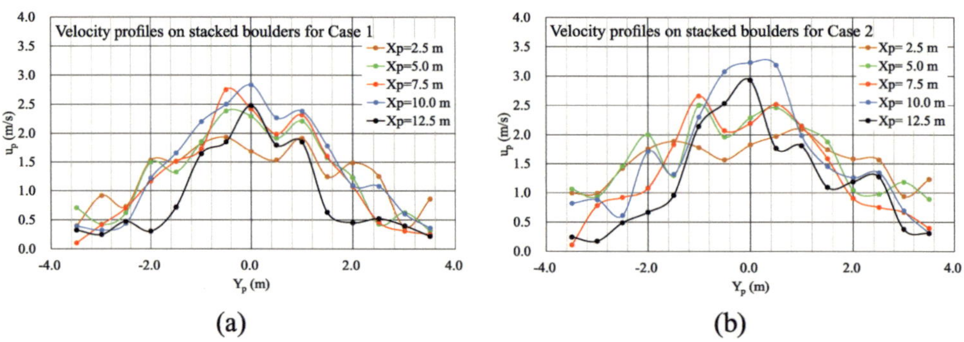

Figure 10.
Velocity profiles for a 1.12.5 downward slope and a 1/11 transverse slope. (a) Velocity profiles for Case 1. (b) Velocity profiles for Case 2.

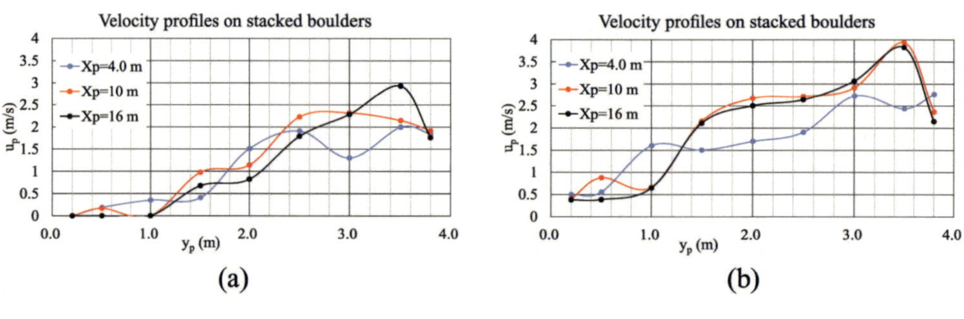

Figure 11.
Velocity profiles for a 1/8.5 downward slope and a 1/11 transverse slope. (a) Velocity profiles for Case 1. (b) Velocity profiles for Case 2.

Figure 15 shows the velocity distributions in the assembled boulders under both normal and flooding conditions. The values of the velocity are expressed in the prototype. This experiment was an additional experiment with a 1/4 scale model; the velocity was measured using a two-dimensional electric magnetic current meter of type I with a diameter of 3 mm (sampling time: 30 s; sampling frequency: 50 ms). As shown in **Figure 15**, the mean velocity u_p and the standard deviation u_p were always low in the assembled boulders under both the normal and flooding conditions.

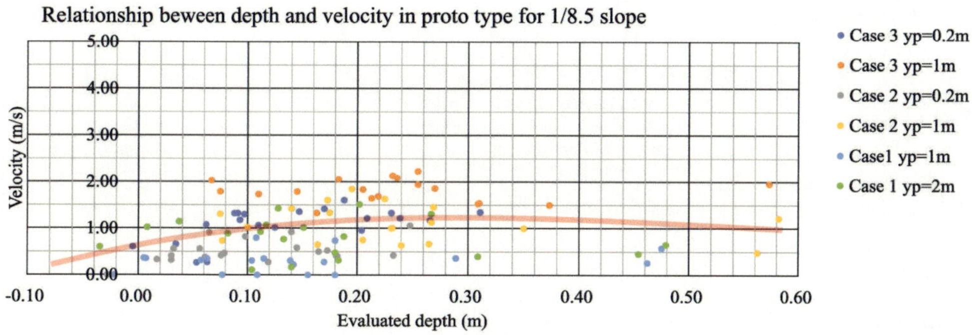

Figure 12.
Relationship between evaluated depth and mean velocity for the 1/8.5 slope.

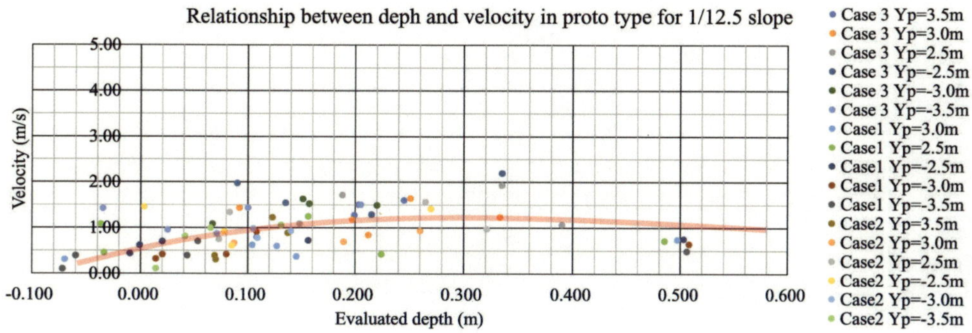

Figure 13.
Relationship between evaluated depth and mean velocity for the 1/12.5 slope.

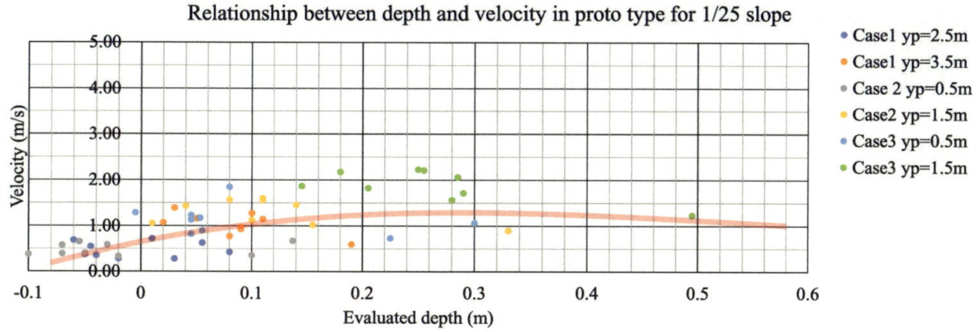

Figure 14.
Relationship between evaluated depth and mean velocity for the 1/25 slope.

As the flow concentrated toward the lowest part of the assembled boulders, the velocity reached its maximum value. The main flow was always located near the water surface. **Figure 16** shows the maximum velocity decay downstream of the consecutively assembled boulders for downward slopes of 1/12.5 and 1/8.5.

As shown in these figures, the main flow continued all the way downstream to the lowest part of the assembled boulders, and the maximum velocity downstream of the shallow water flow region was lower than 1 m/s. From these results,

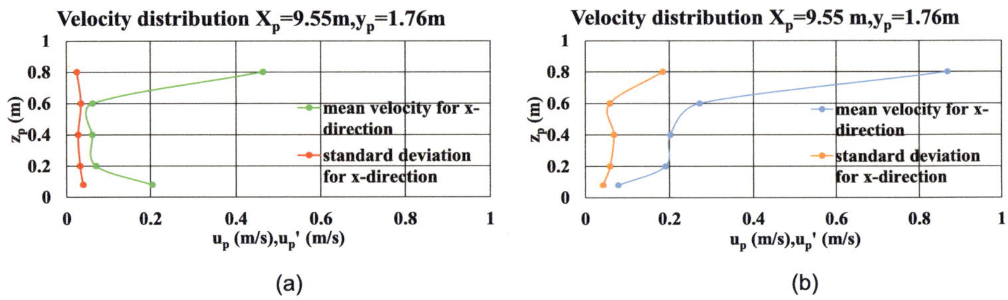

Figure 15.
Velocity distribution in assembled boulders (1/4 scale model, 1/10 downward slope). (a) Normal conditions (h_{cp} = 0.166 m). (b) Flooding conditions (h_{cp} = 0.453 m).

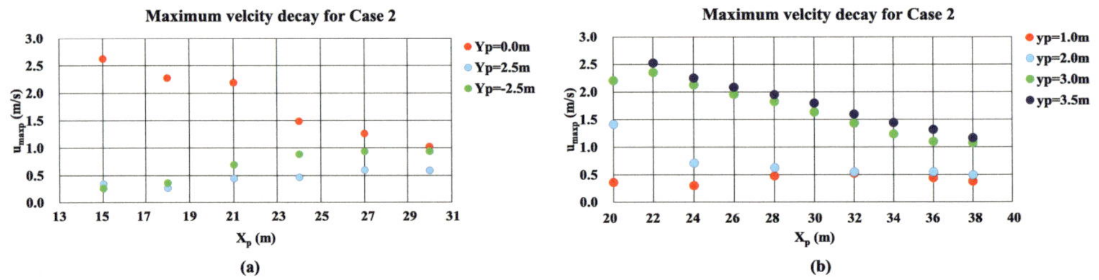

Figure 16.
Maximum velocity decay downstream of consecutively assembled boulders. (a) 12.5 slope. (b) 1/8.5 slope.

fish and crustaceans could easily migrate upstream through the consecutively assembled boulders.

3.3 Velocity fields during floods

Figures 17 and **18** show the streamwise change in the maximum velocity around the assembled boulders amidst a large flood. The maximum velocity was defined at each vertical measurement section. In the prototype, the critical flow depth was set to about 1.1–1.5 m, and the maximum velocity reached 6.0 and 6.5 m/s for the 1/8.5 and 1/12.5 slopes respectively. In the experiments shown in **Figures 6–8**, the boulders were assembled only on each step, and the material for fixing was not used. The velocity near the boulders reached 4.0–5.0 m/s during floods, but the assembled boulders were not destroyed at all. In the case of the imbrication of boulders in a gravel river (**Figure 1**), the stacked boulders might be fragile during large floods because the boulders might not support each other against the fluid force. Regarding the maximum velocity decay downstream of the consecutively assembled boulders, as the main flow was located near the water surface, it was possible to prevent bed erosion. Moreover, the maximum velocity decay changed transversely. If the consecutively assembled boulders were arranged symmetrically, it was possible to prevent side bank erosion.

To evaluate the flow resistance of the assembled boulders in a flood flow, the depth of the water in a pseudo-uniform flow was evaluated according to the shape of the assembled boulders and the water surface profile. The friction coefficient was evaluated using the Darcy–Weisbach equation; it was found to vary with the slope in the downstream direction. For the 1/8.5 slope, the friction coefficient was 0.180, and

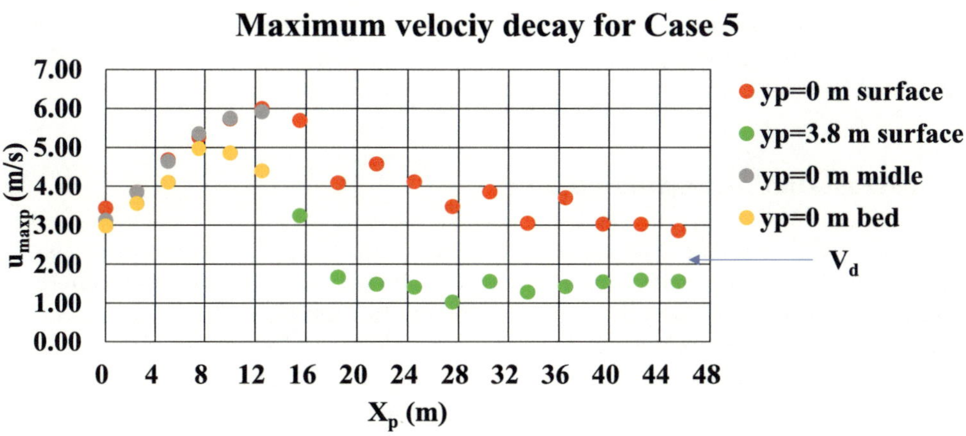

Figure 17.
Streamwise change in velocity for Case 5 (q_p = 5.75 m^2/s) under the 1/12.5 slope. V_d = averaged velocity at downstream subcritical flow (X_p = 46 m).

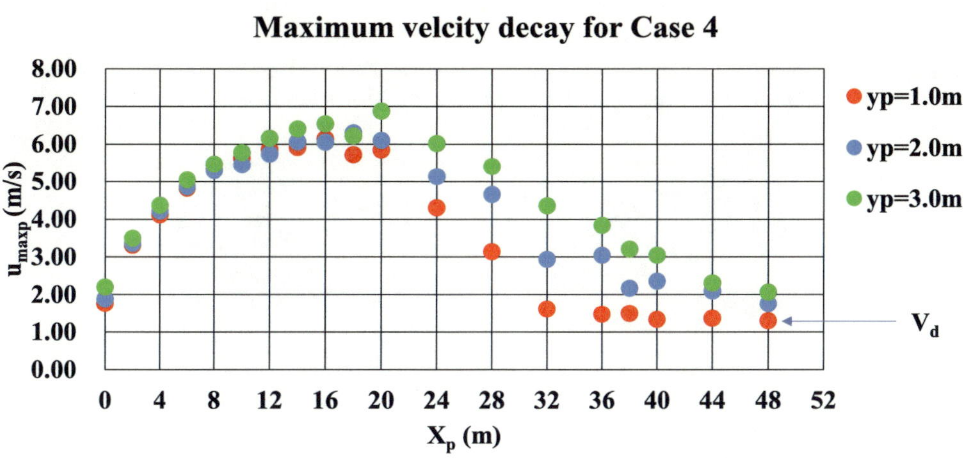

Figure 18.
Streamwise change in velocity for Case 4 (q_p = 3.61 m^2/s) under the 1/8.5 slope. V_d = averaged velocity at downstream subcritical flow (X_p = 46 m).

for the 1/12.5 slope, it was 0.252. The measured results obtained by Rickenmann and Recking [13] were converted into a friction coefficient corresponding to the relative roughness height, which is shown in Eq. (1).

$$\sqrt{\frac{1}{f}} = 2.821\left(\frac{d}{D_{84}}\right)^{0.697} \quad 0.646 < \frac{d}{D_{84}} < 5.07 \tag{1}$$

where d is the flow depth, and D84 is used as characteristic grain size (following Ferguson [16]).

If the value of D_{84} is evaluated from the friction factor and the flow depth, D_{84} = 0.204 m for the 1/8.5 slope and D_{84} = 0.445 m for the 1/12.5 slope respectively. Accordingly, the equivalent roughness height, which contributes directly to a flow resistance, might become smaller as the downward slope increases. The estimation of the friction coefficient can be applied to the stable structure as in the case of the assembled boulders.

4. Field installation and monitoring

The artificial installation of assembled boulders can be applied to eliminate discontinuities due to drop structures. In order to maintain the balance between energy dissipation during floods and the migration of aquatic animals under normal conditions, consecutively assembled boulders were installed with 74.4 m in width of a channelized weir with a 2.3 m drop. The weir was located in the Ohmu River, Hokuto city, Yamanashi Prefecture, Japan (**Figure 19**). Every year, many boulders, including ones 0.7 m in size, are transformed during large floods; strength in the structure is therefore required. In order to install consecutively assembled boulders near an apron behind the channelized weir, the downward slope was set to 1/8. The size of the boulders was set to about 0.7 m to ensure water depth between the assembled boulders (**Figure 20**). During floods, the main stream was formed along the water surface downstream of assembled boulders, and the flow velocity above the boulders could be reduced by shape resistance from the assembled boulders. As shown in **Figure 21**, considering the water width of the main stream under normal conditions (in this case, 10 m in width)

Figure 19.
Consecutively assembled boulders.

Figure 20.
Structure of assembled boulders.

Figure 21.
Aerated main flow and shallow water flow.

Figure 22.
Shallow water flow region.

Figure 23.
Transported rocks during floods.

the bed level at the upstream end of the consecutively assembled boulders was adjusted to be 0.20 m lower than both sides of the main stream. The river bed downstream of the consecutively assembled boulders was adjusted by installing rocks in order to form a subcritical flow during floods. As shown in **Figure 22**, a shallow water flow was formed on both sides of the main stream. The flow condition can be confirmed from the physical model (**Figure 23**). After the installation of the consecutively assembled boulders, it was confirmed that aquatic organisms at various stages of growth that are native to the river could migrate upstream through the channelized weir. Moreover, the main flow was located along the water surface during floods; there was thus no local scouring.

5. Discussion

As shown in **Figures 4** and **5**, as the consecutively assembled boulders were installed in transverse steps, it was easy for shallow water to flow to form near the water side for different discharges. Moreover, the gap flow among the assembled boulders had a low velocity with low turbulence (**Figure 15**). The flow condition confirmed these facts at the structure constructed in the Ohmu River (**Figure 22**). The width of the assembled boulders installed on site was 74 m. In this case, the experiment corresponded to the investigation with a width of 8 m, which is a partial reproduction of the original condition. With the assistance of many fishermen, a local government office, and fishery associations, it was confirmed that swimming fishes (e.g. Plecoglossus altivelis, Tribolodon hakonensis, and Oncorhynchus masou ishikawae) and benthic fishes (e.g. Rhinogobius flumineus and Cottus pollux) were able to migrate upstream on the assembled boulders for various discharges. The flow conditions and velocity fields in the assembled boulders can be observed experimentally (**Figures 15** and **24**), but they are difficult to quantitatively evaluate in detail. Therefore, it is important to understand the flow condition in the space from the field construction of assembled boulders. The experimental results shown in **Figures 12–14** are the most important results. Practically, the selection of a downward slope of consecutively assembled boulders is limited because the drop structure must be designed on the basis of the hydraulic design in Japan. As the relationship between the surface velocity and the evaluated depth in the shallow water flow region is independent of downward slopes (at least 1/8.5 to 1/25 slopes), the application of assembled boulders might be flexible. The main flow passing over the consecutively assembled boulders is always located near the water surface, as shown in **Figures 12–14**, and the velocity of the main flow changes not only in a streamwise direction but also in a transverse direction, as shown in **Figure 16**. As the main flow continues far downstream, it is easy for swimming fishes, benthic fishes, and crustaceans to find the upstream migration route. Especially if the fish passage with the assembled boulders is installed in part, the installation of the assembled boulders is effective for guidance toward the upstream migration route.

During floods, the structure of the artificially assembled boulders should be stable. The experiments with a 1/10 scale model confirmed that the artificially assembled boulders were stable during floods. In this case, the boulders were assembled on each step, and the material for fixing was not used. More than 10 assembled boulder structures were installed on-site, all of them were found to be stable even after the floods. **Figure 23** shows the transported rocks during floods. These rocks were the same size as the assembled boulders, but the consecutively assembled boulders were not destroyed. The structure of artificially assembled boulders was constructed on the basis of the naturally assembled rocks and was quite different from that caused by imbrication in the river. The artificially assembled boulders were shaped so that the boulders were stacked on the top of the downstream boulders. There were at least four

Figure 24.
Flow conditions at assembled boulders (1/4 scale model, 1/10 downward slope). (a) Normal condition (h_{cp} = 0.166 m). (b) Flood condition (h_{cp} = 0.453 m).

Figure 25.
Ground sill of assembled boulders.

Figure 26.
Alternative protruding assembled boulders.

points of contact for the boulders (overlapping points, points touching both sides, and one point at the bottom). The shape of the boulders was not spherical, but a flat shape is recommended for the boulders in order to be stabilized them.

From the point of view of the structure of artificially assembled boulders, as shown in **Figures 25** and **26**, the installation of assembled boulders is helpful for the improvement of river environments [17]. After many floods, the structure of the assembled boulders was still stable. Accordingly, stacked boulders can be installed to improve river environments, refuges, and other aquatic habitats and the migration routes of aquatic animals. Limitations for the application of assembled boulders should be discussed after a systematic investigation in the near future.

Rock weirs, cross vanes, and Syvde-type weirs have been proposed as stable structures for river passage [18–21]. These structures consist of densely packed boulders. According to their design manuals and references, these kinds of boulders are not mechanically assembled to prevent them from being destroyed during large floods.

6. Conclusions

The flow conditions, velocity fields, and stability of artificially assembled boulders, in addition to the possibility for upstream migration routes for aquatic animals, were shown experimentally in response to various changes in discharges. It was found that, at down-ward slopes of 1/8.5, 1/12.5, and 1/25, a transverse gradient about 1/10 in the cross-section created a shallow water flow and a gap flow through the assembled boulders with low time-averaged velocity and low turbulence, regardless of the degree of the downward slope, even if discharges changed under normal conditions. The experimental results showed that a shallow water flow and a gap flow of assembled boulders could allow swimming fish, benthic fish, and crustaceans to migrate upstream. The assembled boulders were constructed in such a way to support each other and resist fluid forces, confirming the stability of the assembled boulders during floods. These findings were confirmed from the field construction. In addition, the upstream migration of swimming fishes (e.g. Plecoglossus altivelis, Tribolodon hakonensis, and Oncorhynchus masou ishika-wae) and benthic fishes (e.g. Rhinogobius flumineus and Cottus pollux) were observed at sites where consecutively assembled boulders were constructed in the Ohmu river. This study supports the practical application of assembled boulder installations for river improvement. The installation of assembled boulders can be applied to improve river environments, aquatic habitats such as refugees and aquatic animal migration routes. Research on consecutively assembled boulders is still in its early stages. Practically, six sites were installed in rivers of various sizes, and biological studies have not been carefully conducted. However, it is true that a variety of aquatic animals were observed at the upstream end of the consecutively assembled boulders. Further biological surveys may be needed. In addition, a protruding assemblage of boulders, a pool-type fish passage with assembled boulders, and ground sills of assembled boulders were constructed, but limits to the application of assembled boulders may need to be examined after a systematic investigation in the future.

Acknowledgements

This study was supported by the Eel Food Culture Promotion Foundation.

Author details

Youichi Yasuda[1*] and Nozomi Fuchino[2]

1 Department of Civil Engineering, College of Science and Technology, Nihon University, Japan

2 Department of Civil Engineering, Graduate School of Science and Technology, Nihon University, Japan

*Address all correspondence to: yasuda.youichi@nihon-u.ac.jp

References

[1] Yasuda Y. Contribution due to river engineering in preservation for diadromous aquatic animals. Aquabiology. 2016;**38**(4):387-396

[2] Marwan AH, Bonnie JS, Dan LH, David SL, Andre EZ, Brett CE. Sediment storage and transport in coarse bed streams: Scale considerations. In: Gravel-Bed Rivers VI: From Process Understanding to River Restoration. Vol. 18. Amsterdam: Elsevier; 2008. pp. 473-497

[3] Strom K, Icolaou AN, Papanicolaou AN. Morphological characterization of cluster microforms. Sedimentology. 2008;**55**:137-153

[4] Church M, Hassan MA, Wolcott JF. Stabilizing self-organized structures in gravel-bed stream channels: Field and ex-perimental observations. Water Resources Research. 1998;**34**(11):3169-3179

[5] Lamarre H, Roy AG. A field experiment on the development of sedimentary structures in a gravel-bed river. Earth. 2008;**33**:1064-1081

[6] Wittenberg L, Newson MD. Particle clusters in gravel-bed rivers: an experimental morphological approach to bed mate-rial transport and stability concepts. Earth Surface Processes and Landforms. 2005;**30**(11):1351-1368

[7] Papanicolaou AN, Strom K, Schuyler A, Talebbeydokhti N. The role of sediment specific gravity and availability on cluster evolution. Earth Surface Processes and Landforms. 2003;**28**(1):69-86

[8] Hassan MA, Reid I. The influence of microform bed roughness elements on flow and sediment transport in gravel bed rivers. Earth Surface Processes and Landforms. 1990;**15**(8):739-750

[9] Strom K, Papanicolaou AN, Evangelopoulos N, Odeh M. Microforms in gravel bed rivers: Formation, disintegration, and effects on bedload transport. Journal of Hydraulic Engineering, ASCE. 2004;**130**(6):554-567

[10] Biggs BJF, Duncan MJ, Francoeur SN, Meyer WD. Physical charactersation of microform bed cluster refugia in 12 headwater streams. New Zealand, Journal of Marine and Freshwater Research. 1997;**31**(4):37-41

[11] Yasuda Y, Masui K. Comparison of energy dissipator between roughness slope with stacked boulders and abrupt drop in low drops. Journal of Japan Society of Civil Engineers, Ser. B1 (Hydraulic Engineering). 2020;**76**(2):I-955-I-960

[12] Yasuda Y, Fuchino N. Hydraulics of roughness mild slope with stacked boulders in half trapezoidal section. Journal of Modern Environmental Science and Engineering, Academic Star Publishing Company. 2021;**7**(12):1099-1108

[13] Rickenmann D, Recking A. Evaluation of flow resistance in gravel-bed rivers through a large field data set. Water Re-sources Research. 2011;**47**:22. W07538

[14] Hey RD. Flow resistance in gravel-bed rivers. Journal of Hydraulic Division-ASCE. 1979;**91**(4):365-379

[15] Kenek Co., Ldt. Propeller Current Meter: Detector Model: VR-301/ VR3T-3-20N VR3T-4-20N. VR3T-2-20K

VR3T-3-20K VR3T-4-20K, product information sheet, https://www.sokken.com/img/120406094031.pdf [Accessed on 6th September, 2021]

[16] Ferguson R. Flow resistance equations for gravel and boulder bed streams. Water Resources Research. 2007;**43, W05427**:12

[17] Yasuda Y. Improvement of flow condition in channelized river due to stacked boulders. In: IOP Conf. Series: Earth and Environ-mental Science 626. Beijing, China: ACEER; 2021. p. 012001. DOI: 10.1088/1755-1315/626/1/012001

[18] U.S. Department of Agriculture Portland. Design of Rock Weirs, Technical Notes, Engineering No.24. Oregon; 2000. p. 10

[19] Rosgen DL. The Cross-Vane, W-Weir and J-Hook Vane Structures...Their Description, Design and Application for Stream Stabilization and River Restoration, Wetlands Engineering & River Restoration 2001. ASCE), Sec.26 Chap.3. Reno, Nevada, United States: Conference 2001,August 27-31; 2001. pp. 1-22

[20] Washington State Department of Transportation: Hydraulics. Manual M 23-03.03. 2010, pp.4_23-25

[21] Brittain JE. Weirs as a mitigation measure in regulated rivers-the norwegian experience. Journal of Canadian Water Re-sources. 2003;**28**(2):217-229

Chapter 4

A Hierarchical Approach to Fish Conservation in Semiarid Landscapes: A Need to Understand Multiscale Environmental Relationships

Robert Mollenhauer, Shannon K. Brewer, Desiree Moore, Dusty Swedberg and Maeghen Wedgeworth

Abstract

A multiscale perspective is essential for conservation planning of riverine fishes. Coarse-scale habitat (e.g., basis) can influence both finer-scale habitat characteristics (e.g., reaches and microhabitat) and associated species distributions. Finer-scale management and habitat rehabilitation efforts can fail without the consideration of coarser-scale constraints. We provide a conceptual hierarchical framework for multiscale fish conservation strategies in the semiarid Great Plains. The Great Plains stream network is highly fragmented due to dam construction, water withdrawals, and increased drought severity. Our framework uses relationships with basin-scale connectivity and streamflow and reach-scale physicochemical characteristics in the context of aiding species reintroduction and stream habitat improvements.

Keywords: drought, drying, arid, multiscale, fish conservation

1. Introduction

The importance of multiscale habitat use by aquatic organisms is well recognized and central to the development of meaningful fisheries conservation actions. The distribution of fishes relies on natural features such as the appropriate climate and geology that comprise the physicochemical characteristics that are typically tolerated by species. These ultimate determinants constrain intermediate and proximate determinants on aquatic organisms (e.g., benthic algae, [1]). For example, the pH of a river is dictated, in part, by the underlying lithology of the region [2], and fishes have specific pH tolerances that regulate a variety of life-history attributes (e.g., egg hatching in salmonids [3]). A combination of other physicochemical factors at finer spatial and temporal scales contribute to a heterogeneous riverscape that shape species assemblages [4, 5], where fish use a set of variables that are assumed to maximize fitness [6] or describe behavioral responses (e.g., changes in cover use, [7]). The habitat needs of fishes are often used as the foundation of conservation and recovery plans [8]. For example, priority use areas can be identified and restoration actions planned. As human pressures on fish populations increase, establishing multiscale relationships is more important than ever for guiding conservation actions.

1.1 Landscape change and anthropogenic pressures on fish populations

Human pressures increase the threats on freshwater ecosystems and taxa. The modification of landscapes from historical land cover to agriculture and urban uses has resulted in significant physicochemical changes and water demands on rivers. Urban rivers, for example, are often associated with flashy hydrographs, increased contaminants, and degraded channel morphology [9] including channel incision and erosion [10]. Agriculture land use is also associated with higher sediment and nutrient loads [11, 12], more homogenous substrates and water depths [13], and bank instability [14, 15]. Pressure on water resources needed for human uses has resulted in rivers being dammed, leveed, and pumped thereby disrupting both flow and sediment regimes [16, 17]. The magnitude, duration, timing, frequency, and rate of change of stream discharge (hereafter flow regime) is considered a primary driver of ecosystem processes [18, 19] and biotic integrity [20, 21]. Flow regimes globally have been altered due to human activities [19, 22, 23]. Current and future changes to our climate and associated weather patterns will only exacerbate threats facing freshwater ecosystems.

North American freshwater fishes are experiencing the highest extinction rates among vertebrates [24]. Flow regime alteration and fragmentation due to dam construction (hereafter damming) are often cited as primary causes [22, 25, 26]. Dams alter fish-assemblage structure [27–29] and prey availability [30–32] by creating streamflow conditions and instream habitats favorable to lentic (i.e., lake and reservoir) species. The changes to flow magnitude, in particular, due to damming, have negatively affected many fishes leading to declines in diversity [22, 24, 26]. Altered flow regimes and fragmentation caused by damming are particularly detrimental to lotic (i.e., river and stream) fishes due to their mobility and requirement of multiple habitat types to complete a life cycle [4]. Dams disrupt spawning cues and block migration routes, prevent access to spawning and nursery habitats, and alter nutrient cycles [26, 27, 32]. Damming has largely ceased in North America, but the long-term effects of modified ecosystems result in numerous stream-fish conservation challenges [22, 33, 34]. Existing levels of flow regime alteration and fragmentation are also exacerbated due to climate change and additional anthropogenic pressures, particularly in arid and semiarid regions.

1.2 Climate change and multiscale fish conservation in arid and semiarid landscapes

Flow-regime alteration and stream fragmentation are expected to increase due to climate change and growing human water demands. The combination of extended periods of drought and increased human water demands magnify threats to the long-term persistence of many stream fishes [35–37]. Native fish species in arid and semiarid ecoregions are particularly vulnerable because they have both the highest level of damming [33, 38] and naturally harsh environmental conditions [39–41]. Fishes native to arid and semiarid streams have evolved to tolerate intermittent drying, flashy changes is flow, and extreme physicochemical conditions [38, 41, 42]. However, interactions of dams, loss of base flows due to water withdrawals, and increased drought have intensified already harsh environments [37, 38, 43]. Identifying multiscale stream-fish relationships with fragmentation and flow regimes is essential to developing successful conservation strategies in arid and semiarid ecoregions. Understanding coarse-scale constraints on species distributions is particularly important to guiding finer-scale conservation and recovery efforts. For example, the stream reach is a natural scale for fish conservation and management activities, but effective strategies are dependent on basin-scale relationships [5, 44].

1.3 Great Plains small-bodied minnows

Streams of the semiarid North American Great Plains ecoregion (hereafter Great Plains) are unique, dynamic ecosystems and home to endemic fish species. The Great Plains is one of the most impounded areas of the world [33, 38, 45]. Damming largely ceased in the 1980s, but resulting fragmentation is extensive and flow regimes remain severely altered [33]. Substrate in Great Plains streams is predominately sand and silt, with constantly changing streambed formations [44]. Channel characteristics differ from gravel-bed streams with stable riffle-run-pool formations except during high flows. Environmental conditions are extreme with periods of long drought followed by large flooding and highly variable and wide-ranging water temperatures and salinities [42, 46]. Natural stream drying has been exacerbated by harsher drought periods due to climate change and excessive groundwater pumping [37, 42, 43]. Small-bodied fishes are common in the Great Plains, with brackish (i.e., salt-tolerant) species dominating areas of higher salinity [46, 47]. True minnows (Leuciscidae), particularly smaller-bodied species, also occur in varying abundances throughout the ecoregion. True minnows are a large, diverse family of fishes (~700 species) within the order Cypriniformes [48, 49]. In particular, true minnows display a wide range of life-history traits. This includes a group of species, some endemic to the Great Plains, that have a unique reproductive strategy in stream ecosystems.

Pelagic-broadcast spawning [50] is a common reproductive strategy globally in marine and coastal species, but rare in inland freshwater systems. In inland streams of North America, pelagic-broadcast spawning is restricted to mooneyes (Hiodontidae) and three genera of true minnows (*Hybognathus*, *Notropis*, and *Macrhybopsis*, hereafter pelagophils) that occur in the Great Plains [51]. Pelagophils typically spawn in higher-order streams and release transparent, non-adhesive ova that are semi-buoyant [47]. The downstream displacement of eggs and larvae relies on drift [52]. Thus, both minimal fragmentation and higher flow magnitude are essential to successful recruitment [53–55]. There are similarities between pelagophils and marine pelagic-broadcast spawners in juvenile dispersal strategies to microhabitats with high nutrient concentrations and reduced predation pressure [51, 56]. High-flow events increase nutrient loads and create disconnected temporary slackwater habitats that serve as nurseries for juvenile pelagophils. The pelagophil life cycle is completed through extensive upstream movement by juveniles and adults [53, 57, 58]. Great Plains pelagophils have been strongly negatively affected by damming, water withdrawals, and climate change due to disrupted stream networks, altered flow patterns with reduced magnitude, and loss of floodplain habitats [42, 52, 59]. Numerous studies have reported declines in pelagophil relative abundance and range reductions in the Great Plains (e.g., [43, 54, 60–62]).

2. Modeling at different spatial scales

We modeled occurrence and relative abundance of true minnows (hereafter minnows) in the upper Red River basin. Species occurrence (i.e., distributions) and abundance (i.e., population size) are fundamental ecological state variables used both in research studies and for conservation and management problems [63–65]. Both fish distributions and population sizes may be constrained by coarser-scale characteristics [1, 5, 44]. One analytical approach is to model variation in occurrence or abundance as a function of environmental variables at multiple scales (e.g., [66–68]). However, state variables are quantified differently (e.g., a binary response for occurrence and integers for abundance); thus, it is typically not possible to model multiple states in a single analysis. Different state variables also more naturally align with certain spatial scales. For example, abundance is often not ecologically meaningful at very coarse scales (e.g., basins) and measuring and managing population size at these scales is not realistic. Therefore, occurrence is typically examined at the basin scale, where the stream reach (i.e., a series of representative habitat complexes nested in tributary complexes) is the natural scale for studies and management of abundance [5, 44]. An alternative multiscale analytical approach (our approach here) is to model variation in occurrence and abundance separately at relevant spatial scales.

2.1 Study area

Our study area was in the upper Red River basin of the Great Plains. The area comprised portions of the Central Great Plains and Southwestern Tablelands sub-ecoregions of Oklahoma and Texas (**Figure 1**). The eastern boundary with the Cross Timbers sub-ecoregion corresponded with a transition from sand-bed to gravel-bed streams with lower salinity, increased vegetation, and fish species hybridization zones [69–71]. The western boundary corresponded with the higher-elevation, more-arid High Plains sub-ecoregion. The Central Great Plains is characterized by mixed-grass prairie vegetation, cropland, and landforms that include sand dunes, low mountains, and salt flats [70]. The Southwestern Tablelands has a more rugged terrain, with dissected plain, hill, and canyon landforms, sparse short grass prairie vegetation, and less cropland. Annual precipitation in the study area, though highly variable, increases from east to west eastward (mean rainfall 56–97 cm). In addition to minnows, the brackish plains killifish *Fundulus zebrinus* and Red River pupfish *Cyprinodon fluviatilis* are also common in the study area.

2.2 Small-bodied minnow occurrence among hydraulic response units

We modeled occurrence probability of nine minnows species among hydraulic response units (HRUs) nested in the upper Red River basin (**Figure 1**). HRUs are sub-basins that represent 10-digit hydrologic units with refined boundaries for flow modeling based on local characteristics [72, 73]. The focal species included four pelagophils (emerald shiner *Notropis atherinoides*, plains minnow *Hybognathus placitus*, prairie chub *Macrhybopsis australis*, and Red River shiner *N. bairdi*, [52]), bullhead minnow *Pimephales vigilax*, fathead minnow *P. promelas,* red shiner *Cyprinella lutrensis*, sand shiner *N. stramineus*, and suckermouth minnow *Phenacobius mirabilis*. Most species occur elsewhere in North America east of the Rocky Mountains, with emerald shiner, fathead minnow, and red shiner widely distributed (www.iucnredlist.org).

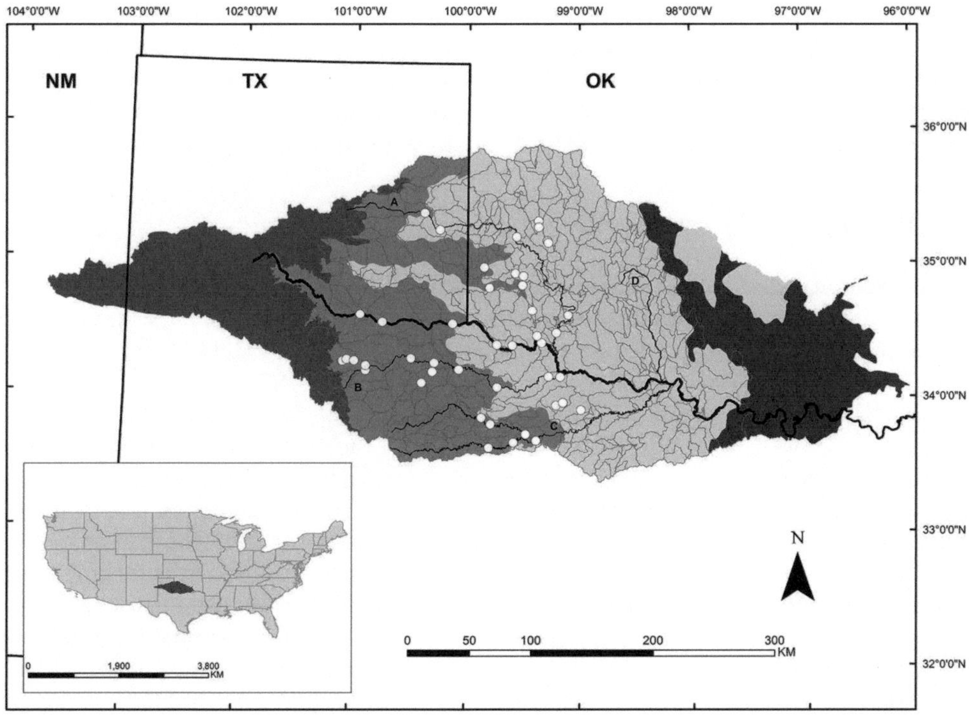

Figure 1.
Study area in the upper Red River basin. The shading in the upper panel denotes level-three sub-ecoregion boundaries (from west to east: High Plains, Southwestern Tablelands, Central Great Plains, and Cross Timbers). Inner borders on polygons show the delineation of hydrologic response units included in the occurrence modeling (Section 2.2). White circles are stream reaches surveyed for prairie chub in 2019 (Section 2.3). The polylines show the Red River mainstem (thicker line) and select major tributaries: A is the North Fork, B is the Pease River, C is the Wichita River, and D is Cache Creek.

Red River shiner is endemic to most of the Red River basin and introduced to other basins of Arkansas, Oklahoma, and Texas [61]. Prairie chub, a species of conservation concern (see Section 2.3), is restricted to the upper Red River basin [69, 71]. We did not include minnows that only occur near the ecotone with the Cross Timbers.

Our study period was 2002–2015. The temporal range encompassed a relatively dry climatic period (2002–2014) and one year of heavy rainfall (**Figure 3** in [74]). We assumed static species occurrence states (i.e., no colonization or extirpations of HRUs) across the study period and at least a one-year lag time for any changes in minnow occurrence states following the end of the dry period.

2.2.1 Fish surveys

We compiled stream-fish surveys from state conservation and management agencies, data repositories, and online databases (Appendix 1, **Table A1**). For online databases, we used the terms "fish" and "fishes" to search all Oklahoma and Texas counties within the study area from 2002–2015. Data were processed to remove duplicate surveys. Each unique survey was spatially referenced to an HRU based on the latitude and longitude. We compiled capture histories for each species (i.e., one for detection and zero for nondetection) at each HRU. Repeat surveys at HRUs were treated as spatial replicates with replacement [76]. We also compiled the date, sampling gear (reported in 92% of surveys), and collector (e.g., agency or individual, Appendix 1, **Table A2**). We assumed seining for surveys that did not report the gear type because it is the most-common stream-fish sampling method [77] and comprised the majority of reported surveys (64%). The additional gear types were backpack electrofishing (18% of surveys) and boat or barge electrofishing (9% of surveys).

2.2.2 Flow regime and fragmentation metrics

We characterized the flow regime of each HRU using mean daily discharge estimates at the outlet. The discharge estimates were obtained from a precipitation-runoff modeling system [72] adapted from the National Hydrologic Model [78] for the Red River basin [79, 80]. We calculated flow regime metrics [81, 82] using EflowStats (version 5.0.1, median option, [83]). Due to the size of the dataset and inherent correlations among metrics, we limited the set to five with each metric representing one flow regime component (**Table 1**). The selected flow metrics were based on expected ecological relationships with minnow occurrence, particularly pelagophils (see Section 1.3), and maintaining the absolute value of Pearson's pairwise correlation coefficient $r < 0.5$.

We used the upstream network density of major dams (UNDR, #/100 rkm) at the outlet of each HRU obtained from an online database [85] to characterize fragmentation. We used UNDR because it was highly correlated with numerous other fragmentation metrics and upstream dams have been shown to be strongly associated with pelagophil distributions in the upper Red River basin [56]. We natural log transformed UNDR prior to modeling due to a right-skewed distribution. Correlation levels were reasonable between UNDR and flow regime metrics ($|r|<0.30$).

2.2.3 Occupancy modeling methods

We modeled minnow occurrence relationships, while accounting for variable detection probability, using the hierarchical framework described by [86]. The latent occurrence state for minnow i at HRU j was treated as partially observed, with $z_{ij} = 1$ if the species was truly present and $z_{ij} = 0$ if the species was truly absent. Each z_{ij} followed a Bernoulli distribution with occurrence probability Ψ:

$$z_{ij} \sim \text{Bernoulli}\left(\Psi_{ij}\right) \quad (1)$$

Metric	Description
Low flow duration variability (DL17, %)	CV of annual Q below 25th PCTL
†Flood pulse count (FH3, # of d/yr)	Median of annual # of days Q is above 3 * median daily Q
†Median daily flow (MA2, m^3/s)	Median daily Q for entire flow record
†Variability of reversals (RA9, %)	CV of flow reversals
Annual maxima variability (TH2, %)	CV of Julian day of annual maximum Q

†natural-log transformed prior to analysis due to a right-skewed distribution.
The alphanumeric codes for metrics correspond to detailed descriptions in Appendix 7 of Kennen et al. [84].
Q: stream discharge.
PCTL: percentile.

Table 1.
Flow-regime metrics used in the species occurrence model.

The detection of species i at HRU j for survey k was conditional on both the true occurrence state and detection probability p (the probability of detecting a species in a single survey if present), where y_{ijk} followed a Bernoulli distribution:

$$y_{ijk} \sim \text{Bernoulli}\left(z_{ij}{}^{*}\ p_{ijk}\right) \tag{2}$$

We modeled variation in Ψ and p as a function of covariates [86]. Detection covariates comprised HRU surface area (km^2, hereafter area) and drainage area. Spatially replicated surveys can result in a violation of the closed-system assumption for occupancy modeling because a species may occur at a site, but not be locally present at the time of the survey [76]. Thus, we used area to account for variation in p associated with patchier species distributions in larger HRUs. Drainage area characterized the stream order of the mainstem for each HRU to account for variation in p associated with species abundance. We natural-log transformed both detection covariates due to right-skewed distributions. Detection relationships with covariates were allowed to vary by species as deflections around the group mean hyperparameter governed by a probability distribution [64, 87, 88]. More common minnows may have inherently higher detection probability. Thus, we modeled the correlation (ρ) between species occurrence probability intercepts α_i and species detection probability intercepts υ_i. The intercepts were jointly distributed as $[\alpha_i , \upsilon_i \mid \Sigma] \sim N(0, \Sigma)$, where Σ is a 2 x 2 matrix comprising variance components $\boldsymbol{\sigma}_{\alpha}^2$ and $\boldsymbol{\sigma}_{\upsilon}^2$ and covariance $\sigma_{\alpha\sigma}$ [88, 89]. We also allowed each species detection intercept to vary by both sampling gear type g (1, seining, 2,backpack electrofishing, 3, boat and barge electrofishing) and collector c (1–6, Appendix 1, **Table A2**) using a grouping factor [90, 91]. The detection component of the occupancy model can be written as:

$$\text{logit}\left(p_{ijkgc}\right) = \upsilon_i + \beta_{1i}X_{1jk} + \beta_{2i}X_{2jk} + \gamma_{ig} + \tau_{ic}, \text{for}\, i = 9, \text{for}\, j = 97, \text{for}\, k = 1,2\ldots K, \tag{3}$$

$$\beta_{ni} \sim \textit{Gaussian}\ (\mu_{\beta n}, \sigma_{\beta n}^2),\ \text{for}\ n = 2$$
$$\gamma_{ig} \sim \textit{Gaussian}\ (0, \sigma_{\gamma}^2),\ \text{for}\ g = 3$$
$$\tau_{ic} \sim \textit{Gaussian}\ (0, \sigma_{\tau}^2),\ \text{for c} = 6$$

where υ is the detection probability intercept, β_1 and β_2 are slopes for associated detection covariates area X_1 and drainage area X_2, γ is the gear type grouping factor, τ is the collector grouping factor, and μ is species group mean. Occurrence covariates comprised UNDR and five flow regime metrics (see Section 2.2.3). We allowed each occurrence covariate to vary by species using the same model structure described for the detection component of the model. The occurrence component of the occupancy model can be written as:

$$\text{logit}(\Psi_{ij}) = \alpha_i + \beta_{ni}X_{nj},\ \text{for}\, i = 9, \text{for}\, j = 97, \text{for}\, n = 6, \beta_{ni} \sim \textit{Gaussian}\left(\mu_{\beta n}, \boldsymbol{\sigma}_{\beta n}^2\right), \tag{4}$$

where α is the detection probability intercept and β_1–β_6 are slopes for associated occurrence covariates X_1–X_6. All detection and occurrence covariates were standardized to a mean of zero and standard deviation (SD) of one. Model coefficients are reported as the mode (most likely value) with a 90% highest density interval (HDI, [92]). Model specifications, diagnostics, and fit tests are provided as supplemental information (Appendix 2).

2.2.4 Occupancy modeling results

There were varying detection relationships among minnows. Detection probability across all minnows (i.e., the group mean) at mean levels of covariates was 0.47 (90% HDI: 0.31, 0.64). Detection probability was lower in larger HRUs for all minnows, with the strength of the relationship similar to the group mean (Appendix 1, **Table A3**). The strength of the detection relationship with drainage area was higher than the group mean for emerald shiner, plains minnow, prairie chub, and Red River shiner and lower for the remaining five minnows. There was more unexplained variation in detection probability attributed to collector (SD = 0.84) than gear type (SD = 0.25). Detection probability and occurrence probability intercepts were moderately positively correlated (ρ = 0.63).

The direction of occurrence relationships with flow-regime characteristics and fragmentation varied both among all minnows and within pelagophils (emerald shiner, plains minnow, prairie chub, and Red River shiner). Occurrence probability across all minnows at mean levels of covariates was 0.76 (90% HDI: 0.60, 0.89).

Occurrence probability for all pelagophils increased sharply with increasing daily streamflow magnitude (MA2, **Table 2** and **Figure 2a**). There was a weak positive occurrence relationship with MA2 for bullhead minnow and suckermouth minnow, a weak negative relationship for sand shiner, and no relationship for fathead minnow

Species	Intercept	UNDR	DL17	FH3	MA2	RA9	TH2
Mean	1.1 (0.4, 1.8)	−1.1 (−3.1, 0.9)	0.5 (0.2, 0.9)	0.4 (0.1, 0.9)	0.8 (−0.1, 1.9)	−1.0 (−1.6, −0.4)	0.3 (−0.4, 0.8)
BUM	2.0 (1.3, 2.8)	1.3 (0.4, 2.2)	0.5 (0.1, 1.0)	0.5 (0.1, 1.0)	0.2 (−0.4, 0.7)	−1.4 (−2.4, −0.6)	0.3 (−0.2, 0.9)
EMS	0.3 (−0.8, 1.4)	−0.1 (−1.9, 1.9)	0.6 (0.1, 1.2)	0.4 (−0.2, 1.0)	1.6 (0.5, 2.7)	−1.5 (−2.9, −0.5)	1.2 (0.3, 2.8)
FAM	0.5 (−0.1, 1.1)	0.4 (−0.2, 0.9)	0.6 (0.2, 1.1)	0.3 (−0.2, 0.8)	0.0 (−0.5, 0.5)	−1.0 (−1.8, −0.3)	0.3 (−0.2, 0.8)
PLM	1.7 (0.8, 3.0)	−4.6 (−7.7, −2.0)	0.6 (−0.1, 1.5)	0.5 (−0.1, 1.1)	1.4 (0.3, 2.6)	−0.6 (−1.4, 0.3)	0.0 (−1.1, 0.8)
PRC	0.8 (−0.4, 2.2)	−2.5 (−5.3, −0.3)	0.5 (−0.2, 1.2)	0.4 (−0.2, 1.1)	2.2 (0.3, 4.2)	−0.8 (−1.6, 0.2)	−0.2 (−1.7, 0.8)
RRS	0.5 (−0.9, 1.8)	−4.7 (−9.3, −1.0)	0.5 (−0.2, 1.2)	0.4 (−0.3, 1.0)	2.4 (0.5, 4.7)	−0.4 (−1.3, 0.7)	−0.2 (−1.9, 0.7)
RES	2.6 (1.9, 3.6)	−0.1 (−0.9, 0.6)	0.4 (−0.1, 0.9)	0.5 (0.1, 1.0)	−0.0 (−0.6, 0.6)	−1.0 (−1.6, −0.4)	0.5 (−0.1, 1.0)
SAS	0.4 (−0.3, 1.1)	−0.7 (−1.4, −0.1)	0.5 (0.1, 1.0)	0.3 (−0.2, 0.8)	−0.6 (−1.4, 0.2)	−0.8 (−1.7, −0.1)	0.3 (−0.3, 0.8)
SUM	1.5 (0.7, 2.4)	1.2 (0.2, 2.2)	0.6 (0.2, 1.2)	0.5 (0.2, 1.2)	0.4 (−0.4, 1.2)	−1.0 (−2.0, −0.2)	0.7 (0.2, 1.4)

BUM, bullhead minnow; EMS, emerald shiner; FAM, fathead minnow; PLM, plains minnow; PRC, prairie chub; RRS, Red River shiner; RES, red shiner; SAS, sand shiner; SUM, Suckermouth minnow; UNDR, upstream network dam density; DL17, low flow duration variability; FH3, flood pulse count; MA2, median daily flow; RA9, variability of reversals; and TH2, annual maxima variability.

Table 2.
Minnow occurrence model coefficients reported on the logit scale as the mode with associated 90% highest density interval (HDI) from the posterior distribution. The intercept is interpreted as estimated occurrence probability at mean levels of covariates. Each covariate coefficient is interpreted with others held constant.

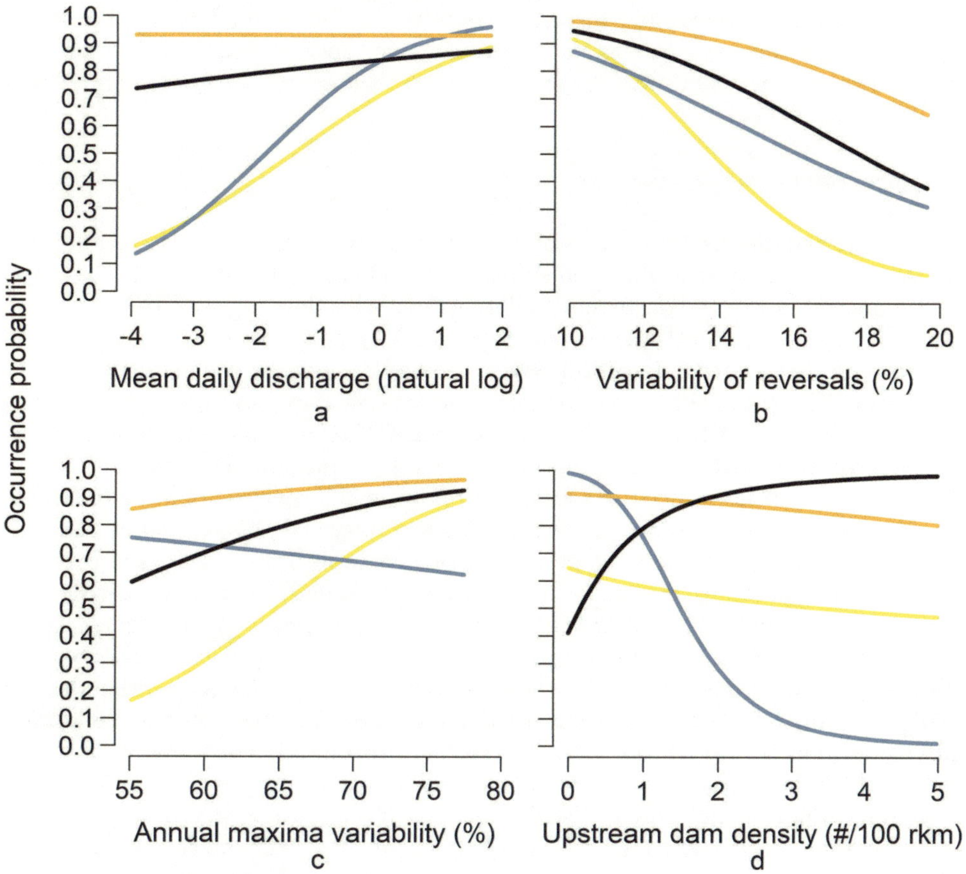

Figure 2.
Line graphs showing relationships between occurrence probability and daily streamflow magnitude (MA2, panel a), variability in reversals of flow (RA9, panel b), variability in timing of annual maximum streamflow (TH2, panel c), and upstream dam density (UNDR, panel d) for four small-bodied minnows in hydraulic response units of the upper Red River basin. The yellow lines represent emerald shiner, the bluelines represent prairie chub, the orange lines represent red shiner, and the black lines represent suckermouth minnow. Mean daily discharge was measured as m^3/s.

and red shiner. Occurrence probability decreased with increasing variability of reversals (RA9) for all minnows, including the cosmopolitan red shiner (**Table 2** and **Figure 2b**). Emerald shiner occurrence probability also increased sharply with increasing variability in annual maxima (TH2, **Table 2** and **Figure 2c**). The positive occurrence relationship with TH2 was weaker for bullhead minnow, fathead minnow, red shiner, sand shiner, and suckermouth minnow. Prairie chub and Red River shiner had a weak negative occurrence relationship with TH2, while plains minnow had no relationship. The group mean and all minnow occurrence relationships were similar with low flow duration variability (DL17) and flood pulse count (FH3). Occurrence probability increased with both increasing DL17 and FH3 (**Table 2**). Plains minnow, prairie chub, and Red River shiner occurrence probability decreased sharply with increasing upstream dam density (UNDR, **Table 2, Figure 2d**). However, there was no occurrence relationship with UNDR for emerald shiner. Sand shiner occurrence was also negatively associated with UNDR, but the strength of the relationship was weaker than that for the three pelagophils. Suckermouth minnow and bullhead minnow occurrence probability increased sharply with increasing UNDR (**Figure 2d**). Fathead minnow occurrence was also positively associated with UNDR, but the strength of the relationship was weaker. There was no occurrence relationship with UNDR for red shiner.

2.2.5 Projected minnow distributions

We projected distributions adjusted for detection probability across the study area for emerald shiner, prairie chub, red shiner, and suckermouth minnow. These minnows represented varying occurrence relationships with flow regimes and fragmentation among focal species (see Section 2.2.4). A species was considered present at all HRUs with a detection. We calculated occurrence probabilities for each species at HRUs where either there were no surveys, or they were not detected using occurrence model coefficients and covariate values (**Table 2**). We emphasize that a high occurrence probability more appropriately represents suitable conditions for occurrence, not an assurance the species is present, and a species might be present at an HRU with a low occurrence probability. Opportunity (i.e., biogeography) and other spatial and environmental factors (i.e., biogeography) not considered here also play a role in aquatic species distributions [1, 5, 93] and, like any modeled relationship, there is inherent error. This is particularly true for HRUs closer to western ecotone with the desert-like High Plains (**Figure 1**). Nevertheless, the projected distributions reflect underlying ecological relationships based on where species were either detected during sampling or likely present but not detected. Further, these modeled relationships can provide an initial step to guide species reintroduction or translocation efforts, which do not depend on present occurrence state (also see Section 3.1).

All four minnows had similar distributions based on naïve occurrence (i.e., where the species was detected) and high occurrence probability along the downstream portion of the Wichita River; however, pelagophils were less likely to occur in HRUs elsewhere (**Figure 3**). As expected, red shiner had a high occurrence probability throughout the study area (**Figure 3a**). The lower red shiner occurrence probability in the southwest portion of the study area reflects the negative relationship with higher variability in flow reversals shared with all upper Red River minnows. Suckermouth minnow had a high occurrence probability along the upstream portion Red River mainstem and in the northern portion of the study area (**Figure 3b**). The higher suckermouth minnow occurrence probability corresponds to HRUs with more dams upstream. Suckermouth minnow occurrence probability was low in the southern portion of the study area with fewer upstream dams. Conversely, prairie chub had a low occurrence probability in the northern portion of the study area and along the upstream mainstem (**Figure 3c**). Prairie chub occurrence probability was higher than both emerald shiner and suckermouth minnow in the southern portion of the study area, particularly along the upper Wichita River. Although emerald shiner is more widespread overall than prairie chub, its projected distribution was narrower in the upper Red River (**Figure 3d**). The higher emerald shiner occurrence probability in the northern portion of the study area and along the upper mainstem is reflective of not sharing the negative upstream dam relationship with other pelagophils.

2.3 Prairie chub relative abundance at the stream reach scale

Historically, the endemic prairie chub was abundant in the upper Red River mainstem and its higher-order tributaries [52]. Suspected population declines and

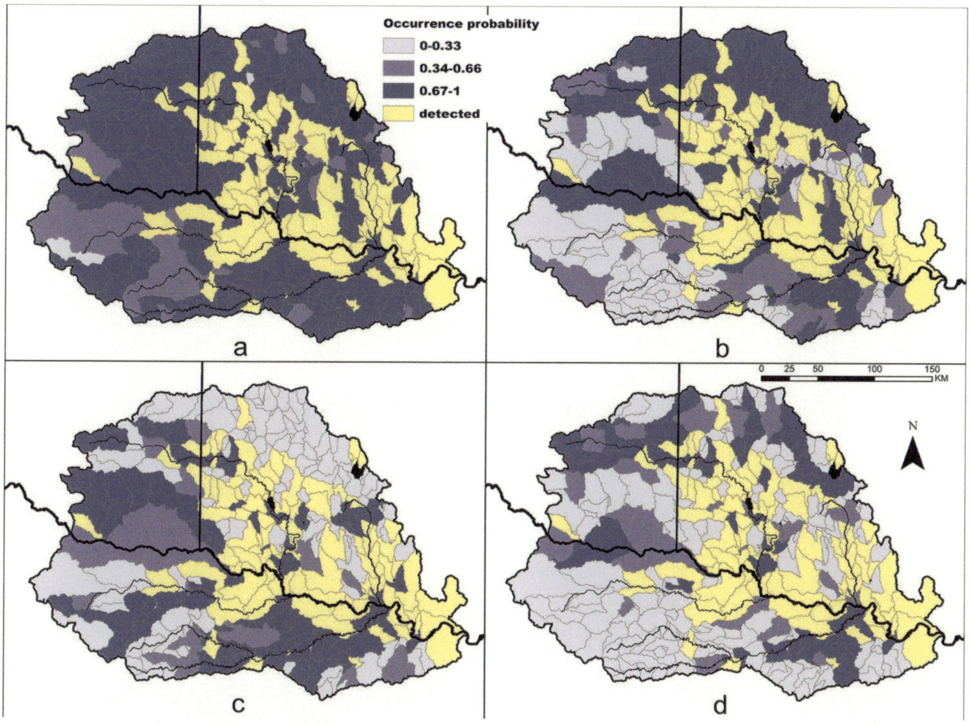

Figure 3.
Predicted species distribution maps for hydraulic response units (HRUs) in the Cross Timbers and Southwestern Tablelands level-three sub-ecoregions in the upper Red River basin (Section 2.1, ***Figure 1****). Occurrence probabilities were estimated using modeled relationships with flow regime metrics and upstream dam density (Section 2.3). Panel a is red shiner, panel b is suckermouth minnow, panel c is prairie chub, and panel d is emerald shiner. Species were assumed to not occur at HRUs filled with black, which are completely under reservoirs.*

poorly understood ecology has resulted in conservation concerns for prairie chub in multiple states [94, 95]. At the federal level, prairie chub is currently threatened and included as a potential endangered species on the 2021–2025 National Domestic Listing Workplan [96].

We show relationships between reach-scale adult prairie chub counts and the predictor variables longitude, stream discharge, and salinity. Prairie chub populations were surveyed at 44 stream reaches of the upper Red River basin in early autumn 2019 (**Figure 1**). A reach was defined as a longitudinal distance of twenty times mean wetted width constrained by a minimum of 100 m and a maximum of 500 m. Adult prairie chub counts were obtained using multi-pass removal sampling with a seine. Sampling occurred west of the overlap and hybridization zone with the morphologically similar shoal chub *Macrhybopsis hyostoma* [71] to minimize misidentified prairie chub (see [97] for a detailed description of sampling methods). Capture probability was fairly constant, which allowed for relative abundance comparisons among reaches. We present relationships as descriptive scatterplots of adult prairie chub counts versus each predictor variable; thus, relationships are independent and not additive. Eastern longitude and stream discharge were more highly correlated (r = 0.40) than salinity and longitude (r = 0.08) and salinity and discharge (r = −0.02).

Adult prairie chub counts of zero were most common (n = 32 reaches), with a count of only one at two reaches (**Figure 4**). All counts >1 were in the eastern half of

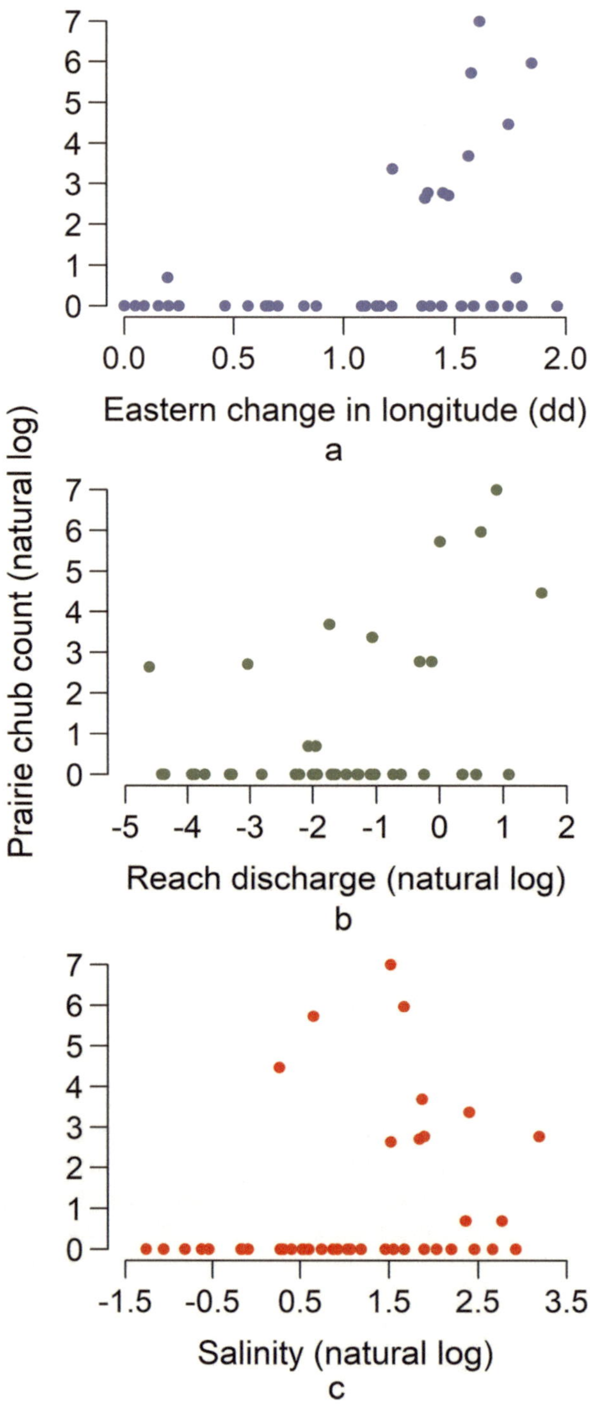

Figure 4.
Scatterplots depicting relationships between reach-scale adult prairie chub counts and longitude (panel a), stream discharge (panel b), and salinity (panel c). A constant of one was added to each count prior to natural-log transforming. Reach discharge was measured as m^3/s and salinity was measured as PPT.

the study area, with the highest counts furthest east (**Figure 4a**). There was a general positive linear trend in relative abundance with increasing reach discharge for counts >1 (**Figure 4b**). Counts >1 also increased with increasing salinity (**Figure 4c**). However, the highest counts were associated with intermediate salinity (~2–6 PPT), which suggested a quadratic relationship.

3. Conservation implications

Our study highlights three important fish conservation aspects in a river basin: (1) the consideration of multiple spatial scales for directing conservation, (2) the tradeoffs of assemblage level (i.e., multiple species) conservation, and (3) the implications of ignoring detection error. Although beyond the scope here, the minnow occurrence relationships can also be used for predictive simulations under different flow regime and fragmentation scenarios. For example, changes in species distributions could be predicted under different levels of dam removal or long-term changes in flow magnitude with increased drought (or both).

3.1 Multiscale fish conservation strategies

The occurrence and relative abundance relationships for prairie chub can be used to identify target areas in the stream network with a higher chance of habitat restoration or reintroduction success. Spatial position has been shown to be strongly associated with the structure of fish populations and assemblages [5, 98, 99]. Prairie chub's distribution in the upper Red River basin is severely constrained by upstream dams (**Figure 2d**). There is essentially no probability of prairie chub occurring below heavily dammed HRUs, presumably due to connectivity requirements for pelagophil reproduction (see Section 1.3). Thus, finer-scale conservation actions (e.g., reintroduction or instream habitat enhancements) in these HRUs would be futile and waste available resources. The most favorable HRUs in the upper Red River stream network for prairie chub occurrence are along the mainstem or higher-order tributaries (i.e., higher long-term flow magnitude), with low upstream dam density and more constancy in rate of change. In particular, HRUs along the Wichita River without prairie chub detections had a high occurrence probability (**Figures 1** and **3c**). If feasible, increasing flow in HRUs with lower dam density could increase the range of the favorable area. Spatial position was also associated with reach-scale prairie chub relative abundance. All high adult counts were associated with reaches in the eastern portion of the study area and higher discharge (**Figure 4a** and **c**). Longitude and discharge were somewhat confounded, and we did not consider the effect of each with the other held constant (see Section 2.3). The number of low-flow days increases further east in the upper Red River basin [75]. However, stream discharge was variable at both the HRU and reach scale. Thus, management actions targeting prairie chub would likely be most effective at reaches in high occurrence probability HRUs in the eastern portion of the study area with higher average flow magnitude. Higher prairie chub relative abundance was also associated with intermediate salinity levels (**Figure 4c**). Salinity has been shown to be strongly associated with fish assemblage structure in Great Plains streams [100], and a quadratic relationship with population size makes sense ecologically for a freshwater species adapted to semiarid streams. However, salinity is highly variable across both space and time in the upper Red River basin [101]. There is also a salinity gradient at the ecotone with the Cross Timbers that constrains prairie chub's eastern distribution and forms a hybrid zone with shoal chub *Macrhybopsis hyostoma* [71]. Thus, to effectively implement a salinity target for prairie chub conservation, improved salinity monitoring would be needed. The findings for our study period are likely reflective of multiscale prairie chub ecological relationships in both wet and dry periods. Occurrence probability among HRUs has been shown to be similar in both wet and dry periods [56], and reach-scale adult counts were collected during a relatively wet period [97].

3.2 Assemblage-level fish conservation

The mixed occurrence relationships among minnows with flow regime and fragmentation have implications for upper Red River basin conservation strategies. It is important that managers consider conservation actions that benefit target species without detrimental effects to other native fishes in the assemblage. Reducing variability in annual maxima timing and removing upstream dams might be beneficial for prairie chub. However, emerald shiner and suckermouth minnow have high occurrence probabilities in the northern portion of the study area where these changes might take place. Unless the mechanism(s) driving the distributions of emerald shiner and suckermouth minnow is better understood, it is possible for conservation actions designed to improve conditions for prairie chub to incidentally harm other species. Fragmentation might prevent prairie chub from successful upstream movement that is important for completing the pelagophil life cycle [58, 102]. However, emerald shiner and suckermouth minnow are more widespread and might be able to adapt to a wider variety of conditions including fragmented river systems (e.g., phenotypic plasticity, [103, 104]). Although prairie chub and emerald shiner are both pelagophils, there is evidence that emerald shiner is less sensitive to flow disturbances than some other pelagophil species [105] and portions of some pelagophil populations are residents that do not make upstream spawning movements [106, 107]. Emerald shiner is also adapted to lentic environments [108] and may benefit competitively in habitats near reservoirs. Because suckermouth minnow is not a pelagophil species, it might not require long unimpeded lengths of river for spawning. It is also possible that another unmeasured or confounding habitat metric is the driver of the emerald shiner and suckermouth minnow distributions. It is prudent to balance conservation efforts to benefit target species while maintaining habitat for other natives. For example, strategic dam removal in the northern portion of the upper Red River basin could benefit prairie chub while preserving habitat favorable to emerald shiner and suckermouth minnow. Also, consideration of flow-regime patterns that benefit numerous minnows (e.g., increased flow constancy) or may not affect other species (e.g., increasing flow magnitude) provide a balanced assemblage-level conservation approach.

3.3 Imperfect and variable detection

Species occurrence is never perfectly observed (i.e., detection is imperfect), and detection probability varies differently among species and sampling methods across sampling conditions [86, 109, 110]. Thus, ignoring detection error results in only naïve occurrence and apparent species distributions (i.e., true distributions are always larger than observed). In addition to underestimating true occurrence, a high detection probability can be misinterpreted as high habitat suitability and lead to misinformed conservation strategies [110–112]. Species-specific fish detection probability varies in relation to numerous environmental characteristics (e.g., water depth, water temperature, conductivity, water clarity, and flow [113, 114]). We show ignoring detection probability at the HRU scale resulted in similar apparent distributions for all minnows in the upper Red River basin (**Figure 3**). There is, of course, uncertainty in the predicted distributions (also see Section 2.2.50). Nevertheless, adjusting for detection error resulted in a clearer picture of true distributions and underlying ecological relationships. Detection probability has also been shown to vary among species and across sampling conditions at finer scales in Great Plains streams [115, 116]. Given the highly variable nature of the Great Plains stream environment, accounting for detection error in fish species distribution studies at all spatial scales is particularly important for sound river basin conservation.

Acknowledgements

This research is a contribution of the Oklahoma Cooperative Fish and Wildlife Research Unit (U.S. Geological Survey, Oklahoma Department of Wildlife Conservation, Oklahoma State University, and Wildlife Management Institute cooperating). Funding was provided by the Oklahoma Department of Wildlife Conservation (F18AF00575) and the U.S. Geological Survey, Water Smart Program. This study was performed under the auspices of Oklahoma State University protocol # AG-19-11. We thank the numerous scientists for their sampling and data collection efforts that made this study possible. We also thank Glena Osban for technical assistance. Any use of trade, firm, or product names is for descriptive purposes only and does not imply endorsement by the U.S. Government. The Red River fish survey dataset compiled by the authors is available at https://doi.org/10.22488/okstate.22.000006.

Appendix 1: supplemental tables

Data source
Fishes of Texas (www.fishesoftexas.org/home/)
iDigBio (www.idigbio.org/)
MARIS (www.sciencebase.gov/catalog/item/51c45ef1e4b03c77dce65a84)
Oklahoma Conservation Commission (www.ok.gov/conservation/)[1]
Oklahoma Museum of Natural History (www.samnoblemuseum.ou.edu/)
Oklahoma Water Resources Board (www.owrb.ok.gov/)[2]
Perkin et al. [75][3]
Texas State University[4]
VertNet (www.vertnet.org/index.html)

Footnotes denote contact(s) or source for datasets not available online.
[1]*Cheryl Cheadle (cheryl.cheadle@conservation.ok.gov) and Jason Ramming (jason.ramming@conservation.ok.gov).*
[2]*Chris Adams (chris.adams@owrb.ok.gov).*
[3]*DOI:10.1890/14-0121.1.*
[4]*David Ruppel (dsruppel@txstate.edu) and Tim Bonner (tbonner@txstate.edu).*

Table A1.
Data sources for stream-fish assemblage surveys compiled from 1983 to 2015.

ID	Collector	Proportion of surveys
1	Oklahoma Conservation Commission	0.55
2	Oklahoma Water Resources Board	0.15
3	Oklahoma Department of Environmental Quality	0.08
4	Perkin et al.	0.06
5	Texas Tech University	0.08
6	Miscellaneous	0.08

Surveys were pooled by predominant collectors (i.e., ≥ 10 surveys, also see ***Table A1****).*

Table A2.
Collector descriptions used for the grouping factor in the detection model and the proportion of surveys.

Species	Intercept	HRU area	Drainage area
Mean	−0.1 (−0.7, 0.5)	−0.3 (−0.5, −0.1)	0.1 (-0.4, 0.5)
BUM	0.9 (0.2, 1.6)	−0.3 (−0.6, −0.1)	−0.2 (−0.5, 0.1)
EMS	−0.8 (−1.5, −0.1)	−0.4 (−0.6, −0.1)	0.6 (0.2, 1.0)
FAM	−0.1 (−0.8, 0.6)	−0.2 (−0.5, 0.1)	−0.3 (−0.7, 0.1)
PLM	−0.0 (−0.7, 0.7)	−0.3 (−0.5, −0.1)	0.4 (0.1, 0.7)
PRC	−1.2 (−2.0, −0.4)	−0.4 (−0.7, −0.2)	1.0 (0.6, 1.4)
RRS	−0.8 (−1.6, −0.1)	−0.3 −0.6, −0.2)	0.5 (0.2, 0.8)
RES	1.9 (1.0, 2.7)	−0.3 (−0.6, 0.1)	−0.1 (−0.4, 0.3)
SAS	−0.5 (−1.3, 0.2)	−0.1 (−0.4, 0.3)	−0.4 (−0.9, −0.1)
SUM	−0.2 (−0.9, 0.4)	−0.2 (−0.4, 0.1)	−0.7 (−1.0, −0.3)

HRU, hydraulic response unit; BUM, bullhead minnow; EMS, emerald shiner; FAM, fathead minnow; PLM, plains minnow; PRC, prairie chub; RRS, Red River shiner; RES, red shiner; SAS, sand shiner; SUM, Suckermouth minnow; UNDR, upstream network dam density; DL17, low flow duration variability; FH3, flood pulse count; MA2, median daily flow; RA9, variability of reversals; and TH2, annual maxima variability.

Table A3.
Minnow detection model coefficients reported on the logit scale as the mode with an associated 90% highest density interval from the posterior distribution. The intercept is interpreted as estimated detection probability at mean levels of covariates. Each Covariate coefficient is interpreted with other covariates held constant.

Appendix 2: occupancy model specifications, diagnostics, and fit test

We fit the occupancy model with the program JAGS [117] called from the statistical software R [118] using the package jagsUI [119]. We used vague truncated normal priors for species coefficients and vague gamma priors for associated standard deviations [64]. Posterior distributions for coefficients were estimated with Markov chain Monte Carlo methods using four chains of 20,000 iterations each run in parallel after a 5,000-iteration burn-in phase (thinning = 10). We considered adequate convergence a potential scale reduction factor ($\hat{R}$) <1.05 [120] and "grassy" trace plots for all parameters [64]. We calculated the 90% highest density intervals using the R package HDInterval [121].

We examined model fit using a posterior predictive check [64] based on the goodness-of-fit test described by [122]. We simulated expected species encounter histories under model parameters to compare discrepancies with observed encounter histories and calculated a Bayesian p-value (0.47). A Bayesian p-value near 0.5 suggests adequate fit and extreme values (i.e., <0.05 or >0.95) indicate a lack of fit [64, 123].

Author details

Robert Mollenhauer[1], Shannon K. Brewer[2]*, Desiree Moore[3], Dusty Swedberg[4] and Maeghen Wedgeworth[5]

1 U.S. Geological Survey, Wetland and Aquatic Research Center, Gainesville, FL, USA

2 U.S. Geological Survey, Alabama Cooperative Fish and Wildlife Research Unit, Auburn University, Auburn, AL, USA

3 U.S. Fish and Wildlife Service, San Marcos Aquatic Resources Center, San Marcos, TX, USA

4 Illinois Natural History Survey, Prairie Research Institute, Champaign, IL, USA

5 South Carolina Department of Natural Resources, Marine Resources Research Institute, Charleston, SC, USA

*Address all correspondence to: skb0064@auburn.edu

References

[1] Stevenson RJ. Scale-dependent determinants and consequences of benthic algal heterogeneity. Journal of the North American Benthological Society. 1997;**16**:248-262. DOI: 10.2307/1468255

[2] Hynes HBN. The stream and its valley. Internationale Vereinigung für Theoretische und Angewandte Limnologie: Verhandlungen. 1975;**19**: 1-15. DOI: 10.1080/03680770.1974.11896033?journalCode=tinw19

[3] Carrick TR. The effect of acid water on the hatching of salmonid eggs. Journal of Fish Biology. 1979;**14**:165-172. DOI: 10.1111/J.1095-8649.1979.TB03506.X

[4] Marsh-Matthews E, Matthews WJ. Geographic, terrestrial and aquatic factors: Which most influence the structure of stream fish assemblages in the midwestern United States? Ecology of Freshwater Fish. 2000;**9**:9-21. DOI: 10.1034/j.1600-0633.2000.90103.x

[5] Fausch KD, Torgersen CE, Baxter CE, Li HW. Landscapes to riverscapes: Bridging the gap between research and conservation of stream fishes. Bioscience. 2002;**52**:483-498. DOI: 10.1641/00063568(2002)052[0483:LTRBTG]2.0.CO;2

[6] Krebs JR, Kacelnik A. Decision-making. In: Krebs JR, Davies NB, editors. Behavioural Ecology: An Evolutionary Approach. 4th ed. Oxford, United Kingdom: Blackwell Scientific Publications; 1997. pp. 105-136

[7] Grabowska J, Kakareko T, Błońska D, Przybylski M, Kobak J, Copp GH. Interspecific competition for a shelter between non-native racer goby and native European bullhead under experimental conditions – Effects of season, fish size and light conditions. Limnologica. 2016;**56**:30-38. DOI: 10.1016/J.LIMNO.2015.11.004

[8] Peterson JT, Rabeni CF. The relation of fish assemblages to channel units in an Ozark stream. Transactions of the American Fisheries Society. 2001;**130**: 911-926. DOI: 10.1577/1548-8659(2001)130,0911:TROFAT.2.0.CO;2

[9] Paul MJ, Meyer JL. Streams in the urban landscape. Annual Review of Ecology, Evolution, and Systematics. 2001;**32**:333-365

[10] O'Driscoll M, Soban J, Lecce S. Stream channel enlargement response to urban land cover in small coastal plain watersheds. North Carolina. Physical Geography. 2013;**30**:528-555. DOI: 10.2747/0272-3646.30.6.528

[11] Osborne LL, Wiley MJ. Empirical relationships between land use/cover and stream water quality in an agricultural watershed. Journal of Environmental Management. 1988;**26**: 9-27

[12] Johnson LB, Richards C, Host GE, Arthur JW. Landscape influences on water chemistry in Midwestern stream ecosystems. Freshwater Biology. 1997;**37**: 193-208. DOI: 10.1046/j.1365-2427.1997.d01-539.x

[13] Walser CA, Bart HL. Influence of agriculture on in-stream habitat and fish community structure in Piedmont watersheds of the Chattahoochee River System. Ecology of Freshwater Fish. 1999;**8**:237-246. DOI: 10.1111/j.1600-0633.1999.tb00075.x

[14] Richards C, Johnson LB, Host GE. Landscape-scale influences on stream habitats and biota. Canadian Journal of

Fisheries and Aquatic Sciences. 1996;**53**: 295-311. DOI: 10.1139/f96-006

[15] Wang L, Lyons J, Kanehl P, Gatti R. Influences of watershed land use on habitat quality and biotic integrity in Wisconsin streams. Fisheries. 1997;**22**: 6-12. DOI: 10.1577/1548-8446(1997)022 <0006:IOWLUO>2.0.CO;2

[16] Kondolf GM. Hungry water: Effects of dams and gravel mining on river channels. Environmental Management. 1997;**21**:533-551. DOI: 10.1007/ s002679900048

[17] Pizzuto J. Effects of dam removal on river form and process. Bioscience. 2002; **52**:683-691. DOI: 10.1641/0006-3568 (2002)052[0683:EODROR]2.0.CO;2

[18] Poff NL, Allan JD, Bain MB, Karr JR, Prestegaard KL, Richter BD, et al. The natural flow regime. Bioscience. 1997;**47**: 769-784. DOI: 10.2307/1313099

[19] Sofi MS, Bhat SU, Rashid I, Kuniyal JC. The natural flow regime: A master variable for maintaining river ecosystem health. Ecohydrology. 2020;**13**:e2247. DOI: 10.1002/eco.2247

[20] Karr JR, Dudley DR. Ecological perspective on water quality goals. Environmental Management. 1981;**5**: 55-68. DOI: 10.1007/BF01866609

[21] Baron JS, Poff NL, Angermeier PL, Dahm C, Gleick PH, Hairston NG, et al. Balancing human and ecological needs for freshwater: The case for equity. Ecological Applications. 2002;**12**: 1247-1260

[22] Xenopoulos MA, Lodge DM, Alcamo J, Märker M, Schulze K, Van Vuuren DP. Scenarios of freshwater fish extinctions from climate change and water withdrawal. Global Change Biology. 2005;**11**:1557-1564. DOI: 10.1111/j.1365-2486.2005.001008.x

[23] Poff NL, Zimmerman JKH. Ecological responses to altered flow regimes: A literature review to inform the science and management of environmental flows. Freshwater Biology. 2010;**55**:194-205. DOI: 10.1111/ j.1365-2427.2009.02272.x

[24] Burkhead NM. Extinction rates in North American freshwater fishes, 1900–2010. Bioscience. 2012;**62**: 798-808. DOI: 10.1525/bio.2012.62.9.5

[25] Carlisle DM, Wolock DM, Meador MR. Alteration of streamflow magnitudes and potential ecological consequences: A multiregional assessment. Frontiers in Ecology and the Environment. 2011;**9**:264-270. DOI: 10.1890/100053

[26] Cooper AR, Infante DM, Daniel WM, Wehrly KE, Wang L, Brenden TO. Assessment of dam effects on streams and fish assemblages of the conterminous USA. The Science of the Total Environment. 2017;**586**: 879-889. DOI: 10.1016/j. scitotenv.2017.02.067

[27] Anderson EP, Freeman MC, Pringle CM. Ecological consequences of hydropower development in Central America: Impacts of small dams and water diversion on neotropical stream fish assemblages. River Research and Applications. 2006;**22**:397-411. DOI: 10.1002/rra.899

[28] Guenther CB, Spacie A. Changes in fish assemblage structure upstream of impoundments within the upper Wabash River catchment, Indiana. Transactions of the American Fisheries Society. 2006;**135**:570-583. DOI: 10.1577/ T05-031.1

[29] Freedman JA, Lorson BD, Taylor RB, Carline RF, Stauffer JR. River of the dammed: Longitudinal changes in fish assemblages in response to dams. Hydrobiologia. 2014;**727**:19-33. DOI: 10.1007/s10750-013-1780-6

[30] Vinson MR. Long-term dynamics of an invertebrate assemblage downstream from a large dam. Ecological Applications. 2001;**11**:711-730. DOI: 10.1890/1051-0761(2001)011[0711: LTDOAI]2.0.CO;2

[31] Martínez A, Larrañaga A, Basaguren A, Pérez J, Mendoza-Lera C, Pozo J. Stream regulation by small dams affects benthic macroinvertebrate communities: From structural changes to functional implications. Hydrobiologia. 2013;**711**:31-42. DOI: 10.1007/ s10750-013-1459-z

[32] Liu X, Hu X, Ao X, Wu X, Ouyang S. Community characteristics of aquatic organisms and management implications after construction of Shihutang Dam in the Gangjiang River. China. Lake and Reservoir Management. 2018;**34**:4257. DOI: 10.1080/10402381.2017.137371

[33] Costigan KH, Daniels MD. Damming the prairie: Human alteration of Great Plains river regimes. Journal of Hydrology. 2012;**444**:90-99. DOI: 10.1016/j.jhydrol.2012.04.008

[34] Liermann CR, Nilsson C, Robertson J, Ng RY. Implications of dam obstruction for global freshwater fish diversity. Bioscience. 2012;**62**:539-548. DOI: 10.1525/bio.2012.62.6.5

[35] Magalhaes MF, Beja P, Schlosser IJ, Collares-Pereira MJ. Effects of multi-year droughts on fish assemblages of seasonally drying Mediterranean streams. Freshwater Biology. 2007;**52**: 1494-1510. DOI: 10.1111/ j.1365-2427.2007.01781.x

[36] Bond NR, Lake PS, Arthington AH. The impacts of drought on freshwater ecosystems: An Australian perspective. Hydrobiologia. 2008;**600**:3-16. DOI: 10.1007/s10750-008-9326-z

[37] Falke JA, Fausch KD, Magelky R, Aldred A, Durnford DS, Riley LK, et al. The role of groundwater pumping and drought in shaping ecological futures for stream fishes in a dryland river basin of the western Great Plains, USA. Ecohydrology. 2011;**4**:682-697. DOI: 10.1002/eco.158

[38] Palmer MA, Reidy Liermann CA, Nilsson C, Flörke M, Alcamo J, Lake PS, et al. Climate change and the world's river basins: Anticipating management options. Frontiers in Ecology and the Environment. 2008;**6**:81-89. DOI: 10.1890/060148

[39] Matthews WJ, Zimmerman EG. Potential effects of global warming on native fishes of the southern Great Plains and the Southwest. Fisheries. 1990;**15**: 26-32. DOI: 10.1577/1548-8446(1990) 015<0026:PEOGWO>2.0.CO;2

[40] Puckridge JT, Sheldon F, Walker KF, Boulton AJ. Flow variability and the ecology of large rivers. Marine and Freshwater Research. 1998;**4**:55-72. DOI: 10.1071/MF94161

[41] Bunn SE, Thoms MC, Hamilton SK, Capon SJ. Flow variability in dryland rivers: Boom, bust and the bits in between. River Research and Applications. 2006;**22**:179-186. DOI: 10.1002/rra.904

[42] Dodds WK, Gido K, Whiles MR, Fritz KM, Matthews WJ. Life on the edge: The ecology of Great Plains prairie streams. Bioscience. 2004;**54**:205-216. DOI: 10.1641/0006-3568(2004)054 [0205:LOTETE]2.0.CO;2

[43] Perkin JS, Gido KB, Costigan KH, Daniels MD, Johnson ER. Fragmentation and drying ratchet down Great Plains stream fish diversity. Aquatic Conservation: Marine and Freshwater Ecosystems. 2015;**25**:639-655. DOI: 10.1002/aqc.2501

[44] Frissell CA, Liss WJ, Warren CE, Hurley MD. A hierarchical framework for stream habitat classification: Viewing streams in a watershed context. Environmental Management. 1986;**10**:199-214. DOI: 10.1007/BF01867358

[45] Lehner B, Liermann CR, Revenga C, Vörösmarty C, Fekete B, Crouzet P, et al. High-resolution mapping of the world's reservoirs and dams for sustainable river-flow management. Frontiers in Ecology and the Environment. 2011;**9**: 494-502. DOI: 10.1890/100125

[46] Matthews WJ. North American prairie streams as systems for ecological study. Journal of the North American Benthological Society. 1988;7:387409. DOI: 10.2307/1467298

[47] Echelle AA, Echelle AF, Hill LG. Interspecific interactions and limiting factors of abundance and distribution in the Red River pupfish, *Cyprinodon rubrofluviatilis*. American Midland Naturalist. 1972;**88**:109-130. DOI: 10.2307/2424492

[48] Schönhuth S, Vukić J, Šanda R, Yang L, Mayden RL. Phylogenetic relationships and classification of the Holarctic family Leuciscidae (Cypriniformes: Cyprinoidei). Molecular Phylogenetics and Evolution. 2018;**127**: 781-799. DOI: 10.1016/j.ympev.2018.06.026

[49] Tan M, Armbruster JW. Phylogenetic classification of extant genera of fishes of the order Cypriniformes (Teleostei: Ostariophysi). Zootaxa. 2018;**13**:4476

[50] Balon EK. Reproductive guilds of fishes: A proposal and definition. Journal of the Fisheries Board of Canada. 1975; **32**:821-864. DOI: 10.1139/f75-110

[51] Hoagstrom CW, Turner TF. Recruitment ecology of pelagic-broadcast spawning minnows: Paradigms from the ocean advance science and conservation of an imperilled freshwater fauna. Fish and Fisheries. 2015;**16**:282-299. DOI: 10.1111/faf.12054

[52] Worthington TA, Echelle AA, Perkin JS, Mollenhauer R, Farless N, Dyer JJ, et al. The emblematic minnows of the North American Great Plains: A synthesis of threats and conservation opportunities. Fish and Fisheries. 2018; **19**:271-307. DOI: 10.1111/faf.12254

[53] Platania SP, Altenbach CS. Reproductive strategies and egg types of seven Rio Grande catchment cyprinids. Copeia. 1998;**1998**:559-569. DOI: 10.2307/1447786

[54] Durham BW, Wilde GR. Effects of streamflow and intermittency on the reproductive success of two broadcast-spawning cyprinid fishes. Copeia. 2009; **2009**:21-28. DOI: 10.1643/CE-07-166

[55] Perkin JS, Gido KB. Stream fragmentation thresholds for a reproductive guild of Great Plains fishes. Fisheries. 2011;**36**:371-383. DOI: 10.1080/03632415.2011.597666

[56] Mollenhauer R, Brewer SK, Perkin JS, Swedberg D, Wedgeworth M, Steffensmeier ZD. Connectivity and flow regime direct conservation priorities for pelagophil fishes. Aquatic Conservation: Marine and Freshwater Ecosystems. 2021; **31**:3215-3227. DOI: 10.1002/aqc.3631

[57] Albers JL, Wildhaber ML. Reproductive strategy, spawning induction, spawning temperatures and early life history of captive sicklefin chub *Macrhybopsis meeki*. Journal of Fish Biology. 2017;**91**:58-79. DOI: 10.1111/jfb.13329

[58] Archdeacon TP, Davenport SR, Grant JD, Henry EB. Mass upstream dispersal of pelagic-broadcast spawning cyprinids in the Rio Grande and Pecos River, New Mexico. Western North American Naturalist. 2018;**78**:100-105. DOI: 10.3398/064.078.0110

[59] Hoagstrom CW, Brooks JE, Davenport SR. A large-scale conservation perspective considering endemic fishes of the North American plains. Biological Conservation. 2011; **144**:21-34. DOI: 10.1016/j.biocon.2010.07.015

[60] Cross FB, Moss RE. Historic changes in fish communities and aquatic habitats in plains streams of Kansas. In: Matthews WJ, Heins DC, editors. Community and Evolutionary Ecology of North American Stream Fishes. Norman: University of Oklahoma Press; 1987. pp. 155-165

[61] Luttrell GR, Echelle AA, Fisher WL, Eisenhour DJ. Declining status of two species of the *Macrhybopsis aestivalis* complex (Teleostei: Cyprinidae) in the Arkansas River basin and related effects of reservoirs as barriers to dispersal. Copeia. 1999;**1999**:981-989. DOI: 10.2307/1447973

[62] Bonner TH, Wilde GR. Changes in the Canadian River fish assemblage associated with reservoir construction. Journal of Freshwater Ecology. 2000;**15**: 189-198. DOI: 10.1080/02705060.2000.9663736

[63] Begon M, Townsend CR, Harper JL. Ecology: From Individuals to Ecosystems. 4th ed. Hoboken, New Jersey: Wiley-Blackwell; 2005

[64] Kéry M, Royle JA. Applied Hierarchical Modeling in Ecology: Analysis of Distribution, Abundance and Species Richness in R and BUGS: Volume 1: Prelude and Static Models. San Diego, CA: Elsevier Science; 2016

[65] MacKenzie DI, Nichols JD, Royle JA, Pollock KH, Bailey LL, Hines JE. Occupancy Estimation and Modeling: Inferring Patterns and Dynamics of Species Occurrence. 2nd ed. San Diego, CA: Elsevier Science; 2017

[66] Sály P, Takács P, Kiss I, Bíró P, Erős T. The relative influence of spatial context and catchment-and site-scale environmental factors on stream fish assemblages in a human-modified landscape. Ecology of Freshwater Fish. 2011;**20**:251-262. DOI: 10.1111/j.1600-0633.2011.00490.x

[67] Marion CA, Scott MC, Kubach KM. Multiscale environmental influences on fish assemblage structure of South Atlantic coastal plain streams. Transactions of the American Fisheries Society. 2015;**144**:1040-1057. DOI: 10.1080/00028487.2015.1059887

[68] Mollenhauer R, Zhou Y, Brewer SK. Multiscale habitat factors explain variability in stream fish occurrence in the Ozark Highlands ecoregion, USA. Copeia. 2019;**2019**:219-231. DOI: 10.1643/CE-18-099

[69] Eisenhour DJ. Systematics, variation, and speciation of the *Macrhybopsis aestivalis* complex west of the Mississippi River. Bulletin of the Alabama Museum of Natural History. 2004;**2004**:9-47

[70] Woods AJ, Omernik JM, Butler DR, Ford JG, Henley JE, Hoagland BW, Arndt DS, Moran BC. Ecoregions of Oklahoma U.S. Geological Survey. 2005. Available from: https://gaftp.epa.gov/EPADataCommons/ORD/Ecoregions/ok/ok_front.pdf [Accessed: July 5, 2022]

[71] Sotola VA, Ruppel DS, Bonner TH, Nice CC, Martin NH. Asymmetric introgression between fishes in the Red River basin of Texas is associated with variation in water quality. Ecology and Evolution. 2019;**9**:2083-2095. DOI: 10.1002/ece3.4901

[72] Markstrom SL, Regan RS, Hay LE, Viger RJ, Webb RMT, Payn RA, et al. PRMS-IV, the precipitation-runoff modeling system, version 4. U.S. Geological Survey Techniques and Methods. 2015;**2015**:6-B7

[73] U.S. Geological Survey. Hydrologic response units (HRUs) used with the precipitation runoff modeling system for hydrologic simulations of the southeastern United States. 2017. Available from: https://www.sciencebase.gov/catalog/item/59f0d966e4b0220bbd9c58ac [Accessed: May 5, 2022]

[74] Smith SJ, Ellis JH, Wagner DL, Peterson SM. Hydrogeology and simulated groundwater flow and availability in the North Fork Red River aquifer, southwest Oklahoma, 1980–2013. U.S. Geological Survey. 2017. Available from: https://pubs.er.usgs.gov/publication/sir20175098 [Accessed: May 5, 2022]

[75] Perkin JS, Gido KB, Cooper AR, Turner TF, Osborne MJ, Johnson ER, et al. Fragmentation and dewatering transform Great Plains stream fish communities. Ecological Monographs. 2015;**85**:73-92. DOI: 10.1890/14-0121.1

[76] Kendall WL, White GC. A cautionary note on substituting spatial subunits for repeated temporal sampling in studies of site occupancy. Journal of Applied Ecology. 2009;**46**:1182-1188. DOI: 10.1111/j.1365-2664.2009.01732.x

[77] Rabeni CF, Lyons J, Mercado-Silva N, Peterson JT. Warmwater fish in wadeable streams. In: Bonar SA, Hubert WA, Willis DW, editors. Standard Methods for Sampling North American Freshwater Fishes. Bethesda, Maryland: American Fisheries Society. pp. 43-58

[78] Viger RJ, Bock A. GIS features of the geospatial fabric for National Hydrologic Modeling, U.S. Geological Survey. 2014. Available from: http://dx.doi.org/doi:10.5066/F7542KMD [Accessed: July 5, 2022]

[79] U.S. Geological Survey. Red River focus area study. 2019. Available from: https://www.usgs.gov/centers/tx-water/science/red-river-focus-area-study?qt-science_center_objects=0#qt-science_center_objects [Accessed: July 5, 2022]

[80] Roland VL. Model input and output from Precipitation Runoff Modeling System (PRMS) simulation of the Red River basin 1981-2016. U.S. Geological Survey data release. 2020. DOI: 10.5066/P9ZI5IVX

[81] Olden JD, Poff NL. Redundancy and the choice of hydrologic indices for characterizing streamflow regimes. River Resources and Applications. 2003; **19**:101-121. DOI: 10.1002/rra.700

[82] Keenen JG, Henriksen JA, Nieswand SP. Development of the hydroecological integrity assessment process for determining environmental flows for New Jersey streams. USGS Scientific Investigations Report. 2007:2007-5206. DOI: 10.3133/sir20075206

[83] U.S. Geological Survey. USGS-R/EflowStats: Hydrologic indicator and alteration stats. 2019. Available from: https://rdrr.io/github/USGS-R/EflowStats/ [Accessed: July 5, 2022]

[84] Kennen JG, Henriksen JA, Nieswand SP. Development of the hydroecological integrity assessment process for determining environmental flows for New Jersey streams. U.S. Geological Survey Scientific Investigations Report. 2007. p. 2007-5206. DOI: 10.3133/sir20075206

[85] Cooper AR, Infante DM. Dam metrics representing stream fragmentation and flow alteration for the conterminous United States linked to the NHDPLUSV1. U.S. Geological Survey data release. 2017. Available from: https://www.sciencebase.gov/catalog/item/58a60b88e4b057081a24f99d [Accessed: July 5, 2022]

[86] MacKenzie DI, Nichols JD, Lachman GB, Droege S, Royle JA, Langtimm CA. Estimating site occupancy rates when detection probabilities are less than one. Ecology. 2002;**83**:2248-2255. DOI: 10.1890/0012-9658(2002)083[2248:ESORWD]2.0.CO;2

[87] Dorazio RM, Royle JA. Estimating size and composition of biological communities by modeling the occurrence of species. Journal of the American Statistical Association. 2005;**100**:389-398. DOI: 10.1198/016214505000000015

[88] Dorazio RM, Royle JA, Söderström B, Glimskär A. Estimating species richness and accumulation by modeling species occurrence and detectability. Ecology. 2006;**87**:842-854

[89] Zipkin EF, DeWan A, Royle JA. Impacts of forest fragmentation on species richness: A hierarchical approach to community modelling. Journal of Applied Ecology. 2009;**46**:815-822. DOI: 10.1111/j.1365-2664.2009.01664.x

[90] Gelman A, Hill J. Data Analysis Using Regression and Multilevel/Hierarchical Models. United Kingdom: Cambridge University Press; 2007

[91] Wagner T, Hayes DB, Bremigan MT. Accounting for multilevel data structures in fisheries data using mixed models. Fisheries. 2006;**31**:180-187. DOI: 10.1577/1548-8446(2006)31[180:AFMDSI]2.0.CO;2

[92] Kruschke JK, Liddell TM. The Bayesian New Statistics: Hypothesis testing, estimation, meta-analysis, and power analysis from a Bayesian perspective. Psychonomic Bulletin & Review. 2018;**25**:178-206. DOI: 10.3758/s13423-016-1221-4

[93] Jackson DA, Peres-Neto P, Olden JD. What controls who is where in freshwater fish communities – The roles of biotic, abiotic, and spatial factors. Canadian Journal of Fisheries and Aquatic Sciences. 2001;**58**:157-170. DOI: 10.1139/f00-239

[94] Texas Parks and Wildlife Department. Connally W, editor. Texas conservation action plan 2012–2016: Overview. 2012. Available from: https://tpwd.texas.gov/huntwild/wild/wildlife_diversity/nongame/tcap/ [Accessed: July 5, 2022]

[95] Oklahoma Department of Wildlife Conservation. Oklahoma comprehensive wildlife conservation strategy: A strategic conservation plan for Oklahoma's rare and declining wildlife. 2016. Available from: https://www.wildlifedepartment.com/sites/default/files/Oklahoma%20Comprehensive%20Wildlife%20Conservation%20Strategy_0.pdf [Accessed: July 5, 2022]

[96] U.S. Fish and Wildlife Service. Endangered species: Listing and critical habitat National Listing Workplan. 2021. Available from: https://www.fws.gov/endangered/what-we-do/listing-workplan.html [Accessed: July 5, 2022]

[97] Wedgeworth M. Variation in Abundance and Hatch Date of prairie chub *Macrhybopsis australis* in the Upper Red River Basin. Stillwater: Oklahoma State University; 2021

[98] Osborne LL, Wiley MJ. Influence of tributary spatial position on the structure of warmwater fish communities. Canadian Journal of Fisheries and Aquatic Sciences. 1992;**49**:671-681. DOI: 10.1139/f92-076

[99] Smith TA. Kraft CE Stream fish assemblages in relation to landscape position and local habitat variables. Transactions of the American Fisheries Society. 2005;**134**:430-440. DOI: 10.1577/T03-051.1

[100] Higgins CL, Wilde GR. The role of salinity in structuring fish assemblages in a prairie stream system. Hydrobiologia. 2005;**549**:197-203. DOI: 10.1007/s10750-005-0844-7

[101] Ruppel DS, Sotola VA, Gurbux OA, Martin NH, Bonner TH. RFP No. 212f for endangered species research projects for the prairie chub. Final Report. 2017. Available from: https://comptroller.texas.gov/programs/naturalresources/docs/reports/Prairie_Chub_TXST.pdf+&cd=1&hl=en&ct=clnk&gl=us [Accessed: July 5, 2022]

[102] Ruppel DS, Sotola VA, Craig CA, Martin NH, Bonner TH. Assessing functions of movement in a Great Plains endemic fish. Environmental Biology of Fishes. 2020;**103**:795-814. DOI: 10.1007/s10641-020-00983-8

[103] Chase NM, Caldwell CA, Carleton SA, Gould WR, Hobbs JA. Movement patterns and dispersal potential of Pecos bluntnose shiner (*Notropis simus pecosensis*) revealed using otolith microchemistry. Canadian Journal of Fisheries and Aquatic Sciences. 2015;**72**: 1575-1583. DOI: 10.1139/cjfas-2014-0574

[104] Lang JJ, Snyder RJ, Clapsadl MD, Michalak P, Kang L, Pérez-Fuentetaja A. Morphometric differentiation and gene flow in emerald shiners (*Notropis atherinoides*) from the lower Great Lakes and the Niagara River. Journal of Great Lakes Research. 2019;**45**:324-332. DOI: 10.1016/j.jglr.2018.12.002

[105] Perkin JS, Starks TA, Pennock CA, Gido KB, Hopper GW, Hedden SC. Extreme drought causes fish recruitment failure in a fragmented Great Plains riverscape. Ecohydrology. 2019;**12**: e2120. DOI: 10.1002/eco.2120

[106] Moore DM. Movement and Flow-ecology Relationship of Great Plains pelagophil Fishes. Stillwater: Oklahoma State University; 2020

[107] Young BA. Intraspecific Variation among Emerald Shiners (*Notropis atherinoides*) of the Missouri River. Brookings: South Dakota State University; 2001

[108] Fischer JR, Quist MC. Characterizing lentic freshwater fish assemblages using multiple sampling methods. Environmental Monitoring and Assessment. 2014;**186**:4461-4474. DOI: 10.1007/s10661-014-3711-z

[109] Tyre AJ, Tenhumberg B, Field SA, Niejalke D, Parris K, Possingham HP. Improving precision and reducing bias in biological surveys: Estimating false-negative error rates. Ecological Applications. 2003;**13**:1790-1801. DOI: 10.1890/02-5078

[110] Kéry M, Schmidt B. Imperfect detection and its consequences for monitoring for conservation. Community Ecology. 2008;**9**:207-216. DOI: 10.1556/comec.9.2008.2.10

[111] Gu W, Swihart RK. Absent or undetected? Effects of non-detection of species occurrence on wildlife–habitat models. Biological Conservation. 2004; **116**:195-203. DOI: 10.1016/S0006-3207(03)00190-3

[112] Guillera-Arroita G, Lahoz-Monfort JJ, Elith J, Gordon A, Kujala H, Lentini PE, et al. Is my species distribution model fit for purpose? Matching data and models to applications. Global Ecology and Biogeography. 2015;**24**:276-292. DOI: 10.1111/geb.12268

[113] Gwinn DC, Beesley LS, Close P, Gawne B, Davies PM. Imperfect detection and the determination of environmental flows for fish: Challenges, implications and solutions. Freshwater Biology. 2016; **61**:172-180. DOI: 10.1111/fwb.12684

[114] Peterson JT, Paukert CP. Converting nonstandard fish sampling data to standardized data. In: Bonar SA, Hubert WA, Willis DW, editors. Standard Methods for Sampling North American Freshwater Fishes. Bethesda, Maryland: American Fisheries Society. pp. 195-215

[115] Falke JA, Bailey LL, Fausch KD, Bestgen KR. Colonization and extinction in dynamic habitats: An occupancy approach for a Great Plains stream fish assemblage. Ecology. 2012;**93**:858-867. DOI: 10.1890/11-1515.1

[116] Mollenhauer R, Logue D, Brewer SK. Quantifying seining detection probability for fishes of Great Plains sand-bed rivers. Transactions of the American Fisheries Society. 2018;**147**: 329-341. DOI: 10.1002/tafs.10030

[117] Plummer M. JAGS: A program for analysis of Bayesian graphical models using Gibbs sampling. In: Proceedings of the 3rd International Workshop on Distributed Statistical Computing. Vienna: Austrian Science Foundation; 2003. pp. 1-10

[118] R Core Team. R: A Language and Environment for Statistical Computing. Vienna: R Foundation for Statistical Computing;

[119] Jags KK. UI: A wrapper around 'rjags' to streamline JAGS analyses. 2021. Available from: https://cran.case.edu/web/packages/jagsUI/jagsUI.pdf [Accessed: May 5, 2002]

[120] Brooks SP, Gelman A. General methods for monitoring convergence of iterative simulations. Journal of Computational and Graphical Statistics. 1998;**7**:434-455

[121] Meredith M, Kruschke JK. HDInterval: Highest (posterior) density intervals. 2020. Available from: https://cran.rstudio.com/web/packages/HDInterval/index.html [Accessed: May 5, 2002]

[122] MacKenzie DI, Bailey LL. Assessing the fit of site-occupancy models. Journal of Agricultural, Biological, and Environmental Statistics. 2004;**9**: 300-318

[123] Conn PB, Johnson DS, Williams PJ, Melin SR, Hooten MB. A guide to Bayesian model checking for ecologists. Ecological Monographs. 2018;**88**:526-542

Section 3

Water Resources Management

Chapter 5

Trend Analysis of Streamflow and Rainfall in the Kosi River Basin of Mid-Himalaya of Kumaon Region, Uttarakhand

Utkarsh Kumar, Rashmi, Dhirendra Kumar Singh, Suresh Chandra Panday, Manoj Parihar, Jaideep Kumar Bisht and Lakshmi Kant

Abstract

Due to climate change phenomenon and substantial decrease in water resources, analyzing the streamflow trend is of significant importance. In the present study, investigation was carried out to find rainfall and streamflow trends in the Kosi river watershed at different timescales from 1986 to 2016. Kosi river is one of the principal rivers in the Kumaon region. The different methods employed for trend detection of streamflow and rainfall were the Mann–Kendall (MK) test and the Sen's slope (SS) estimator. Results showed a statistically significant decreasing trend in pre-monsoonal and annual rainfall with a Sen's slope of -2.27 and -1.49 mm/year, respectively. The decreasing trends in pre-monsoon, post-monsoon, and winter streamflow were found during 1986–2016, which were not statistically significant. The results of the study help in understanding the variation and availability of rainfall and streamflow in different seasons of the year and motivate to adopt effective water management and agricultural practices for rainfed hills.

Keywords: Himalayas, climate change, streamflow, trend analysis, statistical test

1. Introduction

The prominent challenge being faced by the Indian Himalayan region (IHR) is climate change [1]. Study in connection with climate change is of great importance for the Kumaon region of Uttarakhand [2]. Changing temperature and precipitation patterns and their impact on water resources, glaciers, ecology, and agriculture are the results of changing climate over the Himalayas region [3]. Several researchers have studied the impact of climate change on Himalayan region and found that temperature is showing an increasing trend in the western Himalayas, while precipitation is showing a decreasing trend during winter and summer periods [4–8]. The nonuniform distribution of rainfall in the mountains results in differential rainfall trends within small distances [9]. For example, the central Himalayas receives 80% of annual precipitation due to the Indian summer monsoon (ISM), while the western Himalayas receives ~30% due to western disturbances [10–12].

It has been observed that low rainfall or shift in rainfall patterns at different altitudes had resulted in crop failure, declining in food grain yield. It was reported that traditional crops will soon be replaced with cash crops in the Kumaon region [13].

The impact of change on land use and land cover (LULC) plays an important role in climate change on local to regional scale. Increased urbanization or changes in LULC is known to alter changes in LULC directly affecting the rainfall and mesoscale convective system [14–19]. Therefore, understanding rainfall variability in the Himalayan region becomes extremely critical for holistic Himalayan spatial planning for water resource management.

A growing literature suggests that the Western Himalayas region is witnessing above normal increasing temporal trend in temperature and decreasing trend in rainfall. The worst case is the large-scale devastation owing to the Nanda Devi glacier burst in Uttarakhand's Chamoli district, which triggered a mass of snow, ice, and rocks falling speedily down a mountainside known as an avalanche that led to the water level rising in the river Rishiganga and heavy flood in Dhauliganga [19].

Finding variability and trend in long-term historical streamflow is of crucial importance for the appropriate management and planning of water resources. Some of the important reasons for trend analysis of streamflow are to understand the design flow rate for hydraulic structure and assigning water rights beyond the capacity of river supply.

This study applies Mann–Kendall (MK) test, the Sen's slope (SS) estimator, and nonparametric tests for evaluating the trends in streamflow time series. The MK checks whether the trend increases or decreases with time by examining whether to reject the null hypothesis or accept the alternative hypothesis. SS indicates the median of all pairwise slope values of a set of observed data. These two parameters have been employed in several studies for hydrological assessment of trends over various regions of globe for last decade.

There are several factors that impact the hydrology of a river basin, such as land use, climate change, and hydraulic infrastructure management [20, 21]. Therefore, investigating the hydrological characteristics from the historical time series of discharge data is considered one of the most important objectives in the field of water resource planning [22]. Salarijazi [23] reported that in several published research studies, the hydrological time series from different regions describe significant non-consistency or non-stationary. Due to this concern, trend analysis and change point detection in streamflow time series and other climatic variables (rainfall, evapotranspiration, and temperature) have been studied by many researchers in different watersheds or river basins at different time scales throughout the globe. Hyvärinen [24] analyzed streamflow data from 1913 to 2008 using Mann–Kendall test. He reported a significant decrease in high flow trend in Hawaii. Trend analysis and change point detection are two important tests, which have been popularly used at same time as mentioned by [23]. There is loss of spatial information of the hydro-climatic variable at large scale. Hence, it is recommended to analyze hydro-climatic variables at a small scale [24]. In the present study, the Kosi river watershed was undertaken to investigate the trend in measured streamflow data and the possible linkage for the observed changes in streamflow with rainfall and anthropogenic factors.

According to the published literature, no research study has addressed the trend in stream flow and rainfall, and their association with each other in the Kosi river watershed. Therefore, this study investigates trend of stream flow and rainfall data of this watershed at Ramnagar station during the last 31 years from 1986 to 2016.

2. Study area

Kosi River watershed extending between 29° 18′ N–79° 02′ E and 29° 51′ N–79° 51′ E is located in the Kumaon Lesser Himalaya (**Figure 1**). The Kosi River flows North–South in the northern and southern parts of the watershed while in the middle part, it follows the East–West trend incising the bedrocks and forming broad valleys, strath as well as unpaired terraces. The Kosi River mainly receives its water from several springs, aquifers, and tributaries in its course. Kosi river watershed falls within Almora and Nainital districts of Uttarakhand. The word Kosi refers to "river." Kosi is a Himalayan river that originates from Koshimool near Kausani and flows in the central part of Almora and the western part of Nainital district. River Kosi has the total catchment area of 3,420 sq. km. Kumaon is a mountainous region of eastern Uttarakhand in India. This region consists of the great Himalayan tract. Many rivers and their tributaries got their course from Kumaon. Four major rivers Kali, western Ramganga, Kosi, and Gaula make the surface drainage of Kumaon Himalaya. Kosi is the main river of Almora and Nainital districts. It is an important river flowing in the hills of the Kumaon region and drains central part of Almora and western part of Nainital district. The soil of the watershed falls under the loamy to clay categories. The major agricultural crops in the watershed are wheat, paddy, barley, pulses, and vegetables. The study area is also rich in the temperate horticulture fruit crops.

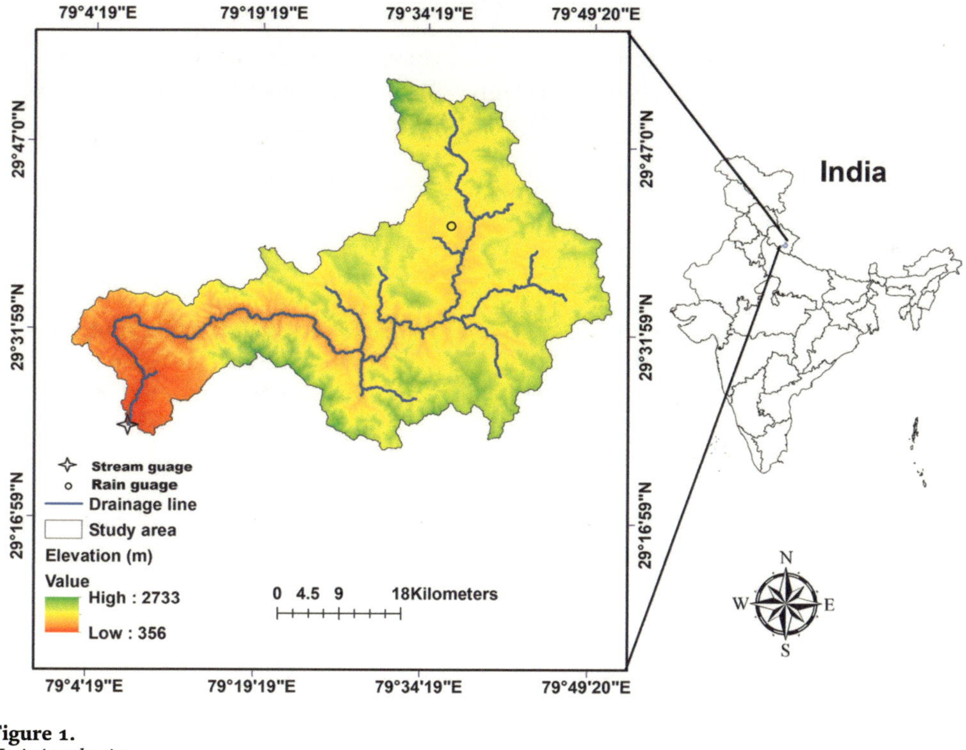

Figure 1.
Kosi river basin.

The average annual rainfall at different locations of the watershed ranges from 850 to 1100 mm. About 75% of annual rainfall is received between June and September due to southwest monsoon. The river Kosi is the major source of water supply to cities of Almora, Pithoragarh, and other cities in the Kumaon region of Uttarakhand.

3. Methodology

Streamflow data were collected from the executive engineer office, Ramnagar Irrigation Division, Nainital, Government of Uttarakhand. The 31 years of streamflow data were obtained for Ramnagar station near the outlet of the basin, which represents the entire river basin. The meteorological data were collected from ICAR–VPKAS, experimental farm Hawalbagh observatory (1986–2016). The limitation of this study is only this station has long-term data of rainfall. The watershed was delineated using 90-m-resolution SRTM data set in ArcGIS.

3.1 Parameters analyzed

The following parameters were analyzed from the data:

- Interannual monthly mean streamflow: monthly mean streamflow of the same month over the years (Jan 1986, Jan 1987, ———).
- Annual mean streamflow: mean of 12 monthly mean streamflow values from January to December for the gauging station.
- Seasonal streamflow: mean or monthly mean streamflow values for the pre-monsoon (March–May), monsoon (June–September), post-monsoon (October–November), and Winter (December–February) season.
- Annual seasonal and monthly rainfall.

3.2 Trend analysis

In this study, monthly streamflow trend analysis was evaluated using nonparametric approach namely Mann–Kendall (MK) [25–33] and Sen's slope estimator (magnitude of change) [34, 35]. MK test is a robust and widely accepted method in different hydro-climatic studies. Although the MK test is robust and widely accepted, it does not account for serial autocorrelation that usually occurs in a hydro-climatic variable time series. The presence of serial correlation in a time series may lead to wrong information because it enhances the probability of finding a significance when actually there is an absence of a significant trend. The trend of different hydro-climatic variables was evaluated at 5% and 10% significant levels (p value) as an indicator of trend strength

3.3 Mann–Kendall (MK) test

The MK test [29–30] computes statistics as Eq. (1)

$$S = \sum_{i=1}^{n-1} \sum_{j=i+1}^{n} \text{sgn}\left(x_j - x_k\right) \tag{1}$$

where S = normal distribution with the mean, n = number of observations (≥ 10), x_j is the jth observation, and sgn () is the sign function defined as sgn (α) = 1 if α >0; sgn (α) = 0 if α = 0; and sgn (α) = -1 if α<0.

$$Var(S) = \frac{n(n-1)(2n+5) - \sum_{i=1}^{m} t_i(t_i - 1)(2t_i + 5)}{18} \tag{2}$$

where n = number of tied groups having similar value for a data group and t_i = number of data in the ith tied group. The actual MK statistics are given as Eqs. (3–4)

$$Z = \frac{S+1}{\sqrt{V(S)}}, ifS < 0 \tag{3}$$

$$Z = 0, ifS = 0 \tag{4}$$

$$Z = \frac{S+1}{\sqrt{V(S)}}, ifS < 0 \tag{5}$$

Two hypotheses are made, that is, H_o (null hypothesis) and H_1 (alternative hypothesis). H_o indicates no statistically significant trend, while H_1 indicates a statistically significant trend.

3.4 Sen's slope

Computation of the magnitude of change in a dataset is done by Sen's slope [36, 38]. This is a simple linear regression method, which can estimate the slope of the median of two different variables (dependent and independent). It can be estimated using Eq. (6)

$$d_{ijk} = \frac{X_{ij} - x_{ik}}{j-k} \tag{6}$$

where X_{ij} and x_{ik} are data values and j and k are the time series.

3.5 Coefficient of variation

The coefficient of variation (CV) is defined as the standard deviation divided by the mean. It was used in the study to reveal the interannual variation of an annual average of rainfall. It is calculated using Eq. (7)

$$CV = \frac{\sum_{i=1}^{n} \left(ARF_i - \overline{ARF}\right)^2}{\overline{ARF}} \tag{7}$$

where ARF_i is the annual rainfall in the year i and $\overline{ARF}$ is the average annual rainfall from 1986 to 2016 (n = 31).

4. Results and discussion

The descriptive statistics of 31-year monthly and seasonal streamflow data are shown in **Table 1** and 2, respectively.

4.1 Statistical analysis of streamflow

To understand the hydrological characteristics of the Kosi river watershed, the overall behavior of streamflow and rainfall of the watershed, mean monthly rainfall, and mean monthly streamflow were analyzed over the periods from 1986 to 2016 (**Figure 2**). The mean monthly streamflow varies from 9.31 m^3/sec (May) to 92.37 m^3/sec (August). The month of high streamflow generally matches with the monsoon season, which clearly demarcates that streamflow in this area is largely dependent on rainfall. The maximum mean monthly streamflow occurs in August,

Months	Mean ± SD	Minimum (m^3/sec)	Maximum (m^3/sec)	CV (%)
January	14.33 ± 8.30	3.68	32.17	0.58
February	15.80 ± 8.14	4.14	33.11	0.52
March	14.03 ± 7.53	3.25	28.89	0.54
April	10.13 ± 6.19	2.37	23.76	0.61
May	9.31 ± 6.12	2.33	23.17	0.66
June	22.84 ± 31.25	1.91	153.30	1.37
July	66.47 ± 38.36	10.67	174.80	0.58
August	92.37 ± 44.70	23.00	235.86	0.48
September	78.36 ± 66.10	17.96	306.40	0.84
October	26.75 ± 11.21	9.23	54.84	0.42
November	14.55 ± 6.12	5.69	28.81	0.42
December	13.39 ± 7.14	4.13	29.54	0.53

Table 1.
Mean monthly streamflow dynamics for the period from 1986 to 2016.

Season	Mean ± SD	Minimum (m^3/sec)	Maximum (m^3/sec)	CV (%)
MAM@	11.16 ± 6.18	2.79	24.22	0.55
JJAS@	52.01 ± 20.01	19.39	102.03	0.38
ON@	20.65 ± 7.76	7.52	35.46	0.38
Annual	31.53 ± 10.11	12.43	55.07	0.32
DJF@	14.51 ± 6.92	4.67	27.87	0.48

@MAM = March, April, and May; JJAS = June, July, August, and September; ON = October, November; DJF = December, January, and February

Table 2.
Mean seasonal streamflow dynamics for the period from 1986 to 2016.

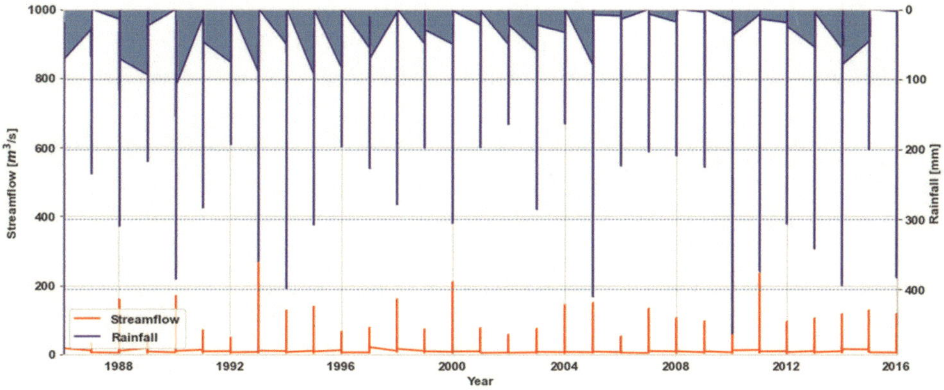

Figure 2.
Mean monthly streamflow and mean monthly rainfall of Kosi watershed from 1986–2016 at Ramnagar gauging station.

while the maximum mean monthly rainfall of the catchment occurs in August in most of the years. This concludes that most of the rainfall occurring in August may be the majority contributing to streamflow.

4.2 Temporal variation of annual mean streamflow and rainfall

The variation of streamflow is shown in **Figure 3**. The annual mean streamflow varies from 12.43 m^3/sec to 55.07 m^3/sec with annual mean value of 31.53 m^3/sec, standard deviation of 10.11 m^3/sec, and coefficient of variation of 0.32.

4.3 Variation of seasonal streamflow

The variation of seasonal streamflow and monthly rainfall was represented in **Figures 4** and **5**, respectively. It clearly shows that the highest value of streamflow was found in the June, July, August, and September (JJAS) seasons, while the March, April, and May (MAM) seasons exhibited the lowest streamflow value.

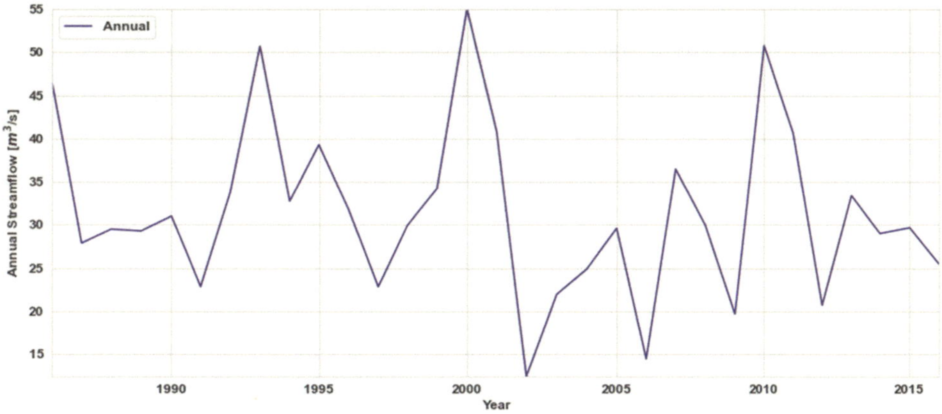

Figure 3.
The annual streamflow analysis (1986 and 2016).

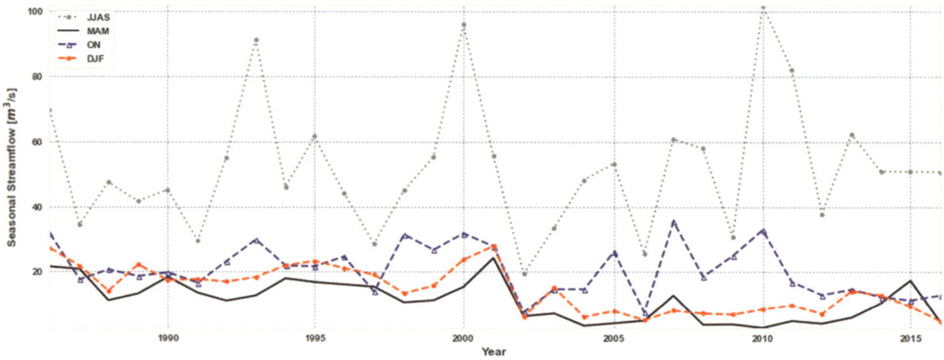

Figure 4.
The seasonal streamflow analysis (1986 and 2016).

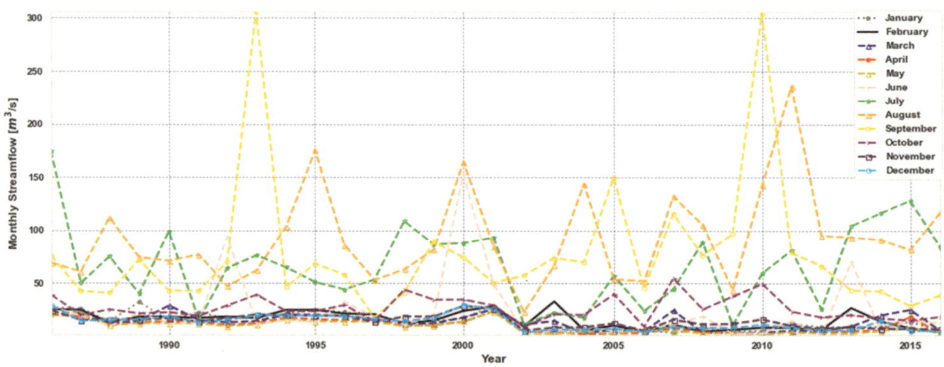

Figure 5.
Time series of monthly streamflow.

These results again corroborate the association between streamflow and rainfall. In this study area, most of the rainfall is received during June and September.

4.4 Trend analysis of monthly streamflow and rainfall

To understand the variations of monthly streamflow behavior, interannual monthly mean streamflow and rainfall were tested using the MK test and Sen's slope (**Table 3**). Similarly, seasonal rainfall and streamflow were analyzed using the MK test and Sen's slope (**Table 4**). The streamflow for all the months shows a decreasing trend at Ramnagar station, although all were not significant.

4.5 Characterizing the factors of trends in streamflow

In this study, we have tried to quantify different drivers for detecting the trend of streamflow. As most of this watershed is under forest land use, we considered urbanization and population growth as the most important factors. It is important to mention that the urban population in Uttarakhand increased from 16.36% of the total in 1971 to 20.7% in 1981, 22.97% in 1991, and 25.59% in 2001 (**Table 5**). The state registered the highest growth of urban population during 1971–1981 (56.38%);

Months	1986–2016 (Streamflow, Ramnagar)			1986–2016 (Rainfall, Hawalbagh)		
	Z value	Sen's slope	p value	Z value	Sen' s slope	p value
January	−0.76	−1.49	0.44	−1.08	−0.52	0.27
February	0.00	−0.02	1.0	−0.34	−0.40	0.73
March	−0.71	−0.92	0.47	−1.44	−0.92	0.14
April	−0.81	−0.79	0.41	−0.85	−0.51	0.39
May	−0.13	−0.20	0.89	−1.75	−1.45	0.07
June	0.30	0.82	0.75	−0.30	−0.62	0.75
July	1.93	48.14	0.05	−0.69	−1.25	0.48
August	0.88	29.35	0.37	−0.88	−1.34	0.37
September	0.61	11.11	0.54	−0.10	−0.17	0.91
October	1.08	6.87	0.27	0.78	0.15	0.43
November	−0.54	−0.75	0.58	−1.61	0.00	0.10
December	−0.95	−1.81	0.34	−0.64	0.00	0.52

** Significant at 5% level, ** significant at 10% level.*

Table 3.
Overall trend analysis of monthly streamflow and rainfall.

Season	1986–2016 (Streamflow)			1986–2016 (Hawalbagh)		
	Z value	Sen' s slope	p value	Z value	Sen's slope	p value
MAM	−0.50	−0.63	0.61	**−2.27***	−3.32	0.02
JJAS	1.49	20.72	0.13	−0.88	−3.01	0.37
ON	−0.74	2.56	0.45	−1.49	−7.23	0.13
Annual	1.39	8.99	0.16	−1.49	−7.23	0.13
DJF	−0.64	−1.18	0.51	−1.17	−1.99	0.24

**Significant at 5% level.*

Table 4.
Overall trend analysis of annual and seasonal streamflow and rainfall.

however, decadal urban population growth declined slightly during 1981–1991 (42.20%) and 1991–2001 (32.81%) [39].

5. Conclusions

The time series of rainfall and streamflow data of the Kosi river watershed for the last 31 years (1986–2016) was statistically analyzed to determine the trend and understand the changes in the streamflow regime. Based on this study, following conclusions were drawn:

- The rainfall is assumed to be the dominant component in the streamflow of the Kosi river watershed.

Census years	Total population	Urban population	Urban content (%)	Urban growth(%)
1901	19,79,866	1,54,424	7.8	—
1911	21,42,258	1,79,332	8.37	16.13
1921	21,15,984	1,91,660	9.06	6.87
1931	23,01,019	1,95,797	8.51	2.16
1941	26,14,540	2,70,503	10.35	38.15
1951	29,45,929	4,00,631	13.6	48
1961	36,10,938	4,95,995	13.74	23.8
1971	44,92,724	7,34,856	16.36	48.16
1981	57,25,972	11,49,136	20.07	**56.38**
1991	71,13,483	16,34,084	22.97	42.2
2001	84,79,562	21,70,245	25.59	32.81
2011	1,01,16,752	30,91,169	30.55	42.43

Table 5.
Trends of urban growth in Uttarakhand (1901–2011).

- The Mann–Kendall analysis of mean monthly streamflow data for last 31 years showed a nonsignificant decreasing trend during monsoon with a significance level of 10%.

The opposite trends observed between the streamflow and rainfall in majority of the watershed area suggest that endogenous change in the catchment dominates over exogenous changes. Abeysingha et al. [40] reported that the trend in annual streamflow for different rivers primarily was driven by changes in rainfall. In addition, Tiwari et al. [41] evaluated the actual evapotranspiration, runoff, and potential evapotranspiration for the past century by using monthly water balance model, and their analysis indicated that rainfall has been the primary factor of variability in the runoff.

The decreasing trend of streamflow in the downstream area of the river may be partly caused by the variations in rainfall and partly by other anthropogenic factors. Human activities such as water consumption, land use, and land cover changes caused by forest disturbances, soil and water conservation projects, new drain construction and city expansion, soil water infiltration, and surface evapotranspiration result in significant hydrological alteration [42, 43]. Nune et al. [44] also found a declining trend in streamflow without significant changes in rainfall Himayat Sagar catchment in India over 24 years (1980–2008). They also reported that streamflow trends declined mainly due to anthropogenic factors, such as changes in land use, watershed development, groundwater abstraction, and storage. Regarding the trend in the seasonal distribution of rainfall, we found that the pre-monsoon seasonal rainfall is increasing significantly, particularly in the month of May. In contrast, post-monsoon rainfall is decreasing significantly, especially in the downstream area.

Conflicts of interest

The authors declare no conflicts of interest.

Author details

Utkarsh Kumar[1,2]*, Rashmi[3], Dhirendra Kumar Singh[1], Suresh Chandra Panday[2], Manoj Parihar[2], Jaideep Kumar Bisht[2] and Lakshmi Kant[2]

1 Division of Agricultural Engineering, ICAR-Indian Agricultural Research Institute, New Delhi, India

2 Crop Production Division, ICAR-Vivekananda Parvatiya Krishi Anusandhan Sansthan, Almora, Uttarakhand 263601, India

3 Department of Agricultural Economics, Institute of Agricultural Sciences, Banaras Hindu University, Varanasi, India

*Address all correspondence to: utkarsh.kumar@icar.gov.in

References

[1] Singh SP, Singh V, Skutsch M. Rapid warming in the Himalayas: Ecosystem responses and development options. Climate and Development. 2010;**2**: 221-232

[2] Aggarwal PK. Impact of climate change on Indian agriculture. Journal of Plant Biology. 2003;**30**:189-118

[3] Dimri AP, Dash SK. Winter time climatic trends in the western Himalayas. Climate Change. 2012;**111**: 775-800

[4] Bhutiyani MR, Kale VS, Pawar NJ. Climate change and the precipitation variations in the northwestern Himalaya: 1866–2006. International Journal of Climatology. 2010;**30**:535-548

[5] Islam ZU, Rao LAK, Zargar AH, Ahmad S, Khan MA. Temperature variability in Himalayas and threat to the glaciers in the region: A study aided by remote sensing and GIS. Journal of Environmental Research Development. 2008;**3**:496-505

[6] Kumar U, Singh S, Bisht JK, Kant L. Augmenting spring discharge using Geospatial cum soil and water conservation technique: A case study of Almora. Indian Journal of Hill Farming. 2021;**34**:134-142

[7] Kumar U, Panday SC, Kumar J, Parihar M, Meena VS, Bisht JK, et al. Use of a decision support system to establish the best model for estimating reference evapotranspiration in sub-temperate climate: Almora, Uttarakhand. Agricultural Engineering International: CIGR Journal. 2022;**24**(1):41-50

[8] Kumar U, Panday SC, Kumar J, Meena VS, Parihar M, Singh S, et al. Comparison of recent rainfall trend in complex hilly terrain of sub–temperate region of Uttarakhand. Mausam. 2021; **72**:349-358

[9] Kumar U, Singh S, Bisht JK, Kant L. Use of meteorological data for identification of agricultural drought in Kumaon region of Uttarakhand. Journal of Earth System Science. 2021;**130**:121

[10] Kumar U, Meena VS, Singh S, Bisht JK, Pattanayak A. Evaluation of digital elevation model in hilly region of Uttarakhand: A case study of experimental farm Hawalbagh. Indian Journal of Soil Conservation. 2021;**49**:77-81

[11] Shafiq MUI, Islam ZU, Ahmad WS, Bhat MS, Ahmed P. Recent trends in precipitation regime of Kashmir valley, India. Disaster Advances. 2019;**12**:1-11

[12] Shekhar MS, Chand H, Kumar S, Srinivasan K, Ganju A. Climate change studies in the western Himalaya. Annals of Glaciology. 2010;**51**:105-112

[13] Joshi S, Kumar K, Joshi V, Pande B. Rainfall variability and indices of extreme rainfall-analysis and perception study for two stations over Central Himalaya, India. Natural Hazards. 2014; **72**(2):361-374

[14] Ménégoz M, Gallée H, Jacobi HW. Precipitation and snow cover in the Himalaya: From reanalysis to regional climate simulations. Hydrology and Earth System Sciences. 2013;**17**(10): 3921-3936

[15] Bolch T et al. The state and fate of Himalayan glaciers. Science. 2012; **336**(6079):310-314

[16] Bharti V. Investigation of Extreme Rainfall Events over the North West

Himalaya Region Using Satellite Data. Twente: University of Twente; 2015

[17] Niyogi D, Kishtawal C, Tripathi S, Govindaraju RS. Observational evidence that agricultural intensification and land use change may be reducing the Indian summer monsoon rainfall. Water Resources Research. 2010

[18] Pielke RA, Walko RL, Steyaert LT, et al. The influence of anthropogenic landscape changes on weather in South Florida. Monthly Weather Review. 1999; **127**(7):1663-1673

[19] Picard D. Testing and estimating change-points in time series. Advances in Applied Probability. 1985;**17**(04):841-867

[20] Halder S, Saha SK, Dirmeyer PA, et al. Investigating the impact of land-use land-cover change on Indian summer monsoon daily rainfall and temperature during 1951-2005 using a regional climate model. Hydrology and Earth System Sciences. 2016

[21] Pettitt AN. A non-parametric approach to the change-point problem. Journal of Royal Statistic Society Series C. 1979;**28**(2):126-135

[22] Raje D. Changepoint detection in hydrologic series of the Mahanadi River basin using a Fuzzy Bayesian approach. Journal of Hydrologic Engineering. 2014; **19**(4):687-698

[23] Salarijazi M. Trend and change-point detection for the annual stream-flow series of the Karun River at the Ahvaz hydrometric station. African Journal of Agricultural Research. 2012;7(32):4540-4552

[24] Hyvärinen V. Trends and characteristics of hydrological time series in Finland. Hydrology Research. 2003;**34**(1–2):71-90

[25] Xiong L, Guo S. Trend test and change-point detection for the annual discharge series of the Yangtze river at the Yichang hydrological station [Test de tendance et détection de rupture appliqués aux séries de débit annuel du fleuve Yangtze à la station hydrologique de Y]. Hydrological Sciences Journal. 2004;**49**(1):99-112

[26] Bassiouni M, Oki DS. Trends and shifts in streamflow in Hawai'i, 1913–2008. Hydrological Processes. 2013; **27**(10):1484-1500

[27] Villarini G, Smith JA, Serinaldi F, Ntelekos AA. Analyses of seasonal and annual maximum daily discharge records for central Europe. Journal of Hydrology. 2011;**399**(3–4):299-312

[28] Chen J, Brissette FP, Leconte R. Uncertainty of downscaling method in quantifying the impact of climate change on hydrology. Journal of Hydrology. 2011;**401**(3–4):190-202

[29] Koutsoyiannis D. Hydrology and change. Hydrological Sciences Journal. 2013;**58**(6):1177-1197

[30] Mann HB. Nonparametric tests against trend. Econometrica. 1945; **13**(606):245-259

[31] Kendall MG. Rank Correlation Methods. London: Griffin; 1975

[32] Yue S, Pilon P, Phinney B, Cavadias G. The influence of auto correlation he ability to detect trend in hydrological series. Hydrological Processes. 2002;**16**(9):1807-1829

[33] Sharif M, Archer DR, Fowler HJ, Forsythe N. Trends in timing and magnitude of flow in the Upper Indus basin. Hydrology and Earth System Sciences. 2013;**17**:1503-1516

[34] Zhang S, Lu XX. Hydrological responses to precipitation variation and diverse human activities in a mountainous tributary of lower Xijiang, China. Catena. 2009;**77**:130-142

[35] Gautam MR, Acharya K. Streamflow trends in Nepal. Hydrological Sciences Journal. 2012;**57**(2):344-357

[36] Theil H. A rank-invariant method of linear and polynomial regression analysis. Indagationes Mathematicae. 1950;**12**:85

[37] Hamed KH, Rao AR. A modified Mann-Kendall trend test for auto correlated data. Journal of Hydrology. 1998;**204**(1–4):182-196

[38] Sen PK. Estimates of the regression coefficient based on Kendall's tau. Journal of the American Statistical Association. 1968;**63**:1379-1389

[39] Milliman JD, Farnsworth KL, Jones PD, Xu KN, Smith LC. Climatic and anthropogenic factors affecting river discharge to the global ocean, 1951-2000. Global and Planetary Change. 2008;**62**: 187-194

[40] Abeysingha NS, Singh M, Sehgal VK, et al. Analysis of trends in streamflow and its linkages with rainfall and anthropogenic factors in Gomti River basin of North India. Theoretical Applied Climatology. 2016;**123**785-799

[41] Tiwari PC, Tiwari A, Joshi B. Urban growth in Himalaya: Understanding the process and options for sustainable development. Journal of Urban and Regional Studies on Contemporary India. 2018;**4**(2):15-27

[42] Gregory J, David M. Temporal and spatial variability of the global water balance. Climatic Change. 2012;**20**: 375-387

[43] Wei XH, Zhang MF. Quantifying streamflow change caused by forest disturbance at a large spatial scale: A single watershed study. Water Resources Research. 2010;**46**:W12525

[44] Nune R, George BA, Teluguntla P, Western AW. Relating trends in streamflow to anthropogenic influences: A case study of Himayat Sagar catchment, India. Hydrology and Earth System Sciences Discussions. 2012;**9**: 9295-9336

Chapter 6

Characteristics and Process Interactions in Natural Fluvial–Riparian Ecosystems: A Synopsis of the Watershed-Continuum Model

Lawrence E. Stevens, Raymond R. Johnson and Christopher Estes

Abstract

The watershed-continuum model (WCM) describes fluvial-riparian ecosystems (FREs) as dynamic reach-based ecohydrogeological riverine landscapes linking aquatic, riparian, and upland domains within watersheds. FRE domains include aquatic (channels, hyporheic zones, springs, other groundwater zones and in-channel lakes), riparian, and adjacent upland zones, all of which can interact spatio-temporally. Occupying only a minute proportion of the terrestrial surface, FREs contain and process only a tiny fraction of the Earth's freshwater, but often are highly productive, flood-disturbed, and ecologically interactive, supporting diverse, densely-packed biotic assemblages and socio-cultural resource uses and functions. FRE biodiversity is influenced by hydrogeomorphology, ecotonal transitions, and shifting habitat mosaics across stage elevation. Thus, the WCM integrates physical, biological, and socio-cultural characteristics, elements, and processes of FREs. Here, we summarize and illustrate the WCM, integrating diverse physical and ecological conceptual models to describe natural (unmanipulated) FRE dynamics. We integrate key processes affecting FRE forms and functions, and illustrate reach-based organization across temporal and spatial scales. Such a holistic approach into natural FRE structure and functions provides a baseline against which to measure and calibrate ecosystem alteration, management, and rehabilitation potential. Integration of groundwater, fluvial, and lacustrine ecological interactions within entire basins supports long-term, seasonally-based sustainable river management, which has never been more urgently needed.

Keywords: conceptual model, continuum, ecohydrogeology, ecosystem, fluvial, riparian, rivers, springs, streams, watershed

1. Introduction

Fluvial-riparian ecosystems (FREs) are watershed- (catchment-, drainage area-) based riverine landscape systems that integrate aquatic, riparian, and upland domains within watersheds, linking physical, biological, and cultural-economic processes [1–3]. From the context of a system FREs consist of "*... a structured set of objects and/or attributes... (,) components or variables ...that exhibit discernible relationships with one another and operate together as a complex whole ...*" ([4], 1–2; [5–7]). FREs include all sources of water that contribute to the basin's riverine ecosystem, including springs, surface runoff, lakes, and atmospheric sources such as humidity and fog. Only an average of 2120 km^3 (0.0002 percent) of the world's water exists in river systems at any given time [8] (**Figure 1**). But while rivers process only a tiny fraction of the Earth's fresh water and occupy only a minute proportion of the Earth's terrestrial surface, FREs are highly productive and ecologically interactive, often supporting complex landforms and diverse, densely packed biotic assemblages that change across fine to coarse spatial and

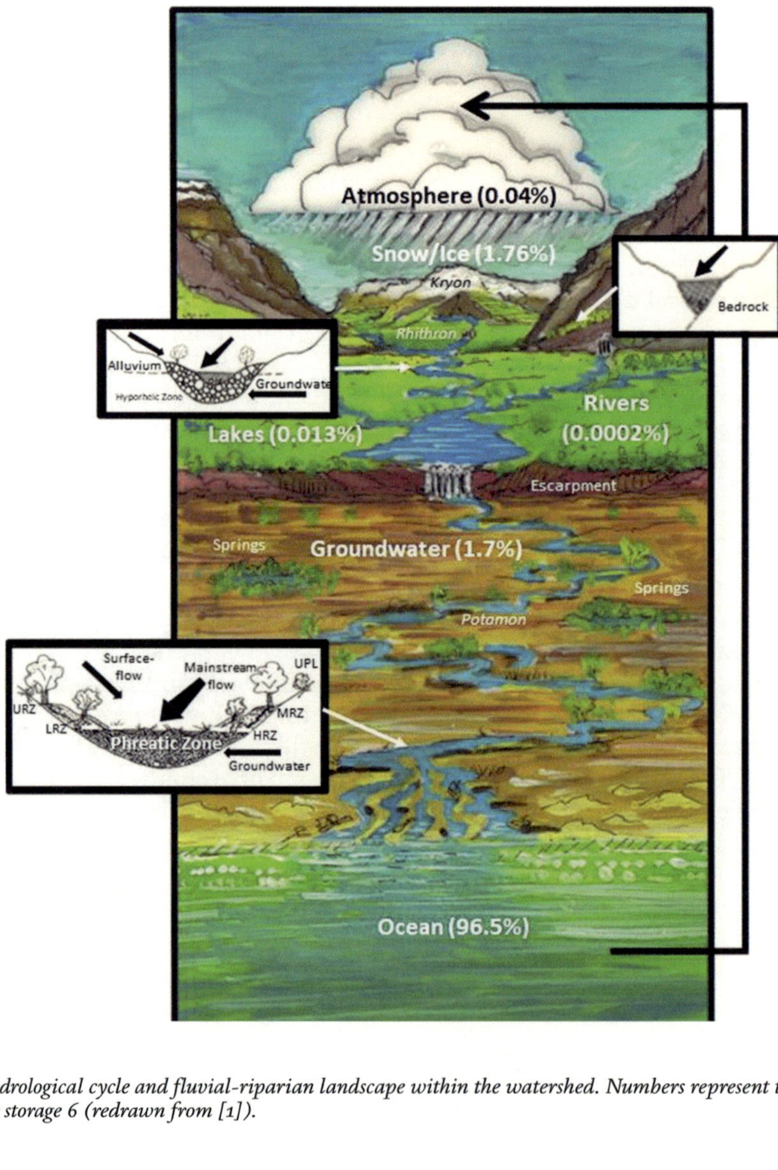

Figure 1.
Surface hydrological cycle and fluvial-riparian landscape within the watershed. Numbers represent the percent of freshwater storage 6 (redrawn from [1]).

temporal scales [9, 10] and burgeoning human populations. FRE physical and biological characteristics and processes among aquatic and riparian domains step, intergrade, and may interact through reaches within the watershed, from the headwaters to the terminus in an endorheic basin or the sea, and can extend far out into the submarine environment (e.g., [11]; **Figure 1**). Physically, FREs are "*complex adaptive process–response system(s) with ...morphological system(s) of channels, floodplains, hillslopes, deltas, ... and cascading system(s) of ...water and sediment*" [12].

Within FREs, the riparian domain is a zone of "*transition between the aquatic ecosystem and the adjacent terrestrial ecosystem* [13]. Riparian zones function as filters that reduce the impacts of flooding and surface runoff, as habitats that support vegetation, fish, and wildlife populations and habitat, and often provide critically important ecosystem goods and services [13–19]. Elevated FRE biodiversity is linked to, and influenced by factors including tectonics, geology, climate, hydrology, geomorphology, and latitude, in the context of shifting habitat mosaic heterogeneity and ecotonal dynamics [20–23]. Human reliance on FREs, and our species' evolutionary history and modern demography clearly demonstrate that reliance. As human domination of the Earth has progressed, rivers have been subjected to a host of anthropogenic alterations, including resource extraction, groundwater withdrawal, flow diversion and regulation, water quality degradation, and introduction of non-native species. The natural dimensions and human impacts on FREs have stimulated deep interest, concern, and much basic and applied research, generating a vast literature and prompting development of a suite of interrelated, but not necessarily integrated ecohydrogeological models. Focus on particular aspects of FRE channel development, geomorphology, ecology, or sociology has sometimes diminished wholistic integration. Also, graphic representation of FRE ecology can be improved to enhance conceptualization, and improve educational outreach.

Here, we provide an overview description and illustrated summary of the watershed-continuum model (WCM) [1], which couples interdisciplinary physical and ecological conceptualization of FRE ecology. The WCM links conceptual models of fluvial spatio-temporal development and geomorphology across stream order and reaches FRE ecology, trophic energy and matter dynamics, biodiversity, and evolutionary interactions from the river's source to its mouth. We provide a chronological analysis of major concepts in FRE ecohydrogeology (**Table 1**) and illustrate the WCM with an improved spatio-temporal, reach-linked conceptual diagram that integrates "bottom-up" physical factors, including geology, hydrology, geochemistry, geomorphology, sedimentology, and fluvial climate, with aquatic and riparian biotic assemblages and ecosystem structure within the watershed, and the potential for trophic cascade effects [26, 27, 36, 37, 99–102]. Due to the brevity of this manuscript, we emphasize here integration of physical and ecological conceptual elements and processes in natural, unmanipulated FREs, recognizing that such an integration is needed as a basis to improve watershed stewardship.

We reserve more detailed discussions on the details of riparian and aquatic community ecology related to the WCM for subsequent summaries [1] but focus on Integration and clear depiction of FRE domain interactions among reaches and over time within the basin. We discuss understudied issues and opportunities, the resolution of which will help advance FRE ecology in the future. Such an objective is essential for sustainable management of rivers.

Model	Description
Trophic-dynamic aspect of ecology [24, 25]	Consistency of energy dynamics across trophic levels within ecosystems (e.g., Cedar Bog, Silver Springs)
Stream order classification [26, 27]	Classification of dendritic hierarchy
Lentic ecosystem ecology (e.g. [28])	Limnology of fresh waters
Stream channel development [29–33]	Depth, width, velocity, slope, discharge, and sediment load interactively control channel geometry
Dynamic equilibrium concept [34]	Channel geomorphology and energy moves toward equilibrium, never reaching it due to subsequent perturbation
Perpetual Riparian Succession [35]	Regular flood scouring of floodplains keeps riparian vegetation in a state of perpetual or suspended succession
Lotic ecosystem ecology [36, 37]	Limnology of moving fresh waters
Riparian ecosystem ecology [13–21, 23, 38–44]	Riparian ecosystems are biologically diverse, structured, and highly ecologically interactive landscapes
River Continuum Concept [6, 45]	Rivers as flow-integrated ecosystems; invertebrate feeding guilds vary longitudinally
Dynamic Equilibrium Model (species richness) [46–49]	Intermediate levels of disturbance intensity and productivity maximize the biodiversity of passively dispersing organisms
Nutrient spiraling concept [50, 51]	Autochthonous (endogenic) and allochthonous (exogenic) nutrients and matter are transported through helical ecological pathways through FREs
Serial discontinuity concept [52, 53]	Relationships between natural as well as anthropogenic dams and tributaries regulate downstream FRE structure and function
Ecological and land use history [54, 55]	FRE ecology requires detailed and long-term understanding of geologic, hydrographic, biota, and land use history of the basin
Flood Pulse Concept [56, 57], e.g. [58]	Flooding regulates developmental cyclicity of rivers
Stream channel classification [59–64]	Systematic analysis of reach-based channel geometry
River Productivity Model [65, 66]	Fluvial productivity is spatially heterogeneous, affecting FRE ecological function
Process domain concept [67, 68]	Tributary and/or bedrock-controlled reaches generate fluvial geomorphic discontinuities
Telescoping material spiral model [69]	Material spirals tend to lengthen with stream order
Link discontinuity concept [70, 71]	Large tributaries create abrupt discontinuities, generating multi-reach alteration downstream
Riparian eco-hydrogeomorphology [13–23, 72–76]	Channel geomorphology and stage shape riparian zonation and ecology
Network Dynamics Hypothesis [77–79]	The complexity of the overall basin shapes tributary contributions to the mainstream
Variation in solar radiation influences FRE ecology [80, 81]	Cliff shading in canyon-bound FRE reaches strongly influences aquatic and riparian productivity
Top predators affect fluvial geomorphology [82]	Trophic cascades can influence channel geomorphology at a reach scale

Riparian vegetation Niche Construction Perspective and Niche-Box Model [83, 84]	Riparian plant species life history traits interact successionally with fluvial landform dynamics
Biogeochemical retention and processing network [85–87]	All parts of the fluvial system (mainstreams, floodplains, lakes, and wetlands) form a fluvial bio-geochemical retention and processing network
FRE species life history strategies interactions [86–90]	FRE guilds of plant and fish include competitive, ruderal, and stress-tolerant species
River wave concept [91]	FRE aquatic domain processes can be viewed as waves that determine or regulate production and transport of organic matter
Ephemeral stream ecology [42, 43, 92–93]	Seasonally ephemeral streams are punctuated, rapidly functioning biogeochemical systems
Biological stream width theory [94]	Resource subsidies from the FRE aquatic domain extend well into the surrounding upland terrain
Least Action Principle (LAP) [95]	FRE teleomatic change through the LAP to achieve maximum energy efficiency
Spring ecosystem classification and ecology [96, 97]	As "zero-order streams", spring contributions to FREs vary by geomorphic type
Integrated Metasystems Theory [98]	Regional processes act across spatial scales to control FRE form and function

Table 1.
Concepts in physical and biological FRE ecology, presented in approximate chronological order of publication. Rows in gray indicate concepts that are primarily focused on physical processes, whereas unshaded rows indicate concepts that are more strongly focused on eco-biological issues.

2. FRE elements, processes, and interactions

2.1 Overview

Natural FREs are terrestrial, dendritic surface-water flow paths transporting matter and energy both downslope and upslope through their channels within watersheds, with flow contributed by groundwater and multiple surface water, as well as atmospheric sources (**Figures 1** and **2**). Rivers play a disproportionately large role in geochemical cycling, biodiversity, human culture, esthetics, and socio-economic appropriation [98, 103]. Collectively, physical state variables regulate flow, hydrography, water quality, and sediment transport in a "bottom-up" ecosystem fashion, generating FRE habitat templates at a local scale and within reaches that are created by parent rock geology and geologic structure, as well as tributary influences [6, 67, 69, 98]. FREs are structured in relation to latitude, the extent of geological constraint, flooding, sediment transport, potential productivity, and many other factors. FRE aquatic and riparian habitats are colonized through both active and passive biogeographic processes and change across stage and over time through disequilibrial or successional processes [34, 46–48, 74, 75]. We present a rough chronology of advances in physical and biological conceptual models (**Table 1**), with additional references in the WCM [1], and illustrate this complex, "bottom-up" array of physical influences and responses on FRE structure (**Figure 2**).

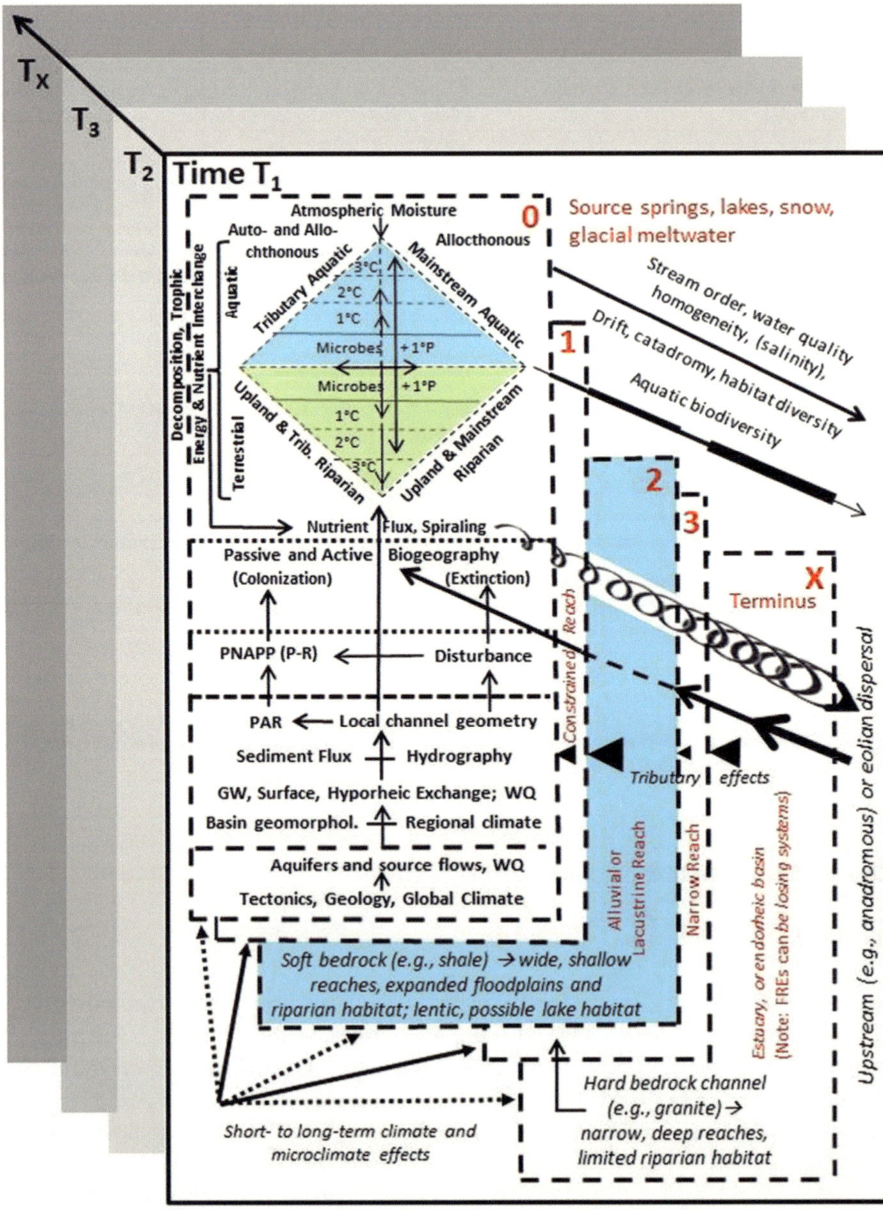

Figure 2.
Conceptual FRE model depicting interactions among independent and dependent physical and ecological variables and processes, across stream order (zero headwaters to X- mouth) and time (T_1 to T_X). Reaches depicted reflect general patterns of width. Line thickness indicates strength of effect; black arrow points indicate relative impacts of tributaries of different sizes. C – Consumers, P – Productivity or producers, PAR – Photosynthetically-active radiation, PNAPP – Potential net annual primary productivity R – Respiration, WQ – Water quality. Redrawn from [1].

2.2 Physical conceptual models

Basin and consequent FRE development vary in relation to complex interactions among tectonic terrain, geologic structure, parent bedrock, and climate over time

(**Table 1**; **Figure 1**). Tectonic setting and geologic history ultimately control regional aquifer and surface watershed development (e.g., [104, 105]). Large, high-order rivers develop over geologic time frames, particularly in interior continental settings. Basin integration across complex landscapes for many large rivers has largely occurred during Neogene time (<23.5 million yr.; e.g., the Amazon, Colorado, Mississippi, Nile, Rhine, Rio Grande, and Yangtze rivers) [106–111], although some river basins are far older (e.g., [112]).

River basins exist in a continuous state of development, expanding and capturing adjacent drainage basins at both gradual and punctuated rates (e.g., through periods of relatively rapid crustal deformation or vulcanism) and achieving transverse integration through at least five mechanisms (antecedence, superposition, anteposition, spillover, and piracy) [113, 114]. River channels respond to perturbations such as flood events by moving toward an equilibrium state that, due to recurring perturbation, may never be reached [101]. Although poorly synthesized, the three main tectonic interactions (convergence, divergence, and transform) are likely to produce different groundwater and surface water basin configurations. Aquifers that source rivers in these landscapes often are assumed to be constrained by surface catchment boundaries (e.g., [115]); however, such is not always the case (e.g., [116]). Regional climate also affects landscape geomorphology, and sometimes reciprocally influences tectonic processes (e.g., [117]). With tectonism as a driver, river course integration within a basin has been analogized to the process of organic evolution, in which drainage-head erosion allows individual tributaries to "competitively" integrate increasingly larger catchment areas, ultimately securing throughflow to the terminus [118].

Drainage network complexity is related to stream order, which increases when two channels of the same magnitude meet [26, 27]. Many of the world's major rivers arise from discrete springs, spring-fed marshes, or groundwater-fed lakes. Headwater springs function as "zero order" streams and springbrooks are commonly regarded as first-order channels. Rivers with spring-sourced baseflows include the Amazon, Colorado, Mississippi, Rhine, Volga, Murray, and many others [119]. Springs often exhibit strikingly different temperature and geochemical characteristics from those in the adjacent higher order streams into which they are confluent (e.g., [120]). Their unique water quality may influence mainstream processes, such as imprinting among larval fish [121] but see [122]. The ecological transition from headwater springs into first-order streams is highly individualistic, often occurring at a chemically and thermally discrete distance from the source [123]. The quantity and quality of riverside or in-stream springs, as well as seasonal flow changes driven by precipitation, also can affect stream channel geomorphology through travertine deposition (e.g., [124]), persistence of in-channel woody vegetation [125], and river water quality (e.g., geothermal spring influences; e.g., [126]).

Stream order increases erratically downstream, with middle-order streams often having the greatest productivity and species richness. FRE lacustrine reaches can occur at any order within a basin (**Figures 1** and **2**) and lake limnological processes like thermal or chemical stratification can influence downstream flow and geochemistry. The largest rivers include the Amazon and Volga, which are regarded as 12th or 13th order streams. Such large rivers buffer temporal and spatial changes in water temperature, geochemistry, and the timing of dynamic equilibration.

River drainage networks often are subdivided into segments and reaches, which function as organizational units within FREs. Reaches lie within segments and are best distinguished geomorphologically on the basis of differences in parent rock geology, shoreline erodibility, slope (gradient), and thalweg position (e.g., [127]). River segments (*sensu* [128]) include one or more river reaches that are collectively subject to distinctive changes in one or more ecosystem characteristics (e.g., water temperature, geochemistry, suspended sediment load, or gamma diversity in an assemblage). Such changes often are introduced by a tributary, thereby affecting downstream FRE ecology [70, 71, 79, 129, 130].

The frequency, duration, magnitude, and timing of both high and low flows are critical determinants of FRE ecology. Discharge and flood frequency and magnitude are increasingly monitored and evaluated in relation to human activities within watershed. Magilligan [131] described variation in channel boundary shear stress and unit stream power on an array of stream channels across 2- to 500-year floods. She noted three-fold order of magnitude variation in flood power through the basin due to valley width, with broad, alluvial channels in wide valleys subject to lower flood power. In contrast, she reported increased stream power in narrow valleys with constrained channels, a pattern influenced both by basin size and by local controls. She also suggested that maximum flood impacts on channel geomorphology occur at discrete points within reaches. Such focal points are likely to shift over time, suggesting that drainage evolution may occur most intensively at the local scale. Antecedent events are critically important, as prior high flows exert long-lasting impacts on FRE structure and ecology (e.g., [34, 132, 133], many others). Detection of such events through dendrochronology is becoming more frequently used, helping to determine long-term drought frequency and duration, and providing insights into adaptive strategic options for FRE management in the face of climate change (e.g., [134–139]).

The impacts of natural, regular, short-term stage fluctuations in rivers are generally poorly known, but are of great consequence in management of rivers impounded for hydroelectric power production (e.g., [140, 141]). Natural semi-daily tidal bores are common in the lowermost reaches of low-gradient rivers that reach the sea. Daily variation in flow stage in such settings may desiccate or freeze macrophytes, macroinvertebrate habitats and eggs, or interrupt aquatic and riparian faunal feeding and other behaviors, leading to reduced or fluctuating primary and secondary consumer production. Understanding the effects of natural fluctuating flows remains a potentially important topic for future natural and regulated FRE research.

River water quality varies across lithology, latitude, elevation, humidity province, season, and stream order within basins and among reaches, and springs, lakes, and glacial melt influence river waters. Water quality characteristics can transition markedly over stream order and are important determinants of macrophyte structure and composition, and life history and feeding guild distributions of aquatic macroinvertebrates, fish, and amphibians, in turn influencing food web linkage ([141], but see [142]) and riparian groundwater quality and quantity. Surface flow geochemical and sedimentological changes occur at tributary confluences (e.g., [129, 130]), abruptly as discontinuities, or more gradually and to a lesser extent in side channels and other shallow, low-velocity shoreline habitats. Limnologically-influenced water quality dominates lake-sourced rivers, but we know of little research on natural downriver responses to such alteration. River water generally trends toward a universal quality across stream order, generating relatively similar geochemistry among the world's major rivers at their mouths; however, the contributions and evolution of FRE water quality depend in large measure on subbasin geology and the relative contributions of tributaries (e.g., [143, 144]), as well as anthropogenic impacts.

The erosion, and deposition of bed, suspended, dissolved loads, and flotsam sediments are related to watershed geology, aquifer properties, flow dynamics, channel configuration, and other factors like glacial influences [29, 32, 33, 59, 127, 145, 146].

Cumulatively, including anthropogenic materials, the world's rivers deposit about 20 billion mt of solids into the sea each year [147]. Rather than being a sole function of basin area, this deposition is largely the result of discharge from many thousands of small basins (<10,000 km^2) with relatively high-gradient rivers that mouth directly into the seas [148]. Large rivers deposit proportionally less sediment due to subaqueous storage in deltas.

Recent research and stewardship attention in fluvial geomorphology has shifted to temporally based based spatial scales of reaches in the continuum of alluvial to constrained rivers (e.g., [10, 53, 67, 68]). Alluvial reaches often have relatively uniform bed materials and channel landform configuration, and often are closer to equilibrium than geologically constrained channels. Models of sediment deposition and erosion are diverse (reviewed in [149], among many others) and can provide adequate two-dimensional prediction of suspended sediment transport through channels with varying bed roughness, channel steepness, and sediment transport. However, most rivers have insufficient historical flow, sedimentological, and hydrographic data to permit high-precision modeling [150]. Variation in turbulence, sheer stress, transport capacity, and bed and suspended sediment loading are likely to increase channel landform and FRE habitat diversity among reaches. In comparison with constrained reaches, seasonally dynamic alluvial reaches often support broader riparian zones, with increasing filtering, storage, and processing of matter from upstream and local sources [75, 76, 151].

Many watershed factors Influence FRE functions. Upland wildfire, forest pest insect outbreaks, coarse wood debris loading, and overgrazing can affect fluvial FRE sedimentation, geomorphology, and nutrient and nutrient transport (e.g., [152, 153]). Ice-related impacts on FREs in temperate and boreal streams involve ice formation, "shoving", and black-ice melting, resulting in severe scouring of shorelines and bed surfaces, damming of channels, uplifting/redeposition of fine to coarse substrata (including boulders and coarse woody debris) [153, 154].

Fluvial climate is influenced by global- to local-scale conditions, the latter including nocturnal cool air subsidence and upriver mountain valley wind patterns (e.g., [155, 156]). However, few meso-scale studies of river influences on basin, reach, or local microclimate have been conducted. Riparian and in-stream interception of photosynthetically-active radiation (PAR) varies temporally and by reach in canyon-bound rivers, influencing in-canyon air temperature, relative humidity, and aquatic and riparian production (e.g., [80, 81]). Although not yet studied, variation in PAR flux also may influence erosion in temperate cliff-dominated channels through increased slope failure frequency, cliff retreat, and canyon landform evolution. At local, cross-sectional scales, discharge, cliff shading, and channel aspect influence microclimate and fluvial solar energy flux, in turn influencing FRE air temperature and relative humidity [157], which has been positively associated with temperate avian species diversity [158]. Thus, along with riparian soil ecology, fluvial microclimate may contribute to the biodiversity and productivity of riparian zones.

Habitat complexity at tributary confluences increases ecological productivity and biodiversity, and sustains habitat spatio-temporal connectivity [10, 39, 65, 70, 71, 159]. Tributary impacts on mainstream water quality are greatest when flows of the former exceed those of the latter; however, differences in biota may follow the opposite pattern. Aquatic macroinvertebrate assemblages in small, spring-fed tributaries may substantially differ from those in the large, adjacent mainstream (e.g., [130]). The WCM depicts the magnitude of tributary influences as dark triangles of varying size (**Figure 2**), but we note that large or influential tributaries can exist and interact with the mainstream anywhere in the watershed. The process domain concept (PDC) posits that variation among such geophysical processes at the reach scale shapes channel form, disturbance responses, and ecosystem structure and dynamics [68].

Many FRE physical models have focused on particular aspects of FRE channel development, form, and function. Leopold summarized much of his research on alluvial channels, through which he was able to describe the negative relationship between sinuosity and slope [160]. He lamented [33] that no comprehensive channel model adequately encompassed the self-regulating nature of alluvial channels. Existing alluvial channel studies and models have been criticized for generating: (1) untested qualitative unifying theories; (2) empirical and statistical analyses that support focused semi-quantitative models; (3) reductionist applications of Newtonian theory; and (4) theoretical resolutions of primary flow eqs. [95]. Based on Morisawa's [34] articulation of the dynamic equilibrium concept, Nanson and Huang proposed focus on the least action principle, as represented in alluvial channels through the trend toward maximum flow efficiency [95]. While the WCM does not presume that all FRE component models (particularly ecological models) will have clear, predictive mathematical solutions, placing these many concepts in a logical order and visually representing them is an important step forward (**Table 1**; **Figure 2**).

2.3 Ecological conceptual models

Ecosystem ecology developed through the diverse contributions of Linnaeus, von Humboldt, Mobius, Darwin, Forbes, Warming, Cowles, Elton, and many others. Tansley defined the ecosystem concept as involving interrelated physical and biological elements and processes [161], and Lindeman [24] and Odum [25] initiated analysis of trophic-dynamic aspects of ecology (**Table 1**). FREs are primarily driven by physical factors, generating "bottom-up" ecosystem structure, with dependent biotic composition, structure, function, and trophic interactions (**Figures 2** and **3**). Lower stream order FRE changes often occur in a punctuated, stepped, or reach-bounded fashion as FREs receive tributary contributions of sediment, water temperature, and nutrients. As with fluvial water quality, the ecology of higher-order streams generally changes more gradually, both spatially and over time.

Hutchinson emphasized lake ecosystem limnology (e.g., [28]), while Hynes described stream limnology [36, 37], including the spatial scale and groundwater influences on the watershed, but with somewhat less attention to the FRE riparian domain. Hynsian (lotic) versus Hutchinsonian (lentic) emphases created long-standing differences in interpretation of the roles of habitat and biotic factors on FRE research [162]. Nonetheless, combining these lines of inquiry initiated a plethora of subsequent integrative research on FRE ecology, which continues today.

The most prominent post-Hynesian FRE conceptual advance was the river continuum concept (RCC) [6, 45, 163]. The RCC described a river ecosystem as "...a continuum of biotic adjustments and consistent pattern of loading, transport, utilization, and storage of organic matter along the (ir) length" ([6], 130). The RCC regarded "...the entire fluvial system as a continuously integrated series of physical gradients and associated biotic adjustments as the river flows from headwater to mouth", with "...maintenance of longitudinal, lateral, and vertical pathways for biological, hydrological, and physical processes" ([99], 9–10); see [12]. The RCC lent support many patterns observed in studies of low-medium order streams and ichthyological studies, primarily in mesic regions [159, 164–166]. However, it has been criticized for not fully recognizing the roles of: (1) fluvial discontinuities; (2) groundwater and spring

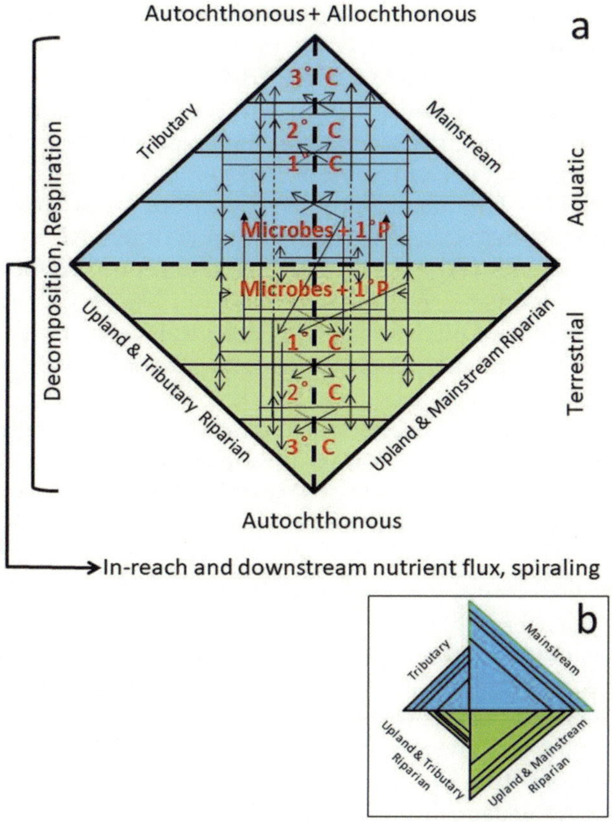

Figure 3.
(a) Expanded detail of foodweb linkages in FREs, contrasting allochthonous (uplands and tributary) vs. autochthonous (mainstream) ecosystem energy inputs with aquatic vs. riparian domain interactions. Arrows indicate common energy pathways among trophic levels in the four FRE arenas. Not all interactions occur in every FRE, and other trophic interactions not depicted here may exist in some FREs. (b) Differential spatial or functional change in reach-based FRE structure and function can occur in response to watershed changes. For example, upland fire can result in sediment, ash, and nutrient loading through tributaries, processes that may diminish FRE productivity and ecological role in the watershed. Similarly, reduction in precipitation or groundwater alterations through climate change or aquifer depletion may reduce mainstream and riparian function. Redrawn from. [1].

sourcing [167]; (3) river-sourced lakes, lentic zones, and productivity hot spots (e.g., [65]); (4) hyporheic refugia [99, 101, 168, 169]; (5) riparian ecology, except as subservient to the aquatic domain; (6) the role of temporal scale in FRE development and function, including dynamic seasonal and interannual geomorphic perturbation and adjustment [34]; (7) ephemeral and intermittent stream FRE ecology [170]; as well as (8) its applicability to higher order streams [171, 172].

Subsequent to the RCC, many FRE syntheses have been undertaken, including comprehensive edited volumes and reviews (including but not limited to [1, 23, 41–44, 66, 69, 73, 75, 91, 99, 101, 102, 151, 164–167, 171, 173–178]). Below we briefly describe some of the major biologically-based conceptual components of the WCM.

Ward clarified four dimensions of spatial and temporal scale operating in most lotic ecosystems, including (1) the "longitudinal" dimension up- and downstream through rivers; (2) across-channel, riparian-aquatic domain interactions; (3) vertical interactions with hyporheic habitats and groundwater; and (4) a broad temporal dimension [102]. Dynamic interactions among these dimensions contribute to the individuality in the character of FREs. Focusing on riparian stage zones and related to Ward's considerations, Nilsson and Svedmark [76] recognized four major processes or characteristics interactively functioning in FREs. (a) Flow regime (hydrographic) dynamics regulate FRE ecological and geomorphological processes (including riparian succession through Connell and Slayters' three modes – facilitation, inhibition, and tolerance [179]). (b) The channel provides a corridor for organic and organic transport, primarily downstream but also upstream via anemochorous and zoochorous dispersal of propagules (**Figure 2**). (c) The riparian zone functions as a filter and boundary between upland and riverine processes (e.g., [85, 180]). (d) Many have recognized the high levels of biodiversity and ecological interactivity of FREs, related to elevated productivity and disturbance and high levels of habitat heterogeneity (e.g., [38, 39, 181]).

The flood pulse concept emphasized the importance of high flow pulses to FRE ecology by regularly restructuring channel and riparian landscapes [56, 57]. Regular seasonal flooding accounts for the state of suspended (or perpetual) succession in natural riparian vegetation zones, particularly in constrained channels [35], with riparian vegetation zonation in belts parallel to the mainstream [182] and with composition controlled by physiological and life history characteristics (e.g., riparian response guilds models [83, 183]).

Advances in nearshore marine ecology patch and disturbance dynamics concepts (e.g., [184–187]) contributed to Thorp and DeLong's [70] riverine productivity model (RPM). The RPM posited that production, as well as decomposition, recruitment, and other important river processes are related to niche diversity, occurring at specific points within the channel, such as at tributary confluences, along shorelines, or in specific depositional settings. Thus, FREs function as microhabitat mosaics.

The serial discontinuity concept (SDC) initially was developed to describe the impacts of impoundments on regulated rivers [52, 53], but also by extension to the roles and impacts of natural dams that form lacustrine reaches, and affect natural FRE channel geomorphology, flow, and population dynamics, both upstream and downstream. Lacustrine reaches can occur anywhere in a basin as a result of tectonism, lava dams, slope failure, or glacier development, and natural dams may persist for evolutionarily significant durations. The SDC posits that the location and size of a dam reset and influence downstream recovery of FRE characteristics through tributary contributions of flow, water quality, and biota [129], and through link discontinuity [70, 71] and network dynamics [77–79]. Examples of natural dams include Lake Victoria in Uganda, which formed as a result of tectonic rifting and interruption of flow in the Kagera and other Nile River headwater streams; Lago de Nicaragua (L. Cocibolca) in Central America, which formed as a result of tectonic uplift in the lower Tipitapa and San Juan River basins; and many basalt dams ni the southwestern USA [188–191]. Six types of slope failure dams that can affect rivers are globally recognized, ranging from relatively common single events with partial valley impoundment to rare, simultaneous impoundment of multiple valleys that create multiple natural lakes [191] (e.g., [192–194]). Ice dam failure also is a well-known phenomenon (e.g., the collapse of Pleistocene Lake Missoula [195], and fjord ice dam failures [196]). These natural impoundments and their failures change downstream channel geomorphology, water quality and flow, hydrography, stage relations, velocity, habitat quality and distribution, and FRE biogeography.

The RCC did not adequately integrate the ecology of ephemeral and intermittent FREs or groundwater-surface water interactions [101]. Colloquially known as dry washes, arroyos, or wadis, ephemeral channels are extremely abundant, comprising far more than half of the global stream channel network [170], and are becoming increasingly abundant as humans and climate change dewater rivers (e.g., [197]). Flooding releases CO_2 sequestered by seasonal or erratic burial of organic matter and invertebrates, such as leaf litter or clams [198]. Benthic invertebrates that shred, graze, or collect organic debris often are generally absent or rare in ephemeral FREs, reducing decomposition rates and transferring those functions to microbial and physical molar actions when the stream floods. Terrestrially, ephemeral versus intermittent riparian zones are bordered by distinctive suites of xeroriparian (dry riparian) to mesoriparian perennial plant species that provide cover and food resources [43]. Analysis of an ephemeral stream in Pakistan revealed deeply rooted woody perennial shrubs in the channel, and a bed dominated by weeds after winter rains, with drought-resistant species occurring on terraces [199]. Aquatic productivity and trophic energetics of arid-land ephemeral streams are reduced and interrupted during dry seasons (e.g., [92]), generating temporal discontinuities of stream processes. Nonetheless, ephemeral channels commonly provide essential wildlife habitat connectivity, and function as punctuated, rapidly changing biogeochemical reactors [93], and warranting additional research.

FRE productivity (P) and disturbance (D) intensity interactively influence aquatic and riparian domain biodiversity through habitat and resource availability, organism size distributions, niche specialization, assemblage composition, competition, and other factors. P and D are related to colonization (C) and extinction (E) processes in insular biogeographic models of species richness, with high levels of P related to C, and high levels of D related to E (**Figure 2**) [46–48, 200–207]. Riparian and shallow aquatic domain P and D, as well as the depth, velocity, and transparency of water, and soil moisture and nutrient content are generally negatively related to stage. FRE riparian biodiversity tends to be maximal at intermediate levels of P and D, at intermediate flood return frequencies, and terrace stages with the maximal "ecological hospitability" to potential colonists. Our observations suggest that this pattern appears to be reversed in the aquatic domain, where shallow shorelines are most biologically diverse; however, more research is needed to understand this FRE "mirror effect" between the two domains. Both the intermediate disturbance hypothesis and insular biogeographic theory [46–48, 207] were developed to describe the biodiversity of sessile taxa, such as plants and corals, not vagile species like many larger macroinvertebrates, fish, and other vertebrates, which often actively cue on hydrographic disturbances (e.g., flood avoidance by belostomatid giant water bugs [208], or hydrograph-cued spawning among fish species). Such relationships may result, in some cases, in FRE riparian trophic cascades in which top predators can reciprocally influence channel geomorphology at reach scales [82]. Such cascades are regularly observed in fish-dominated ecosystems and in some low-order fishless systems, but often are limited in FRE aquatic domains by bottom-up physical processes (e.g., hydrology, sediment transport, ice impacts), with average sheer stress/unit area often negatively related to stream order [131]. Nonetheless, the commonly observed phenomenon of elevated FRE biodiversity, particularly at intermediate stream orders, is at least somewhat related to these coupled gradient interactions and convergent life history strategies.

FREs receive and transmit multidimensional exchanges of ecological matter and energy subsidies in the watershed. Muehlbauer et al. [94] reported that avenues of exchange may be relatively narrow (50% of exchanges occurred within 1.5 m of the stream edge), and 10% of the exchanges occurred 0.5 km into the adjacent uplands. Exchanges are reported to disproportionally influence primary producers and predators, potentially affecting both bottom-up and trophic cascades (e.g., [209]). FRE subsidy exchange involves five spatially directional processes over time, including (a) gravity-driven downslope flow and material allochthonous transport; (b) downstream flow; (c) river-to-uplands eolian and zoochorous transport; (d) lateral and downward surface to hyporheic transport; and (e) upwelling artesian groundwater influences. In addition, downstream main channel or tributary flooding in very low gradient reaches can initiate upstream-directed flow [102, 210]. Like channel geometry, these FRE ecological processes respond dynamically and temporally to climate and other factors, moving toward, but never achieving equilibrium in form, function, boundary conditions, matter transport, or trophic energy dynamics described for channel adjustment by [1, 34, 93].

In the riparian domain, FRE riparian vegetation is "*...a complex of vegetation units along the river network that is functionally related to the other components of the fluvial system and surrounding area*" [211, 212], which interacts in a reach- and segment-dependent fashion with the aquatic domain and its associated processes)" [213]. Like the aquatic domain, the riparian domain is interactively influenced by regional climate (e.g., [214–216]) through direct forcing effects on channel roughness, bedform morphology, and sediment transport during peak flows and seasonally changing rainfall and snowmelt [217–219], and by drought [220, 221]. Groundwater availability also can affect channel and floodplain stability and riparian vegetation [222–226]; (e.g., [227]), [228]. Climate influences on FRE groundwater vary spatiotemporally but can provide recharge that affects reach and segment scales through precipitation and infiltration, with potentially strong seasonal variation, as demonstrated through isotopic studies [229–236].

Coupling the calculation of the standardized precipitation index (SPI; [237]) with the standardized groundwater level index (SGI; [238, 239]) can be used to relate precipitation to groundwater recharge [240, 241]. These metrics affect FRE riparian productivity [242, 243] (e.g., [244–248]) through groundwater recharge in relation to river stage [249], and are affected by air temperature [250–252] and extreme precipitation events [253–255], which in turn affect stream discharge [256], groundwater recharge and availability [257, 258], and phreatic zone and riparian rooting depth [259]. Inorganic sediment transport and turbidity generally (but not always) increase with stream order, reducing downstream PAR availability and 1° through 3° aquatic production [81] and strongly influencing aquatic macroinvertebrate feeding guild structure [6] and riparian nutrient availability. Complex trophic relationships can develop in riparian zones, directly and indirectly influencing primary producer structure and composition. For example, leaf-beetles, grasshoppers, beavers, and ungulates all can strongly influence riparian vegetation composition, structure, and decomposition/soil formation [260–262].

FRE biogeography involves colonization, recruitment, and population establishment overland by volant and other highly vagile species, as well as passive dispersal through gravity, aerial drift, or zoochorous transport of propagules through both overland and dendritic stream corridors (**Figure 2**). Regardless of the pathway, FRE population persistence and assemblage resilience are predicated on the ability of a species to remain in, or disperse-recover their position in the watershed. Therefore, persistence of all FRE species requires some form of upstream dispersal, with eviction or extirpation the inevitable consequence of failed *in situ* or headwater recruitment strategies. Larval aquatic macroinvertebrates may drift downstream, while adult aquatic insects often fly or are blown upstream as aerial drift. Dragonflies, salmonids, and many other fish taxa migrate upstream to spawn, against the dominant current direction, and some fish transport larval unionid mussel larvae upstream. Migratory western North American warblers and other passerine birds intensively use arid-land riparian habitat as stop-over habitat during migration [263, 264], a pattern not strongly evident in mesic eastern North America [265]. Although front-based migration also occurs among some shorebirds, many waterbird species follow FRE corridors, particularly through complex landscapes [80, 266]. In addition, many non-volant vertebrate species follow river corridors as dendritic pathways, although terrestrial faunal movements can be thwarted by steep cliffs, perilous crossings, and anthropogenic landscape interruptions [80].

The riparian plant niche construction perspective [83] and niche-box model (NBM) [84, 183] classified guilds of riparian plants in relation to similarities among life history traits. The NBM incorporates and compares autecological elements for each plant species to improve prediction of vegetation assemblage development in relation to hydrography and riparian conditions. While successfully grouping some species, the large amount of variation in the NBM multivariate plots is a reminder that life history strategies vary tremendously among species, variance that is highly adaptive but which does not readily lend itself to simple classification systems.

Trajectories of vegetation succession through modes of facilitation, inhibition, or tolerance [179] differ temporally among reaches, between humidity provinces and in relation to stream order, fluvial hydrodynamics (disturbance frequency and across stage), geomorphic setting, grain-size distribution, depth to water table [267], and biological effects, such as mycorrhizal succession [268], selective vertebrate [260] or invertebrate herbivory [261, 262], in relation to plant diseases and the presence of some bird species [204]. Surrounding upland assemblages also strongly interact with upper riparian terrace vegetation (e.g., [182, 269]). Due to increased riparian soil water availability and regular flood disturbance, FRE riparian vegetation structure is not well represented by the upland-centric Holdridge [270] global vegetation model [1].

3. Conclusions and research recommendations

Fluvial-riparian ecosystems are hierarchically and dynamically influenced by physical and biotic processes that vary spatially among reaches, over stream order and time within the watershed, approaching but rarely achieving equilibrium in channel geometry, fluid and matter transport, biotic composition, and ecosystem energy dynamics and structure. A wide array of conceptual physical and ecological models has described aspects of FRE ecology and responses to natural and anthropogenic perturbations. However, most models have focused on single or a reduced suite of variables at site-specific, within-reach, alluvial or constrained channels, other watershed scales, and most often on anthropogenically-altered streams. The WCM emphasizes the importance of understanding temporal and spatial scaling across the entire basin in natural systems to provide guidance for improving FRE stewardship.

Despite much progress, a wide array of important ecohydrological processes, questions, and issues remain to be addressed or more fully integrated. Not presented in prioritized order, this list of additional research topics includes but is not limited to: (1) corresponding convergence toward dynamic equilibrium of physical and biological processes; (2) extent of self-similarity among physical and biological processes across reach and stream order, space, and time; (3) groundwater-surface water interactions and connectivity under changing climates; (4) the significance, extent, and roles of groundwater and headwater springs as zero order streams in FRE ecology [120, 271]; (5) natural inter-relationships among lentic and downstream lotic reaches; (6) ephemeral stream ecosystem ecology; (7) the limiting effects of photosynthetically-active radiation in canyon-bound streams; (8) stream microclimate ecology; (9) the interactive effects of flooding, ice, and glacial effects in boreal and high elevation streams; (10) multidirectional subsidy and gene exchange in dendritic pathways; (11) the role of plant physiology in riparian vegetation zonation; (12) the significance of corridors, barriers/filter, and refugial biogeographic effects in dendritic river ecosystems [80]; (13) stream order-driven and cross-sectional spatial impacts on biodiversity; (14) population and successional models among FRE biota across trophic levels; (15) FRE ecosystem genetics and the evolution of endemism across latitude, longitude, and among tectonic landscapes; and (16) the role of noise on FRE faunal assemblages. Adequately examining and incorporating these and other topics will more fully expand the WCM model through future research and will enhance collaborative discussion among hydrogeological, ecological, and socio-cultural disciplines [272]. Such data and integration efforts will improve understanding, modeling, and stewardship of FREs at local, regional, and global spatial and temporal scales.

FREs are complex, continually changing, and vital to life on Earth. Although informative and elegant, all FRE models remain incomplete, and even the most comprehensive FRE conceptual syntheses fall short of adequately representing these remarkable, important, and dynamic ecosystems. Furthermore, discipline-based or regional specificity has often limited the applicability of some models (e.g., [273–278]). Here we present and illustrate a synthesis of FRE knowledge through the WCM, and suggest topics for further investigation. However, FREs cannot be readily, adequately, or usefully reduced to a single suite of equations or simple illustrations. For example, non-Judeo-Christian-Islamic cultures commonly view rivers as living entities, supporting divine spirits, and essential to cultural well-being. Integrating indigenous traditional ecological concepts and knowledge into improved stewardship has rarely been achieved. We suggest that improved comprehension of FREs may require consideration of other socio-cultural dimensions. Enhanced understanding of the complex, multidimensional inter-relationships among Physical, biological, cultural, and socio-economic elements and processes within watersheds is essential to improving FRE stewardship and sustaining ecological functions vital to nature, human life, and societal integrity.

Acknowledgements

We thank the Springs Stewardship Institute for institutional support.

Conflicts of interest

We declare no conflicts of interest related to this manuscript.

Author details

Lawrence E. Stevens[1*], Raymond R. Johnson[2] and Christopher Estes[3]

1 Springs Stewardship Institute, Flagstaff, AZ, USA

2 Hunters Run, Tucson, AZ, USA

3 Chalk Board Enterprises, LLC, Anchorage, AK, USA

*Address all correspondence to: larry@springstewardship.org

References

[1] Stevens LE, Johnson RR, Estes C. The watershed continuum: A conceptual fluvial-riparian ecosystem model. In: Johnson RR, Carothers SW, Finch DM, Kingsley KJ, Stanley JT, editors. Riparian Research and Management: Past, Present, Future. Ft. Collins: U.S. Department of Agriculture, Forest Service; 2020. pp. 80-137

[2] Allan JD, Castillo MM, Capps KA. Fluvial Geomorphology: Stream Ecology. Cham: Springer; 2021. pp. 45-73

[3] Tockner K, Ward JV, Edwards PJ, Kollmann J. Riverine landscapes: An introduction. Freshwater Biology. 2002;**47**:497-500

[4] Chorley RJ, Kennedy BA. Physical Geography: A Systems Approach. London: Prentice-Hall; 1971

[5] Huggett R. A history of the systems approach in geomorphology Une histoire de l' approche systémique en géomorphologie. Géomorphologie Reli. Process. Environ. 2007;**2**:145-158. DOI: 10.4000/geomorphologie.1031

[6] Vannote RL, Minshall GW, Cummins KW, Sedell JR, Cushing CE. The river continuum concept. Canadian Journal of Fisheries and Aquatic Sciences. 1980;**37**:130-137

[7] Ward ND, Sawakuchi HO, Richey JE. The Amazon River's ecosystem: Where land meets the sea. Eos. 2018;**99**. DOI: 10.1029/2018EO088573

[8] Shiklomanov I. World fresh water resources. In: Gleick PH, editor. Water in Crisis: A Guide to the World's Fresh Water Resources. New York: Oxford University Press; 1993. pp. 13-24

[9] Behrensmeyer AK, Damuth JA, Dimichele WA, Potts R, Sues H-D, Wing SL. Terrestrial Ecosystems through Time: Evolutionary Paleoecology of Terrestrial Plants and Animals. Chicago, IL: University of Chicago Press; 1992. pp. 205-325

[10] Sabo JL, Hagen EM. A network theory for resource exchange between rivers and their watersheds. Water Resources Research. 2012;**48**:W04515

[11] Stanford JA, Ward JV. An ecosystem perspective of alluvial rivers: Connectivity and the hyporheic corridor. Journal of North American Benthological Society. 1993;**12**(1):48-60. DOI: 10.2307/1467685

[12] Piégay H. System approaches in fluvial geomorphology. In: Kondolf H, Piegay GM, editors. Tools in Fluvial Geomorphology. West Sussex, UK: Wiley; 2003. pp. 79-101

[13] Verry ES, Dolloff CA, Manning ME. Riparian ecotone: A functional definition and delineation for resource assessment. Water, Air, Soil Pollution Focus. 2004;**4**(1):67-94. DOI: 10.1023/B:WAFO.0000012825.77300.08

[14] Zarnott DH, Tavares VEQ, Giardina C, Alba JMF. The influence of riparian areas on direct surface runoff of precipitation events. Floresta e Ambient. 2020;**27**(4)

[15] Vogel HF, Zawadzki CH, Metri R. Florestas ripárias: importância e principais ameaças. Rev. Saúde e Biol. 2009;**4**(1):24-30

[16] Pusey B, Arthington A. Importance of the riparian zone to the conservation and management of freshwater fish: a review. Marine Freshwater Research. 2003;**54**:1-6

[17] Meehan WR, Swanson FJ, Sedell JR. Influences of riparian vegetation on aquatic ecosystems with particular references to salmonid fishes and their food supply. In: Johnson RR, Jones DA, editors. Importance Preservation and Management of Riparian Habitat: A Symposium. Rocky Mountain Forest and Range Experiment Station. Fort Collins, CO, USA: USDA Forest Service; 1977. pp. 137-145

[18] Swanson FJ, Gregory SV, Sedell JR, Campbell AG. Land-Water Interaction: The Riparian Zone. In: Analysis of Coniferous Forest Ecosystems in the Western United States US/IBP Synthesis Series 14. Stroudsburg, PA, USA: Hutchinson Ross Publishing Co; 1982. pp. 267-291

[19] Dwire KA, Mellmann-Brown S, Gurrieri JT. Potential effects of climate change on riparian areas, wetlands, and groundwater-dependent ecosystems in the Blue Mountains, Oregon, USA. Climate. 2018;**10**:44-52

[20] Gregory SV, Swanson FJ, McKee WA, Cummins KW. An ecosystem perspective of riparian zones. BioScience. 1991;**41**:540-551

[21] Naiman RJ, Décamps H. The ecology of interfaces: Riparian zones. Annual Review of Ecology and Systematics. 1997;**28**:621-658

[22] Naiman RJ, Melillo JM, Lock MA, Ford TE, Reice SR. Longitudinal patterns of ecosystem processes and community structure in a subarctic river continuum. Ecology. 1987;**68**:1139-1156

[23] Naiman RJ, DéCamps H, McClaim ME. Riparia: Ecology, Conservation, and Management of Streamside Communities. New York: Wiley; 2005

[24] Lindeman RL. The tropic-dynamic aspect of ecology. Ecology. 1942;**23**:399-417

[25] Odum HT. Trophic structure and productivity of Silver Springs, Florida. Ecological Monographs. 1957;**27**:55-112

[26] Horton RE. Erosional development of streams and their drainage basins: Hydrophysical approach to quantitative morphology. Bulletin of the Geological Society of America. 1945;**56**:275-370

[27] Strahler AN. Quantitative analysis of watershed geomorphology. Transactions of the American Geophysical Union. 1957;**38**:913-920

[28] Hutchinson GE. A Treatise on Limnology, Volume II. Introduction to Lake Biology and the Limnoplankton. New York: John Wiley & Sons; 1967

[29] Leopold LB, Maddock R Jr. The Hydraulic Geometry of Stream Channels and Some Physiographic Implications. Washington: U.S. Geological Survey; 1953

[30] Schumm SA. Geomorphic thresholds and complex response of drainage systems. In: Morisawa M, editor. Fluvial Geomorphology. Binghamton: SUNY Binghamton Publication in Geomorphology; 1973. pp. 299-310

[31] Schumm SA. Patterns of alluvial rivers. Annual Review of Earth and Planetary Sciences. 1985;**13**:5-27

[32] Schumm SA, Chorley RJ, Sugden DE. Geomorphology. New York: Methuen; 1985. p. 413

[33] Leopold LB. A View of the River. Cambridge: Harvard University Press; 1994

[34] Morisawa M. Streams: Their Dynamics and Morphology. New York: McGraw-Hill; 1968

[35] Campbell CJ, Green W. Perpetual succession of stream-channel vegetation in a semiarid region. Journal of the Arizona Academy of Sciences. 1968;**5**:86-98

[36] Hynes HBN. The Ecology of Running Waters. Toronto: University of Toronto Press; 1970

[37] Hynes HBN. The stream and its valley. Verhandlunger der Internationalen Vereinigung für Theoretische und Angewandte Limnologie. 1975;**91**:1-15

[38] Carothers SW, Johnson RR, Aitchison SW. Population structure and social organization in southwestern riparian birds. American Zoologist. 1974;**14**:97-108

[39] Naiman RJ, DéCamps H, Pollock MM. The role of riparian corridors in maintaining regional biodiversity. Ecological Applications. 1993;**3**:209-212

[40] Nilsson C, Grelsson G, Johansson ME, Sperens U. Patterns of plant species richness along riverbanks. Ecology. 1989;**70**:77-84

[41] Johnson RR, Jones DA. Importance, Preservation and Management of Riparian Habitat: A Symposium. Fort Collins, CO; USA: Department of Agriculture Forest Service Rocky Mountain Research Station; 1977

[42] Johnson RR, Carothers SW, Finch DM, Kingsley KJ, Stanley JT, editors. Riparian Research and Management: Past, Present, Future. Ft. Collins: Department of Agriculture, Forest Service; 2019-2020

[43] Johnson RR, Carothers SW, Simpson JM. A riparian classification system. In: Warner RE, Hendrix KM, editors. California Riparian Systems. Berkeley, CA: University of California Press; 1984. pp. 375-382

[44] Johnson RR, Siebel CD, Patton DR, Ffolliott PF, Hamre RH. Riparian ecosystems and their management: Reconciling conflicting uses. In: First North American Riparian Conference. Tucson, AZ: U.S. Department of Agriculture Forest Service Rocky Mountain Research Station; 1985

[45] Cummins KW. The ecology of running water: Theory and practice. In: Baker DB, Prater BL, editors. Proceedings of the Sandusky River Basin Symposium. Washington, DC: U.S. Government Printing Office; 1976. pp. 277-293

[46] Connell JH. Diversity in tropical rain forests and coral reefs. Science. 1978;**199**:1302-1310

[47] Huston MA. A general hypothesis of species diversity. American Naturalist. 1979;**113**:81-101

[48] Huston MA. Biological Diversity: The Coexistence of Species on Changing Landscapes. Cambridge: Cambridge University Press; 1994

[49] Pollock MM, Naiman RJ, Hanley TA. Plant species richness in riparian wetlands—A test of biodiversity theory. Ecology. 1998;**79**:94-105

[50] Newbold JD, Elwood JW, O'Neill RV, Winkle WV. Measuring nutrient spiraling in streams. Canadian Journal of Fisheries and Aquatic Sciences. 1981;**38**:860-863

[51] Ensign SH, Doyle MW. Nutrient spiraling in streams and river networks. Journal of Geophysical Research. 2006;**111**:G04009

[52] Ward JV, Stanford JA. The serial discontinuity concept of river

ecosystems. In: Fontaine TD, Bartell SM, editors. Dynamics of Lotic Ecosystems. Ann Arbor: Science Publications; 1983. pp. 29-43

[53] Ward JV, Stanford JA. The serial discontinuity concept: Extending the model to floodplain rivers. Regulated Rivers: Research & Management. 1995;**10**:159-168

[54] Décamps H, Fortuné M, Gazelle F, Pautou G. Historical influence of man in the riparian dynamics of a fluvial landscape. Landscape Ecology. 1988;**1**:163-173

[55] Petts GE, Möller H, Roux AL, editors. Historical Change in Large Alluvial Rivers. Chichester: John Wiley; 1989

[56] Junk WJ, Bayley PB, Sparks RE. The flood pulse concept in river continuum systems. Canadian Special Publication on Fisheries and Aquatic Sciences. 1989;**106**:89-109

[57] Tockner K, Malard F, Ward JV. An extension of the flood pulse concept. Hydrological Processes. 2000;**14**:2861-2883

[58] Stella JC, Hayden M, Battles JJ, Fremier A. The role of abandoned channels as refugia for sustaining pioneer riparian forest ecosystems. Ecosystems. 2011;**14**:776-790

[59] Tinkler KJ, Wohl EE. Rivers over rock: Fluvial processes in bedrock channels. American Geophysical Union Geophysical Monograph Series. 1998;**107**:336

[60] Rosgen DL. Applied River Morphology. Pagosa: Wildland Hydrology; 1996

[61] Rosgen DL. Discussion of Critical evaluation of how the Rosgen classification and associated natural channel design methods fail to integrate and quantify fluvial processes and channel responses. Journal of the American Water Resources Association. 2008;**44**:782-792

[62] Griffiths RE, Anderson DE, Springer AE. The morphology and hydrology of small spring-dominated channels. Geomorphology. 2008;**102**:511-521

[63] Buffington JM, Montgomery DR. Geomorphic classification of rivers. In: Shroder J, Wohl EE, editors. Treatise on Geomorphology. San Diego, CA: Academic Press; 2013. pp. 730-767

[64] McManamay RA, DeRolph CR. A stream classification system for the coterminous United States. Scientific Data. 2019;**190**:17

[65] Thorp JH, DeLong MD. The riverine productivity model: An heuristic view of carbon sources and organic processing in large river ecosystems. Oikos. 1994;**70**:305-308

[66] Thorp JH, Thoms MC, Delong MD, editors. The Riverine Ecosystem Synthesis: Toward Conceptual Cohesiveness in River Science. London: Academic Press; 2008

[67] Montgomery DR. Process domains and the river continuum. Journal of the American Water Resources Association. 1999;**35**:397-410

[68] Bellmore JR, Baxter CV. Effects of geomorphic process domains on river ecosystems: A comparison of floodplain and confined valley segments. River Research and Applications. 2014;**30**:617-630

[69] Fisher SG, Grimm NB, Marti E, Holmes RM, Jones JB. Material spiraling

in stream corridors: A telescoping ecosystem model. Ecosystems. 1998;**1**:19-34

[70] Rice SP, Greenwood MT, Joyce CG. Tributaries, sediment sources and the longitudinal organization of macroinvertebrate fauna along river systems. Canadian Journal of Fisheries and Aquatic Science. 2001;**58**:824-840

[71] Rice SP, Kiffney P, Greene C, Pess GR. The ecological importance of tributaries and confluences. In: Rice SP, Roy AG, Rhoads BL, editors. River Confluences, Tributaries and the Fluvial Network. Hoboken, New Jersey: John Wiley & Sons; 2008. pp. 209-242

[72] Bennett SJ, Simon A. Riparian Vegetation and Fluvial Geomorphology. Washington, DC: American Geophysical Union; 2013

[73] Committee on Riparian Zone Functioning and Strategies for Management (CRZFSM). Riparian Areas: Functions and Strategies for Management. Washington, DC: National Academy Press; 2002

[74] Hupp CR, Osterkamp WR. Bottomland vegetation distribution along Passage Creek, Virginia, in relation to fluvial landforms. Ecology. 1985;**66**:670-681

[75] Hupp CR, Osterkamp WR. Riparian vegetation and fluvial geomorphic processes. Geomorphology. 1996;**14**:277-295

[76] Nilsson C, Svedmark M. Basic principles and ecological consequences of changing water regimes: Riparian plant communities. Environmental Management. 2002;**30**:468-480

[77] Benda L, Andras K, Miller D, Bigelow P. Confluence effects in rivers: Interactions of basin scale, network geometry, and disturbance regimes. Water Resources Research. 2004;**40**:W05402

[78] Benda L, Poff NL, Miller D, Dunne T, Reeves G, Pess G, et al. The network dynamics hypothesis: How channel networks structure riverine habitats. Bioscience. 2004;**4**:413-427

[79] Clay PA, Muehlbauer JD, Doyle MW. Effect of tributary and braided confluences on aquatic macroinvertebrate communities and geomorphology in an alpine river watershed. Freshwater Science. 2015;**34**:845-856

[80] Stevens LE. The biogeographic significance of a large, deep canyon: Grand Canyon of the Colorado River, southwestern USA. In: Stevens LE, editor. Global Advances in Biogeography. Rijeka: InTech Publications; 2012. pp. 169-208

[81] Yard MD, Bennett GE, Mietz SN, Coggins LN, Stevens LE, Hueftle S, et al. Influence of topographic complexity on solar insolation estimates for the Colorado River, Grand Canyon, AZ. Ecological Modeling. 2005;**183**:157-172

[82] Beschta RL, Ripple WJ. River channel dynamics following extirpation of wolves in northwestern Yellowstone National Park, USA. Earth Surface Processes and Landforms. 2006;**31**:1525-1539

[83] Corenblit D, Seiger J, Gurnell AM, Naiman RJ. Plants intertwine fluvial landform dynamics with ecological succession and natural selection: A niche construction perspective for riparian systems. Global Ecology and Biogeography. 2009;**18**:507-520

[84] Merritt DM, Scott ML, Poff NL, Auble GT, Lytle DA. Theory, methods, and tools for determining environmental

flows for riparian vegetation: Riparian vegetation-flow response guilds. Freshwater Biology. 2010;**55**:206-225

[85] Bouwman AF, Giffioen MFP, Hefting MM, Middelburg JJ, Middelkoop H, Slomp CP. Nutrient dynamics, transfer and retention along the aquatic continuum from land to ocean: Toward integration of ecological and biogeochemical models. Biogeosciences. 2013;**10**:1-23

[86] Jacobs SM, Bechtold JS, Biggs HC, Grimm NB, Lorentz S, McClain MR, et al. Nutrient vectors and riparian processing: A review with special reference to African semiarid savanna ecosystems. Ecosystems. 2007;**10**:1231-1249

[87] Meybeck M. Carbon, nitrogen and phosphorous transport by world rivers. American Journal of Science. 1982;**282**:401-450

[88] Bennett MG. Effects of Flow Regime on Fishes and Fisheries: From Life Histories to Ecosystem Services. Carbondale, IL: Southern Illinois University; 2015

[89] Grime JP. Evidence for the existence of three primary strategies in plants and its relevance to ecological and evolutionary theory. American Naturalist. 1977;**111**:1169-1194

[90] Mims MC, Olden JD, Shattuck Z, Poff NL. Life history trait diversity of native freshwater fishes in North America. Ecology of Freshwater Fish. 2010;**19**:390-400

[91] Humphries P, Keckeis H, Finlayson B. The river wave concept: Integrating river ecosystem models. BioScience. 2014;**64**:870-882

[92] Jenkins KM, Boulton AJ. Connectivity in a dryland river: Short-term aquatic microinvertebrate recruitment following floodplain inundation. Ecology. 2003;**84**:2708-2723

[93] Larned ST, Datry T, Arscott DB, Tockner K. Emerging concepts in temporary-river ecology. Freshwater Biology. 2010;**55**:717-738

[94] Muehlbauer JD, Collins SF, Doyle MW, Tockner K. How wide is a stream? Spatial extent of the potential "stream signature" in terrestrial food webs using meta-analysis. Ecology. 2014;**95**:44-55

[95] Nanson GC, Huang H. A philosophy of rivers: Equilibrium states, channel evolution, teleomatic change and least action principle. Geomorphology. 2018;**302**:3-19

[96] Springer AE, Stevens LE. Spheres of discharge of springs. Hydrogeology Journal. 2009;**17**:83-93

[97] Stevens LE, Schenk ER, Springer AE. Springs ecosystem classification. Ecological Applications. 2021;**31**:e2218

[98] Cid N, Erős T, Heino J, Singer G, Thibault D. From meta-system theory to the sustainable management of rivers in the Anthropocene. Frontiers in Ecology and the Environment. 2022;**20**:49-57

[99] Annear T, Chisholm I, Beecher H, Locke A, Aarrestad P, Coomer C, Estes C, Hunt J, Jacobson R, Jöbsis G, Kauffman J, Marshall J, Mayes K, Smith G, Wentworth R, Stalnaker C. Instream Flows for Riverine Resource Stewardship, Revised ed. Cheyenne, WY: Instream Flow Council; 2004. p. 268

[100] Annear T, Lobb D, Coomer C, Woythal M, Hendry C, Estes C, Williams K. International Instream Flow Program Initiative, A Status Report of

State and Provincial Fish and Wildlife Agency Instream Flow Activities and Strategies for the Future, Final Report for Multi-State Conservation Grant Project WY M-7-T. Cheyenne, WY: Instream Flow Council; 2009

[101] Stanford JA. Rivers in the landscape: Introduction to the special issue on riparian and groundwater ecology. Freshwater Biology. 1998;**40**:402-406

[102] Ward JV. The four-dimensional nature of lotic ecosystems. Journal of the North American Benthological Society. 1989;**8**:2-8

[103] Reid AJ, Carlson AK, Creek IF, et al. Emerging threats and persistent conservation challenges for freshwater biodiversity. Biological Reviews. 2019;**94**:849-873

[104] Holbrook J, Schumm SA. Geomorphic and sedimentary response of rivers to tectonic deformation: A brief review and critique of a tool for recognizing subtle epeirogenic deformation in modern and ancient settings. Tectonophysics. 1999;**305**:287-306

[105] Duvall A, Kirby E, Burbank D. Tectonic and lithologic controls on bedrock channel profiles and processes in coastal California. Journal of Geophysical Research. 2004;**109**:F03002

[106] Böhme M, Aiglstorfer M, Uhl D, Kullmer O. The antiquity of the Rhine River: Stratigraphic coverage of the Dinotherien-sande (Eppelsheim Formation) of the Mainz Basin (Germany). PLoS ONE. 2012. DOI: 10.1371/journal.pone.0036817

[107] Figueiredo JP, Hoorn C, van der Ven P, Soards E. Late Miocene onset of the Amazon River and Amazon deep-sea fan: Evidence from the Foz do Amazonas Basin. Geology. 2009;**37**:619-622

[108] Mack GH, Seager WR, et al. Pliocene and Quaternary history of the Rio Grande, the axial river of the southern Rio Grande rift. New Mexico Earth-Science Reviews. 2006;**79**:141-162

[109] Wickert AD, Mitrovica JX, Williams C, Anderson RS. Gradual demise of a thin southern Laurentide ice sheet recorded by Mississippi drainage. Nature. 2013;**502**:668-671

[110] Woodward JC, Macklin MG, Krom MD, Williams MAJ. The Nile: Evolution, Quaternary river environments and material fluxes. In: Gupta A, editor. Large Rivers: Geomorphology and Management. London: John Wiley & Sons, Ltd; 2007. pp. 261-292

[111] Zheng H. Birth of the Yangtze River: Age and tectonic geomorphic implications. National Science Review. 2015;**2**:438-453

[112] Heinsalu H, Bednarczyk W. Tremadoc of the East European Platform: Lithofacies and paleogeography. Proceedings of the Estonian Academy of Sciences Geo!ogy. 1997;**46**:59-74

[113] Larson PH, Dorn RI, Gootee BF, Seong YB. Drainage integration in extensional tectonic settings. Geomorphology. 2022;**399**:108082

[114] Ranney W. Carving Grand Canyon: Evidence, Theories, and Mystery. 2nd ed. Grand Canyon: Grand Canyon Association; 2012

[115] Springer AE, Boldt EM, Junghans KM. Local vs. regional groundwater flow delineationfrom stable isotopes at western North America springs. Groundwater. 2016;**55**:100-109

[116] Winograd IJ, Thordarson W. Hydrogeologic and Hydrochemical Framework, South-central Great Basin, Nevada-California, with Special Reference

to the Nevada Test Site. Washington, DC: U.S. Geological Survey; 1975

[117] Montgomery DR, Balco G, Willett SD. Climate, tectonics, and the morphology of the Andes. Geology. 2001;**29**:579-582

[118] Lucchitta I. History of the Grand Canyon and of the Colorado River in Arizona. In: Beus SS, Morales M, editors. Grand Canyon Geology. Oxford: Oxford University Press; 1990. pp. 311-332

[119] Junghans K. Sources of Perennial Stream Flow in Arizona, USA. Flagstaff: Northern Arizona University; 2016

[120] Lowe WH, Likens GE. Moving headwater streams to the head of the class. BioScience. 2005;**55**:196-197

[121] Teears TD. The Effect of Water Quality on the Survival and Fitness of Brook Trout (*Salvelinus fontinalis*) Eggs, Alevins and Fry in Aquaculture and Deep Springs along the South River in Wayesboro, VA. Harrisonburg: James Madison University; 2016

[122] Muller-Schwarze D. Chemical Ecology of Vertebrates. Cambridge: Cambridge University Press; 2006

[123] Morrison RR, Stone MC, Sada DW. Environmental response of a desert springbrook to incremental discharge reductions, Death Valley National Park, California, USA. Journal of Arid Environments. 2013;**99**:5-13

[124] Cantonati M, Segadelli S, Ogata K, Tran H, Sanders D, Gerecke R, et al. A global review on ambient limestone-precipitating springs (LPS): Hydrogeological setting, ecology, and conservation. Science of the Total Environment. 2016;**15**:568-624

[125] Stevens LE, Jenness J, Ledbetter JD. Springs and springs-dependent taxa in the Colorado River Basin, southwestern North America: Geography, ecology, and human impacts. Water. 2020;**12**:1501

[126] Ruzo A. The Boiling River: Adventure and Discovery in the Amazon. New York: TED Books; 2016

[127] Schmidt JC, Graf JB. Aggradation and Degradation of Alluvial Sand Deposits, 1965-1986, Colorado River, Grand Canyon National Park, Arizona. Washington, DC: United States Geological Survey; 1990

[128] Stevens LE, Shannon JP, Blinn DW. Benthic ecology of the Colorado River in Grand Canyon: Dam and geomorphic influences. Regulated Rivers: Research & Management. 1997;**13**:129-149

[129] Dye A. Contribution of Unregulated Tributaries to the Ecological Functioning of the Main Channel of Rivers. Snowy River Recovery: Snowy Flow Response Monitoring and Modeling. Sydney: New South Wales Office of Water; 2010

[130] Stevens LE, Holway JH, Ellsworth C. Benthic discontinuity between an unregulated tributary and the dam-controlled Colorado River, Grand Canyon, Arizona, USA. Annals of Ecology and Environmental Science. 2020;**4**:33-48

[131] Magilligan FJ. Thresholds and the spatial variability of flood power during extreme floods. Geomorphology. 1992;**5**:373-390

[132] Foster DR, Knight DH, Franklin JF. Landscape patterns and legacies resulting from large, infrequent disturbances. Ecosystems. 1998;**1**:497-510

[133] Parsons M, McLoughlin CA, Kotschy KA, Rogers KH, Rountree MW. The effects of extreme floods on the biophysical heterogeneity of river landscapes. Frontiers in Ecology and the Environment. 2005;**3**:487-494

[134] Akkemik U, D'Arrigo R, Cherubini P, Kose N, Jacoby GC. Tree-ring reconstructions of precipitation and streamflow for northwestern Turkey. International Journal of Climatology. 2004;**28**:173-183

[135] Case RA, MacDonald GM. Tree ring reconstructions of streamflow for three Canadian prairie rivers. Journal of the American Water Resouces Association. 2003;**39**:703-716

[136] D'Arrigo R, Abram N, Ummenhofer C, Palmer J, Mudelsee M. Reconstructed Streamflow for Citarum River. Java, Indonesia: Climate Dynamics; 2009

[137] Melo DM, Woodhouse CA, Morino K. Dendrochronology and links to streamflow. Journal of Hydrology. 2012;**412-413**:200-209

[138] Therrell MD, Bialecki MB. A multi-century tree-ring record of spring flooding on the Mississippi River. Journal of Hydrology. 2015;**529**:490-498

[139] US Bureau of Reclamation. Colorado River Basin Water Supply and Demand Study. Washington, DC: US Department of the Interior; 2012

[140] Kennedy TA, Muehlbauer JD, Yackulic CB, Lytle DA, Miller SW, Dibble KL, et al. Flow management for hydropower extirpates aquatic insects, undermining river food webs. BioScience. 2016;**66**:561-575

[141] Merritt RW, Cummins KW, Berg MB. An Introduction to the Aquatic Insects of North America. 4th ed. Dubuque: Kendall-Hunt; 2008

[142] Heino J. A comparative analysis reveals weak relationships between ecological factors and beta diversity of stream insect metacommunities at two spatial scales. Ecology and Evolution. 2015;**5**:1235-1248

[143] Giller S, Malmqvist B. The Biology of Streams and Rivers. Oxford: Oxford University Press; 1998

[144] Kabede S, Travi Y, Alemayehu T, Ayenew T. Groundwater recharge, circulation and geochemical evolution in the source region of the Blue Nile River, Ethiopia. Applied Geochemistry. 2005;**20**:1658-1676

[145] Hey RD, Bathurst JC, Thorne CR. Gravel-bed Rivers: Fluvial Processes, Engineering, and Management. New York: John Wiley & Sons; 1982

[146] Vogel S. Life in Moving Fluids: The Physical Biology of Flow. Boston: Willard Grant Press; 1981

[147] Gray JR, Simões FJM. Estimating sediment discharge. In: Garcia M, editor. Sedimentation Engineering – Processes, Measurements, Modeling, and Practice. Reston, Virginia: American Society of Civil Engineers; 2008. pp. 1065-1086

[148] Milliman JD, Syvitski JPM. Geomorphic/tectonic control of sediment discharge to the ocean: The importance of small mountainous rivers. Journal of Geology. 1992;**100**:525-544

[149] Merritt WS, Letcher RA, Jakeman AJ. A review of erosion and sediment transport models. Environmental Modeling & Software. 2003;**18**:761-799

[150] Alley WM, Evenson EJ, Barber NL, Bruce BW, Dennehy KF, Freeman MC, et al. Progress toward Establishing a National Assessment of Water Availability and Use. Washington, DC: U.S. Geological Survey; 2013

[151] Naiman RJ, Bilby RE, editors. River Ecology and Management. New York: Springer-Verlag; 1998

[152] Bormann FH, Likens GE. Pattern and Process in a Forested Ecosystem. New York, NY: Springer-Verlag; 1979

[153] Stevens V. The Ecological Role of Coarse Woody Debris: An Overview of the Ecological Importance of CWD in BC Forests. Victoria: British Columbia Ministry of Forests; 1997

[154] Prowse TD, Culp JM. Ice breakup: A neglected factor in river ecology. Canadian Journal of Civil Engineering. 2003;**30**:128-144

[155] Draught AE, Rubin DM. Measurements of Wind, Eolian sand Transport, and Precipitation in the Colorado River Corridor. Grand Canyon, Arizona: U.S. Geological Survey; 2006

[156] Whiteman CD. Observations of thermally developed wind systems in mountainous terrain. Meteorological Monographs. 1990;**45**:5-42

[157] Stanitski-Martin D. Seasonal Energy Balance Relationships over the Colorado River and Adjacent Riparian Habitat: Glen Canyon, Arizona. Tempe: Arizona State University; 1999

[158] Selwood K, Clark RH, McGeoch MA, MacNally R. Green tongues into the arid zone: River floodplains extend the distribution of terrestrial bird species. Ecosystems. 2016;**20**:745-756

[159] Merriam G. Connectivity: A fundamental ecological characteristic of landscape pattern. In: Brandt J, Agger P, editors. Proceedings of the First International Seminar on Methodology in Landscape Ecology Research and Planning. Vol. 1. Roskikde: Roskilde Universitessforlag GeoRue; 1984. pp. 5-15

[160] Leopold LB, Wolman MG, Miller JP. Fluvial Processes in Geomorphology. San Francisco: Freeman; 1964

[161] Tansley AG. The use and abuse of vegetational concepts and terms. Ecology. 1935;**16**:284-307

[162] Wurtsbaugh WA, Heredia NA, Laub BG, Meredith CS, Mohn HE, Null SE, et al. Approaches for studying fish production: Do river and lake researchers have different perspectives? Canadian Journal of Fisheries and Aquatic Sciences. 2015;**72**:149-160

[163] Meehan WR, Swanson FJ, Sedell JR. Influences of riparian vegetation on aquatic ecosystems with particular reference to salmonid fishes and their food supply. In: Johnson RR, Jones DA, editors. Importance, Preservation and Management of Riparian Habitat: A Symposium. Fort Collins, CO: U.S. Department of Agriculture Forest Service Rocky Mountain Research Station; 1977. pp. 137-145

[164] Minshall GW. Stream ecosystem theory: A global perspective. Journal of the North American Benthological Society. 1988;**7**:263-288

[165] Minshall GW, Peterson RC, Cummins KW, Bott TL, Sedell JR, Cushing CE, et al. Interbiome comparison of stream ecosystem dynamics. Ecological Monographs. 1983;**53**:1-25

[166] Minshall GW, Cummins KW, Petersen RC, Cushing CE, Bruns DA, Sedell JR, et al. Developments in stream ecosystem theory. Canadian Journal of Fisheries and Aquatic Sciences. 1985;**42**:1045-1055

[167] National Research Council. Riparian Areas: Functions and Strategies for Management. Washington, DC: National Academy Press; 2002

[168] Palmer MA, Bely AE, Berg KE. Response of invertebrates to lotic disturbance: A test of the hyporheic refuge hypothesis. Oecologia. 1992;**89**:182-194

[169] Stanford JA, Ward JV. The hyporheic habitat of river ecosystems. Nature. 1988;**355**:64-66

[170] Datry T, Larned ST, Tocknerr K. Intermittent rivers: A challenge for freshwater ecology. BioScience. 2014;**64**:229-235

[171] Sedell JR, Richey JE, Swanson RJ. The river continuum concept: A basis for the expected ecosystem behavior of very large rivers? Canadian Special Publications in Fisheries and Aquatic Sciences. 1989;**106**:49-55

[172] Johnson BL, Richardson WB, Naimo TJ. Past, present, and future concepts in large river ecology. BioScience. 1995;**45**:134-141

[173] Johnson RR, Lowe CH. On the development of riparian ecology. In: Johnson RR, Ziebell CD, Patton DR, Ffolliott PF, Hamre RH, editors. Riparian Ecosystems and Their Management: Reconciling Conflicting Uses. Tucson, AZ: Department of Agriculture Forest Service; 1985. pp. 112-116

[174] Karr JR. Biological integrity: A long-neglected aspect of water resource management. Ecological Applications. 1991;**1**:66-84

[175] Karr JR. Defining and measuring river health. Freshwater Biology. 1999;**41**:221-234

[176] Karr JR, Chu EW. Restoring Life in Running Waters: Better Biological Monitoring. Washington, DC: Island Press; 1999

[177] Malanson GP. Riparian Landscapes. Cambridge: Cambridge University Press; 1993

[178] Thorp JH, Thoms MC, DeLong MD. The riverine ecosystem synthesis: Biocomplexity in river networks across space and time. River Research & Applications. 2006;**22**:123-147

[179] Connell JH, Slayter RO. Mechanisms of succession in natural communities and their role in community stability and organization. American Naturalist. 1977;**111**:1119-1144

[180] Hough-Snee N, Laub BG, Merritt DM, Long AL, Nackley LL, Roper BB, et al. Multi-scale environmental filters and niche partitioning govern the distributions of riparian vegetation guilds. Ecosphere. 2015;**6**:1-22

[181] Nilsson C. Frequency distributions of vascular plants in the geolittoral vegetation along two rivers in northern Sweden. Journal of Biogeography. 1983;**10**:351-369

[182] Carothers SW, Aitchison SW, Johnson RR. Natural resources, white-water recreation and river management alternatives on the Colorado River, Grand Canyon National Park, Arizona. Proceedings of the First Conference on Scientific Research in the National Parks. 1979;**1**:253-260

[183] Merritt DM, Scott ML, Poff NL, Auble GT, Lytle DA. Theory, methods and tools for determining environmental flows for riparian vegetation: Riparian vegetation-flow response guilds. Freshwater Biology. 2009;**55**:206-225

[184] Sousa WP. The role of disturbance in natural communities. Annual Review of Ecology & Systematics. 1984;**15**:353-391

[185] Pringle CM, Naiman RJ, Bretschko G, et al. Patch dynamics in lotic systems: The stream as a mosaic. Journal of the North American Benthological Society. 1988;**7**:503-524

[186] Townsend CR. The patch dynamics concept of stream community ecology. Journal of the North American Benthological Society. 1989;**8**:36-50

[187] Townsend CR. The intermediate disturbance hypothesis, refugia, and biodiversity in streams. Limnology and Oceanography. 2003;**42**:938-949

[188] Crow R, Karlstrom KE, McIntosh W, Peters L, Dunbar N. History of Quaternary volcanism and lava dams in western Grand Canyon based on LIDAR analysis, (40)Ar/(39)Ar dating, and field studies: Implications for flow stratigraphy, timing of volcanic events, and lava dams. Geosphere. 2008;**4**:183-206

[189] Dalrymple GB, Hamblin KW. K-Ar ages of Pleistocene lava dams in the Grand Canyon in Arizona. Proceedings of the National Academy of Sciences of the United States of America. 1998;**95**:9744-9749

[190] Fenton CR, Poreda RJ, Nash BP, Webb RH, Cerling TE. Geochemical discrimination of five Pleistocene lava-dam outburst flood deposits, western Grand Canyon, Arizona. Journal of Geology. 2004;**112**:91-110

[191] Costa JE, Schuster RL. The Formation and Failure of Natural Dams. Washington, DC: U.S. Geological Survey; 1987

[192] Rogers JD, Pyles MR. Evidence of catastrophic erosional events in the Grand Canyon. In: Proceedings of the 2nd Conference on Scientific Research in National Parks. Washington, DC: National Park Service; 1980. pp. 392-454

[193] Castleton JJ, Moore JR, Aaron J, Christ M, Ivy-Ochs S. Dynamics and legacy of 4.8 ka rock avalanche that dammed Zion Canyon, Utah, USA. GSA Today. 2016;**26**:4-9

[194] Hereford R, Jacoby GC, McCord VAS. Geomorphic History of the Virgin River in the Zion National Park Area, southwest Utah. Washington, DC: U.S. Geological Survey; 1995

[195] Benito G, O'Connor JE. Number and size of last-glacial Missoula floods in the Columbia River valley between the Pasco Basin, Washington, and Portland. Oregon. Geological Society of America Bulletin. 2003;**115**:624-638

[196] Reeburgh WS, Nebert DL. The Birth and Death of Russell Lake. Fairbanks: University of Alaska; 1987

[197] Webb RH, Betancourt JL, Johnson RR, Turner RM. Requiem for the Santa Cruz River: An Environmental History of an Arizona River. Tucson, AZ: University of Arizona Press; 2014

[198] Smith JA, Auerbach DA, Flessa KW, Flecker AS, Dietl GP. Fossil clam shells reveal unintended carbon cycling consequences of Colorado River management. Royal Society Open Science. 2016;**3**:160170

[199] Chaghtai SM, Khattak H-U-R. Ecology of a dry stream bed in Peshawar, Pakistan. Pakistan Journal of Botany. 1983;**15**:93-98

[200] Fukami T, Morin PJ. Productivity-biodiversity relationships depend on the history of community assembly. Nature. 2003;**424**:423-426

[201] Hooper DU, Chapin FS, Ewel JJ, Hector A, Inchausti P, Lavorel S. Effects of biodiversity on ecosystem functioning: A consensus of current knowledge. Ecological Monographs. 2005;**75**:3-35

[202] Marquard E, Weigelt A, Roscher C, Gubsch M, Lipowsky A, Schmid B. Positive biodiversity-productivity relationship

due to increased plant density. Journal of Ecology. 2009;**97**:696-704

[203] Tilman D, Reibch PB. Diversity and productivity in a long-term grassland experiment. Science. 2001;**294**:843-845

[204] Stevens LE. Mechanisms of Riparian Plant Succession. Flagstaff: Northern Arizona University; 1989

[205] Reice SR. Experimental disturbance and the maintenance of species diversity in a stream community. 1985;**67**:90-97

[206] Wilkinson DM. The disturbing history of intermediate disturbance. Oikos. 1999;**84**:145-147

[207] MacArthur RH, Wilson EO. The Theory of Island Biogeography. Princeton: Princeton University Press; 1967

[208] Lytle DA. Use of rainfall cues by Abedus herberti (Hemiptera: Belostomatidae): A mechanism for avoiding flash floods. Journal of Insect Behavior. 1999;**12**:1-12

[209] Bartels P, Cucherousset J, Steger K, Eklov P, Tranvik LJ, Hillebrand H. Reciprocal subsidies between freshwater and terrestrial ecosystems structure consumer resource dynamics. Ecology. 2012;**93**:1173-1182

[210] Larsen S, Muehlbauer JD, Marti E. Resource subsidies between stream and terrestrial ecosystems under global change. Global Change Biology. 2015;**22**:2489-2504

[211] U.S. Department of Agriculture Forest Service. Watershed and Air Management, Section 2526.05, in Agriculture Forest Service: 2000. Washington, DC: U.S. Department of Agriculture Forest Service; 2000

[212] Riis T, Kelly-Quinn M, Aguiar FC, Manolaki P, Bruno D, Bejarano MD, et al. Global overview of ecosystem services provided by riparian vegetation. Bioscience. 2020;**70**(6):501-514

[213] Dufour S, Rodríguez-González PM, Laslier M. Tracing the scientific trajectory of riparian vegetation studies: Main topics, approaches and needs in a globally changing world. Science Total Environment. 2019;**25**(653):1168-1185. DOI: 10.1016/j.scitotenv.2018.10.383

[214] Tockner K, Stanford JA. Riverine flood plains: Present state and future trends. Environmental Conservation. 2002;**29**:308-330

[215] Rood SB, Pan J, Gill KM, Franks CG, Samuelson A, Shepherd GM. Declining summer flows of Rocky Mountain rivers: Changing seasonal hydrology and probable impacts on floodplain forests. Journal of Hydrology. 2008;**349**:397-410

[216] Perry LG, Andersen DC, Reynolds LV, Mark NS, Shafroth PB. Vulnerability of riparian ecosystems to elevated CO2 and climate change in arid and semiarid western North America. Global Change Biology. 2012;**18**:821-842

[217] Vandenberghe J. Climate forcing of fluvial system development: An evolution of ideas. Quaterinary Science Review. 2003;**22**(20):2053-2060. DOI: 10.1016/S0277-3791(03)00213-0

[218] Liu HH. Impact of climate change on groundwater recharge in dry areas: An ecohydrology approach. Journal of Hydrology. 2011;**407**:175-183

[219] Okkonen J, Kløve B. A conceptual and statistical approach for the analysis of climate impact on ground water table fluctuation patterns in cold conditions. Journal of Hydrology. 2010;**388**(1-2):1-12

[220] Bogaart PW, Van Balen RT, Kasse C, Vandenberghe J. Process-based

modeling of fluvial system response to rapid climate change II. Application to the River Maas (the Netherlands) during the Last Glacial–Interglacial Transition. Quaterly Science Review. 2003;**22**(20):2097-2110. DOI: 10.1016/S0277-3791(03)00144-6

[221] Cecil BC, Dulong FT. Precipitation models for sediment supply in warm climates. In: May C, Cecil B, Edgar NT, editors. Climate Controls on Stratigraphy. Tulsa,OK: SEPM Society for Sedimentary Geology; 2003. pp. 21-27. DOI: 10.2110/pec.03.77.0021

[222] Groeneveld DP, Griepentrog TE. Interdependence of groundwater, riparian vegetation, and streambank stability: A Case Study. In: Riparian ecosystems and their management: Reconciling conflicting uses. First North American riparian conference:1985 April 16-18. Tuscon; 1985. pp. 44-48

[223] Gribovszki Z, Kalicz P, Szilágyi J, Kucsara M. Riparian zone evapotranspiration estimation from diurnal groundwater level fluctuations. Journal of Hydrology. 2008;**349**(1-2):6-17. DOI: 10.1016/j.jhydrol.2007.10.049

[224] Hai-liang XU, Mao YE, Ji-mei LI. Changes in groundwater levels and the response of natural vegetation to- second paper.pdf. Journal of Environmental Science. 2007;**19**:1199-1207

[225] Butturini A, Bernal S, Sabater S, Sabater F. The influence of riparian-hyporheic zone on the hydrological responses in an intermittent stream. Hydrological Earth System. 2002;**6**(3):515-525. DOI: 10.5194/hess-6-515-2002

[226] Keller EA, Kondolf GM, Hagerty DJ. Groundwater and fluvial processes; Selected observations. In: Higgins CG, Coates DR, editors. Groundwater Geomorphology; The Role of Subsurface Water in Earth-Surface Processes and Landforms. Boulder, CO: Geological Society of America; 1990

[227] Kløve B, Ala-Aho P, Bertrand G, Gurdak JJ, Kupfersberger H, Kværner J, et al. Climate change impacts on groundwater and dependent ecosystems. Journal of Hydrology. 2014;**518**:250-266

[228] Hiscock K, Sparkes R, Hodgens A. Evaluation of future climate change impacts on european groundwater resources. In: Treidel H, Martin-Bordes JJ, Gurdak JJ, editors. Climate Change Effects on Groundwater Resources: A Global Synthesis of Findings and Recommendations. Abingdon-on-Thames: Taylor & Francis Publishing; 2012. pp. 351-366

[229] Hu W, Wang YQ, Li HJ, Huang MB, Hou MT, Li Z, et al. Dominant role of climate in determining spatio-temporal distribution of potential groundwater recharge at a regional scale. Journal of Hydrology. 2019;**578**:1016

[230] Changnon SA, Huff FA, Hsu C-F. Relations between Precipitation and Shallow Groundwater in Illinois. Journal of Climate. 1988;**9**:1239-1250

[231] Fu G, Crosbie RS, Barron O, Charles SP, Dawes W, Shi X, et al. Attributing variations of temporal and spatial groundwater recharge: A statistical analysis of climatic and non-climatic factors. Journal of Hydrology. 2019;**568**:816-834

[232] Fritz P, Drimmie RJ, Frape SK, O'Shea K. The isotopic composition of precipitation and groundwater in Canada. In: Isotope Techniques in Water Resources Development. Proc. IAEA Symposium. Vienna; 1987. pp. 539-550

[233] Lee KS, Wenner DB, Lee I. Using H- and O-isotopic data for estimating

the relative contributions of rainy and dry season precipitation to groundwater: Example from Cheju Island, Korea. Journal of Hydrology. 1999;**222**(1-4):65-74

[234] Yin L, Hou G, Su XS, Wang D, Dong J, Hao Y, et al. Isotopes (δD and δ18O) in precipitation, groundwater and surface water in the Ordos Plateau, China: Implications with respect to groundwater recharge and circulation. Hydrogeological Journal. 2011;**19**(2):429-443. DOI: 10.1007/s10040-010-0671-4

[235] Jasechko S, Birks SJ, Gleeson T, Wada Y, Fawcett PJ, Sharp ZD, et al. The pronounced seasonality of global groundwater recharge. Water Resources Research. 2014;**50**(11):8845-8867. DOI: 10.1002/2014WR015809

[236] Eastoe CJ, Wright WE. Hydrology of mountain blocks in Arizona and New Mexico as revealed by isotopes in groundwater and precipitation. Geoscience. 2019;**9**(11):461; DOI: 10.3390/geosciences9110461

[237] Bloomfield JP, Marchant BP. Analysis of groundwater drought building on the standardized precipitation index approach. Hydrological Earth System Science. 2013;**17**(12):4769-4787

[238] Kumar R, Musuuza JL, Van Loon AF, Teuling AJ, Barthel R, Ten Broek J, et al. Multiscale evaluation of the Standardized Precipitation Index as a groundwater drought indicator. Hydrological Earth System Science. 2016;**20**(3):1117-1131. DOI: 10.5194/hess-20-1117-2016

[239] Yeh H-F, Hsu H-L. Using the Markov Chain to analyze precipitation and groundwater drought characteristics and linkage with atmospheric circulation. Sustainability. 2019;**11**(6):1817

[240] Thomas BF, Behrangi A, Famiglietti JS. Precipitation intensity effects on groundwater recharge in the southwestern United States. Water (Switzerland). 2016;**8**(3):12-17. DOI: 10.3390/w8030090

[241] Asoka A, Wada Y, Fishman R, Mishra V. Strong linkage between precipitation intensity and monsoon season groundwater recharge in India. Geophysical Research Letters. 2018;**45**(11):5536-5544. DOI: 10.1029/2018GL078466

[242] Liu J, Ma X, Duan Z, Jiang J, Reichstein M, Jung M. Impact of temporal precipitation variability on ecosystem productivity. Wiley Interdisciplinary Review Water. 2020;**7**(6):1-22. DOI: 10.1002/wat2.1481

[243] Lotsch A, Friedl MA, Anderson BT, Tucker CJ. Coupled vegetation-precipitation variability observed from satellite and climate records. Geophysical Research Letters. 2003;**30**(14):8-11. DOI: 10.1029/2003GL017506

[244] Rasmusson EM, Arkin PA. A global view of large-scale precipitation variability. Journal of Climatology. 1993;**6**(8):1495-1522

[245] Panagoulia D. Assessment of daily catchment precipitation in mountainous regions for climate change interpretation. Hydrological Science. 1995;**40**(3):331-350

[246] Panagoulia D, Bárdossy A, Lourmas G. Diagnostic statistics of daily rainfall variability in an evolving climate. Advance Geoscience. 2006;**7**:349-354

[247] Panagoulia D, Bardossy A, Lourmas G. Multivariate stochastic downscaling models for generating precipitation and temperature scenarios

of climate change based on atmospheric circulation. Global Nest Journal. 2006;**10**:263-272

[248] Gherardi LA, Sala OE. Enhanced precipitation variability decreases grass- and increases shrub-productivity. Proceedings of the National Academy Science USA. 2015;**112**(41):12735-12740. DOI: 10.1073/pnas.1506433112

[249] Dong L, Guo Y, Tang W, Xu W, Fan Z. Statistical evaluation of the influences of precipitation and river level fluctuations on groundwater in Yoshino River Basin, Japan. Water. 2022;**14**(4):12735

[250] Pendergrass AG, Knutti R, Lehner F, Deser C, Sanderson BM. Precipitation variability increases in a warmer climate. Scientific Reports. 2017;**7**(1):1-9. DOI: 10.1038/s41598-017-17,966-y

[251] Zhang W, Furtado K, Wu P, Zhou T, Chadwick R, Marzin C, et al. Increasing precipitation variability on daily-to-multiyear time scales in a warmer world. Scientific Advances. 2021;**7**(31):1-12. DOI: 10.1126/sciadv.abf8021

[252] Panagoulia D. Hydrological modeling of a medium-size mountainous catchment from incomplete meteorological data. Journal of Hydrology. 1992;**135**(1-4):279-310

[253] Panagoulia D, Vlahogianni EI. Nonlinear dynamics and recurrence analysis of extreme precipitation for observed and general circulation model generated climates. Hydrological Process. 2014;**28**(4):2281-2292. DOI: 10.1002/hyp.9802

[254] Panagoulia D, Economou P, Caroni C. Stationary and non-stationary GEV modeling of extreme precipitation over a mountainous area under climate change. Environmetrics. 2014;**25**(1):29-43

[255] Zheng W, Wang S, Sprenger M, Liu B, Cao J. Response of soil water movement and groundwater recharge to extreme precipitation in a headwater catchment in the North China Plain. Journal of Hydrology. 2019;**576**:466-477

[256] Boutt DF, Mabee SB, Yu Q. Multiyear increase in the stable isotopic composition of stream water from groundwater recharge due to extreme precipitation. Geophysical Research Letters. 2019;**46**(10):5323-5330. DOI: 10.1029/2019GL082828

[257] Liu Y, Jiang X, Zhang G, Xu YJ, Wang X, Qi P. Assessment of shallow groundwater recharge from extreme rainfalls in the Sanjiang Plain, northeast china. Water. 2016;**8**(10):5323-5330

[258] Qi P, Zhang G, Xu YJ, Wang L, Ding C, Cheng C. Assessing the influence of precipitation on shallow groundwater table response using a combination of singular value decomposition and cross-wavelet approaches. Water. 2018;**10**(5):23

[259] Shao J, Si B, Jin J. Rooting depth and extreme precipitation regulate groundwater recharge in the thick unsaturated zone: A case study. Water. 2019;**11**(6):3390

[260] Bailey JK, Whitham TG. Interactions between cottonwood and beavers positively affect sawfly abundance. Ecological Entomology. 2006;**31**:294-297

[261] Sacchi CF, Price PW. Pollination of the arroyo willow, Salix lasiolepis: Role of insects and wind. American Journal of Botany. 1988;**75**:1387-1393

[262] Simieon G, Stevens LE. Tamarix (Tamaricaceae), Opsius stactogalus (Cicidellidae), and litter fungi interactions limit riparian plant establishment. Advances in Entomology. 2015;**3**:65-81

[263] Carlisle JD, Skagen SK, Kus BE, Van Riper CIII, Paxton KL, Kelly JF. Landbird migration in the American West: Recent progress and future research directions. The Condor. 2009;**111**:211-225

[264] Skagen SK, Kelly JF, Van Riper CIII, Hutto RL, Finch DM, Krueper DJ, et al. Geography of spring landbird migration through riparian habitats in southwestern North America. The Condor. 2005;**107**:212-227

[265] Kelly JF, Hutto RL. An East–West comparison of migration in North American wood warblers. The Condor. 2005;**107**:197-211

[266] Stevens LE, Buck KA, Brown BT, Kline N. Dam and geomorphic influences on Colorado River waterbird distribution, Grand Canyon, Arizona. Regulated Rivers: Research & Management. 1997;**13**:151-169

[267] Geerling GW, Ragas AMJ, Leuven RSEW, van den Berg JH, Breedveld M, Liefhebber D, et al. Succession and rejuvenation in floodplains along the River Allier (France). Hydrobiologia. 2006;**565**:71-86

[268] Piotrowski JS, Lekberg Y, Harner MJ, Ramsey PW, Rillig MC. Dynamics of mycorrhizae during development of riparian forests along an unregulated river. Ecography. 2008;**31**:245-253

[269] Woodbury AM, editor. Ecological Studies of Flora and Fauna in GlenCanyon. Salt Lake City: University of Utah; 1959

[270] Holdridge LR. Determination of world plant formations from simple climatic data. Science. 1947;**105**:367-368

[271] Gomi T, Sidle RC, Richardson JS. Understanding process and downstream linkages of headwater systems. BioScience. 2002;**52**:905-916

[272] Cantonati M, Stevens LE, Segadelli S, Springer AE, Goldscheider N, Celico F, et al. Ecohydrogeology: The interdisciplinary convergence needed to improve the study and stewardship of springs and other groundwater-dependent habitats, biota, and ecosystems. Ecological Indicators. 2020;**110**. DOI: 10.1016/j.ecolind.2019.105803

[273] Lubinski KS. A Conceptual Model of the Upper Mississippi River System. Onalaska: U.S. Fish and Wildlife Service; 1993

[274] Stromberg J, Briggs M, Gourley C, Scott M, Shafroth P, Stevens L. Human alterations of riparian ecosystems. In: Baker MB Jr, Ffolliott PF, DeBano L, Neary DG, editors. Riparian Areas of the Southwestern United States: Hydrology, Ecology, and Management. Boca Raton: Lewis Publishers; 2004. pp. 99-126

[275] Steiger J, Tabacchi E, Dufour S, Corenblit D, Peiry J-L. Hydrogeomorphic processes affecting riparian habitat within alluvial channel-floodplain river systems: A review for the temperate zone. River Research and Applications. 2005;**21**:719-737

[276] Harvey J, Gooseff M. River corridor science: Hydrologic exchange and ecological consequences from bedforms to basins. Water Resources Research. 2015;**51**:6893-6922

[277] Ward JV. Ecology of alpine streams. Freshwater Biology. 1994;**32**:277-294

[278] Scrimgeour GJ, Prowse TD, Culp JM, Chambers PA. Ecological effects of river ice breakup: A review and perspective. Freshwater Biology. 1994;**32**:261-275

Chapter 7

Water Resources Management and Hydraulic Infrastructures in the Senegal River Basin: The Case Study of Senegal, Mali and Mauritania

Cheikh Faye

Abstract

The water resources of the Sahelian countries bordering the Senegal River basin (Senegal, Mali and Mauritania) are limited and unevenly distributed. To overcome the unequal distribution of water resources and to manage floods and droughts in the Senegal River Basin, hydraulic infrastructures have been built in the Senegal River Basin, starting with the Manantali dam. The paper reviews the current water storage capacity in the Senegalese, Malian and Mauritanian parts of the Senegal River Basin from a sustainable water resources management perspective. Data from the Manantali dam and from the water resources of the downstream countries (Mali, Senegal and Mauritania) in the Senegal River basin were employed to assess water storage capacity at country level in this basin. Water storage capacity was found to be lowest in the Mauritanian part and highest in the Malian part. These results led to the conclusion that despite the OMVS based heavy investment in the infrastructure of water storage capacity there is both need and potential for infrastructure increase. As the Senegal River Basin is a transboundary case the riparian countries sharing in order to promote integrated water resources management at the basin level, need to continue to develop additional storage to underpin and modernize the responsible use of water resources through the construction of other multifunctional water infrastructure.

Keywords: water storage, water infrastructure, dams, water use, water management

1. Introduction

Infrastructures are defined as "networks that enable the movement of goods, people or ideas and allow their exchange in space" [1]. The speed and direction of movement is influenced by their topology and physical form and from this point of view infrastructures are technological objects. Water distribution systems can be defined "as networks connecting water from rivers, lakes and storage sites to homes, farmers' fields irrigation systems and factory outlets empowering water's economic and social functions" [2].

The construction of dams is often linked to state policies seeking to meet the needs of populations. Among these needs, the multiplication of the number of structures is caused by regional development, increased access to drinking water and electricity, flood-fighting and irrigation development. Half of the world's rivers have at least one dam, and hydroelectric power plants produce more than 50% of the electricity consumed in a third of the world's countries [3].

Dams are works that block a section of a valley over the entire width and create a basin in a geological way and thus they usually are considered to be barriers. Its origin may be natural or catastrophic e.g., land slides or avalanches, or it can be the result of a disorganization of the river network with a change in the geomorphological system e.g., moranic or glacial dam. A dam is a project, and hence it has a pre-determined lifecycle after which it may end up being filled, or else by yielding, undermined by the infiltration waters.

Dams are the subject of many claims by members of civil society. The latter often criticize manufacturers and decision-makers for a lack of consideration towards them, a lack of transparency, and unfulfilled promises. From these concerns that animate the populations arise demands, forces to fight against dam projects accused of the flooding of forests, the acidity of water, the sterilization of agricultural land, and expropriations. These infrastructures also include the links between these machines that allow them to function as a system, as well as techniques of organization—companies, accounting, bureaucracies, etc. [4]. These infrastructures, of course, exist in society, and often embody, reflect and, in turn, shape their political, economic and social environment [2].

However, water is considered to be an economic good and is classified as a necessity [5, 6] and the economic motives for increase of agricultural/meat production via bulding dams are as below.

As seen above, if in-country production increases substantially Mauritania an Mali have a chance at a zero/positive BoP and Senegal could halve its negative BoP (**Figures 1–4**).

The first dams in the world date back to antiquity. Their objectives were to meet the water needs of the populations and for irrigation. "They are located in the Nile Valley, Mesopotamia, China and South Asia." The oldest known remains come from the Sadd-el-Karafa dam made in Egypt between 2950 and 2750 BC. Even in ancient Rome, more than adequately supplied with water and believing that flowing water was a sign of a high standard of living, water-saving devices (such as taps and storage tanks) were widely deployed [8]. Studies provide information on the mode of operation of dams, their history, as well as their consequences on the fragmentation of watercourses. These impacts are analyzed from several angles, in particular through the transformations of the landscapes, the displacement of populations, the changes of identity, the images projected on the disruptive dam, the economic contributions, the usefulness and the beneficiaries. These analyzes relate to cases of dams already built or projects.

Information also exists on the types of construction. They can be arch dams, gravity dams or buttress dams. This information provides knowledge on the world classification and according to the International Commission on Large Dams (ICOLD). The qualification of "large dam" is attributed to those that rise more than 15 meters above the foundations, according to the said commission.

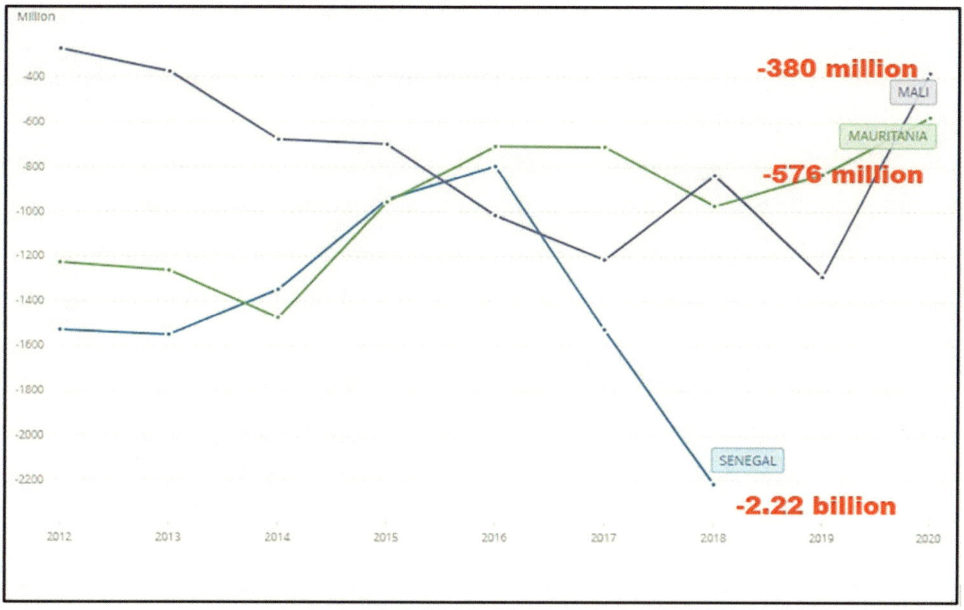

Figure 1.
Current Account (BoP) in Current USD [7].

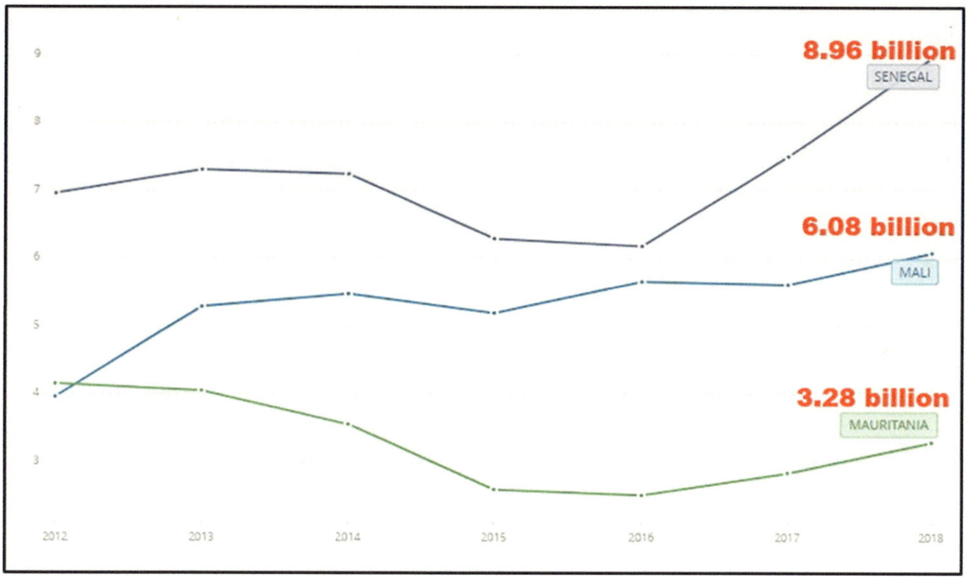

Figure 2.
Imports in Current USD [7].

Dams are installed with a run-of-river or pumped storage reservoir. They are classified into two categories according to the type of material: concrete dams (gravity dams, buttress dams, arch dams) and embankment dams (earth dams, rockfill dams). A third type combining the first two is called hybrid or compound. ICOLD considers that there are 24,395 large earth dams, 3065 rockfill dams, 6688 gravity

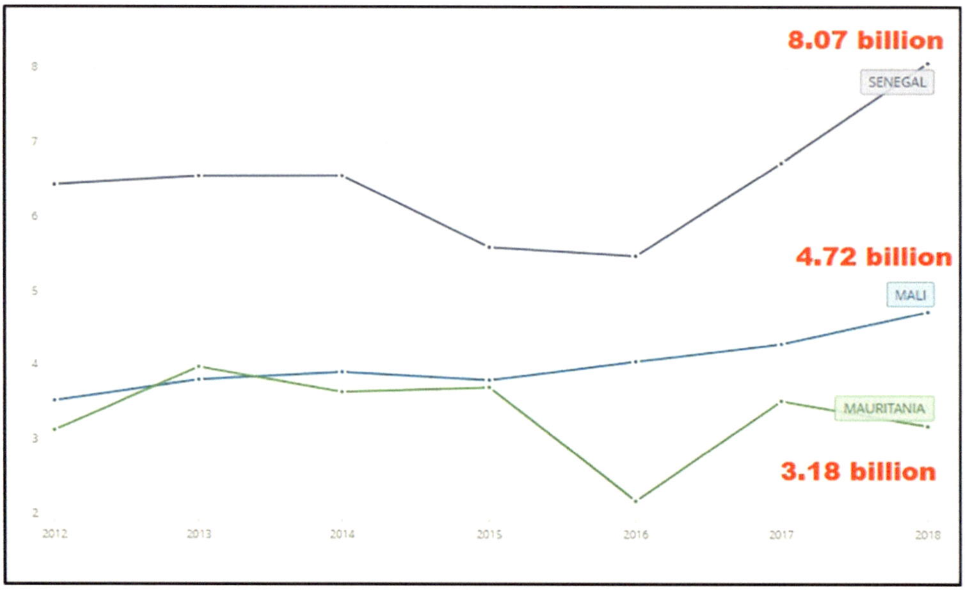

Figure 3.
Merchandise Imports Current USD [7].

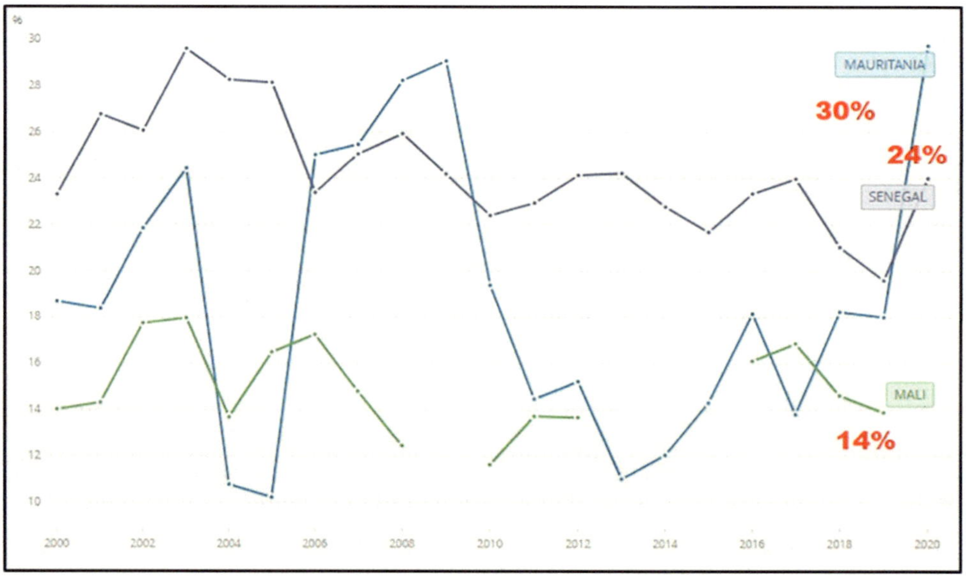

Figure 4.
Food imports as percentage of Merchandise Imports [7].

dams, 426 buttress dams, 1839 arch dams, 172 multi-arch dams and 2603 of another type. Embankment dams are in the majority and constitute nearly 63% of the total number of dams recorded. It is obviously the oldest type of dam and there are traces of embankment dams dating from the oldest civilizations. In addition, this type of dam can be adapted with many types of foundations.

The Senegal River, which is the second largest river in West Africa, owes its formation to the joining of the Bafing and the Bakoye rivers at Bafoulabé, Mali. Its 300,000 km^2 watershed is divided into three subsets [9]: the upper basin, the valley and the delta as seen in (**Figure 5**). The Senegal River runs through four distinguishable climatic zones:

- Guinean (very humid)
- South Sudanian (humid)
- North Sudanian (semi-humid)
- Sahelian (semi-arid).

At 1500 mm/year in the Guinean part the rainfall gradient remains poweful compared to 200–250 mm/year in the northern part leading to an annual average of 550 mm/year, a pluviometric contrast that is a a main basin characteristic which is attenuated to a ceratin extent as billions of m^3 of water are transferred annually by the river from the the upper basin wet regions to the the valley and the delta arid Sahelian regions [10, 11], which explains the great wealth of biophysical environments in the basin and their great diversity.

The impacts of the dam are often examined without taking into account the forms of knowledge of the inhabitants who experience them. The weakness of the

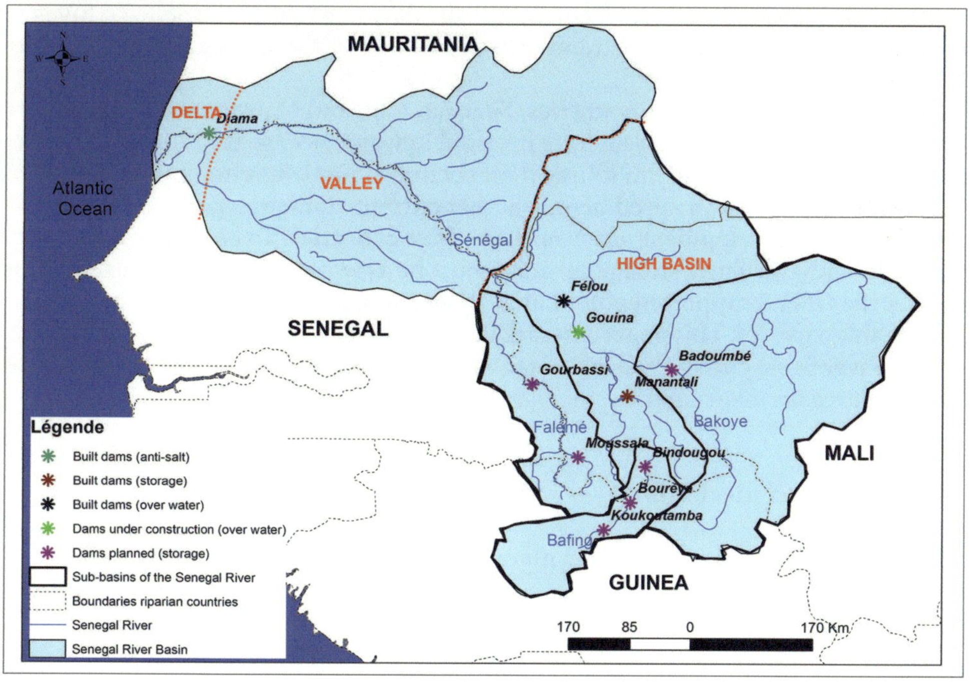

Figure 5.
Situation of the Senegal River watershed and the dams built and planned.

approach to social questions in a context of research into sustainable development is denounced by sociologists, anthropologists, agrogeographers, historians, economists and geographers. On the Manantali dam, the first studies date back to the 1970s. They focused on calculations and simulations before and after the construction of the dam and the hydroelectric power station. The creation of the reservoir, the fish population, the quality of the water, the flows, and the regime of the river were thus questioned. Legally, the status of the Manantali dam is qualified as co-ownership. The specific role of the Organization for the Development of the Senegal River (OMVS) in the operation of the Manantali dam is underlined in certain studies, and the success of the bet on regional integration around the Senegal River is particularly mentioned.

Water withdrawals vary by country. In Senegal, in 2000, withdrawals from water resources amounted to 1591 million m^3 including 1435 million for agriculture (93%), 98 million for communities (4%) and 58 million for industry (3%). In Mauritania, in 2000, water withdrawals were estimated at 1698 million m^3 including 1.5 billion for agriculture (88%), 150 million for domestic use (9%) and 48 million for industry (3%). In Mali, the current withdrawals of the irrigation sector are of the order of 5.0 km^3 in 2006, or 96.4 percent of the total withdrawal [12], and come almost entirely from water resources, surface and almost entirely over a period of 6 months. In the Senegal River Basin water resource availability and distribution are influenced by factors such as: population dynamics, extremely variable climatic conditions ecosystem maintentenance water-affecting environmental issues, and political/socio-cultural issues e.g., food security, and the problem of economic development [13].

Based on the OMVS agreements framework it is seen that via the Manantali dam management a minimum low water flow at the Mali/Senegal border is guaranteed and the Mali, Senegal and Mauritania agreed sharing of stored water is ensured. Thus, a sectoral plan declining master frameset for the development of the Senegal River (SDAGE) is applied in order to attempt the promotion of the watershed's sustainable and concerted development [9].

The water-scarce Sahelian countries, Senegal, Mali and Mauritania, lying on the border of the Senegal River basin have invested in water storage for a long time so as to increase water availability to satisfy their socio-economic/environmental needs. This leads to the question raised here, whether current storage capacity is sufficient to cover the future development needs of these three countries. To be more precise, the Senegal River basin water storage capacity must be assessed taking into consideration whether the OMVS supported potential of these countries is realistically capable to increase this capacity. This paper assesses the the Sahelian countries bordering the Senegal River Basin current water storage capacity from the poin of view of integrated water resources management.

2. Materials and methods

This article is based on data from literature searches and secondary data collection. The main data on water resources were collected from the FAO global database AQUASTAT available at: https://www.fao.org/aquastat/statistics/query/index.html?lang=fr. For the case study on the Senegal River Basin, the databases of the Direction de la Gestion et de la Planification des Ressources en Eau (DGPRE) and of the Organization pour la Mise en Valeur du fleuve Sénégal (OMVS) were used. The combined approach to data collection (on dam issues) is favored here. It consisted first of all of a consultation of unpublished documents (books, reports, dissertations, theses, articles, etc.) which are of great interest for the present. This in-depth review of the literature allowed us to collect different data and information available on the impact of dams in areas where similar studies have been conducted.

3. Results and discussion

3.1 Water availability

Table 1 shows the quantity of wavailable water as seen in FAO [6]. The 2014 three renewable water resource classesi.e., surface water, groundwater, internal water and external water vary according to country. Te estimation results for Mali are at 120 km^3/year compare to those for Senegal, 38.97 km^3/year and for Mauritania, 11.4 km^3/year. For Mali there is a dependency index of 50% between surface renewable water resources and internal renewable water resources estimated at 110 km^3/year and 60 km^3/year repspectively which is explained, in the case of the importance of surface renewable water resources, by the water availability of the Niger River. In Senegal, surface renewable water resources are estimated at 36.97 km^3/year and internal renewable water resources of around 25.8 km^3/year i.e., a dependency index of 33.8%. In the particular case of the the weakest availability country, Mauritania, 97.4% of all water resources is represented by 11.1 km^3/year renewable surface water resources, which essentially are comprised of reservoirs dams distributed widely in the southern and central parts of the territory and of the Senegal River along with its tributaries.

Renewable water resources	Mali		Senegal		Mauritania	
	Quantity	%	Quantity	%	Quantity	%
Total inland water resources (km3/year)	60	50.0	25.8	66.2	0.4	3.5
Total external water resources (km3/year)	60	50.0	13.17	33.8	11	96.5
Surface water resources: total (km3/year)	110	91.7	36.97	94.9	11.1	97.4
Total groundwater resources (km3/year)	20	16.7	3.5	9.0	0.3	2.6
Total renewable water resources (km3/year)	120	100	38.97	100	11.4	100
Dependency index (%)	50		33.8		96.5	
Total water resources per capita (m3/year/capita)	6290		2458		2589	
Total withdrawals (km3/year)	5186		2221		1.3502	
Exploitation index (in %)	4.32		5.70		11.84	
Total capacity of dams (km3)	13,795		0.25		0.5	
Total capacity of dams per inhabitant (m3/inhabitant)	723.09		15.77		113.55	

Source: [6].

Table 1.
Renewable water resources available in 2018 in the three countries.

Internal renewable water resources coorespong to 0.4 km^3/year, which leads to a very high dependency index at 96.5% (**Table 1**) while the total withdrawal estimation regarding Senegal is 2.22 km^3. The exploitation index is relatively low at 5.70% if it is justactaposed to potential of water reserves in the country.

Rural areas and many cities in Mali have to rely exclusively on groundwater as the main source of reliable and safe drinking water while Senegal and Mauritania have to rely on the same for huge tracts of arable land irrigation and livestock watering as well as for the supply of many mines and industries. In terms of total volume of available renewable groundwater, the contrast is sharp between Mali (20 km^3/year) and Senegal (3.5 km^3/year) and Mauritania (0.3 km^3/year).

Regarding the Senegal River the inflows are significant, variable, interannually irregular and in an average year around 20 km^3 while in the wet year of 1924 they reached 41 km^3 in the dry year of 1987 went down to 6.15 km^3 [14] and due to the Sahelian climatic deterioration the average inflow went down to 13 km^3/year. However, user requirements are met due to the water draining by the ricers cris-crossing these countries e.g., the Senegal River. In **Table 1** it is seen that total withdrawals estimation is: in Mali 5.19 km^3/year, in Senegal 2.22 km^3/year in Senegal and in Mauritania 1.70 km^3/year. The exploitation index is relatively low taking into consideration the great potential of water reserves and amounts to 4.33, 5.75 and 14.9% correspondingly.

At the existing socioeconomic circumstances in these three countries, social development invitably causes an increasing demand for water as most national planning initiatives e.g., mining, industry, agricultural development, municipal water supply, energy security, tourism and recreation, and municipal water supply [13]. The demographic and urban growth of these countries exerts strong pressure on the often limited available water resources in these countries.

According to the AQUASTAT database [6], renewable freshwater resources per capita (in m^3) continued to decrease between 1958 and 1962 and 2018–2022 at the level of the three countries. They thus fell from 22,301 m^3 in Mali, 12,538 m^3 in Mauritania and 11,612 m^3 in Senegal in 1958–1962 to only 6290 m^3 in Mali, 2589 m^3 in Mauritania and 2458 m^3 in Senegal in 2017–2022. These results show the tendency of these differences towards a situation first of water stress (below 1700 m^3/inhabitant/

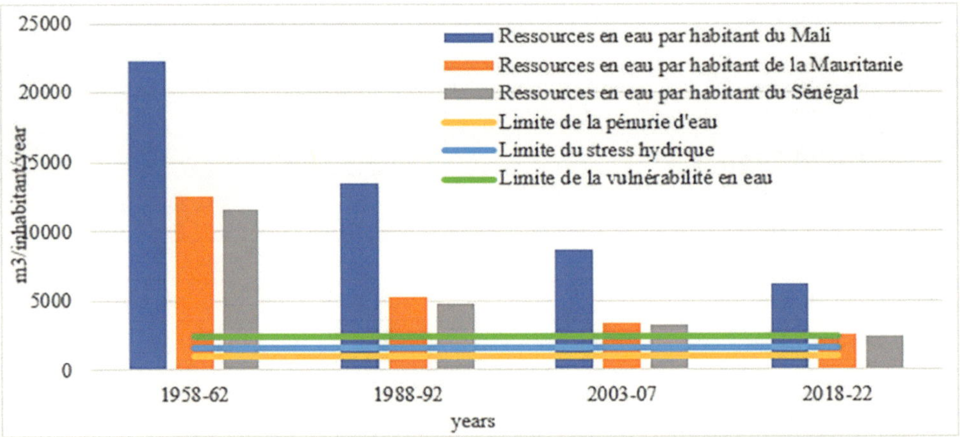

Figure 6.
Evolution of total renewable water resources per capita between 1958 and 1962 and 2013–2017 in the three countries (Source: [6]).

year) and then of water shortage (below 1000 m^3 /inhabitant/year). A country like Senegal is already in a situation of water vulnerability (below 2500 m^3 /inhabitant/year), while Mauritania is not far from such a situation.

As can be seen in the **Figure 6** above in the countries bordering the Senegal river water consumption is increasing at an exponential rate due to population increase leading to the creation of a state of competition for water [15]. According to this these countries will target water uncertainty reduction in-border river flow regulation via dams which is detrimental to the other contesting countries. This leads to peace fracture as high transnational river water import dependent countries e.g., such as Mauritania which is 96.49% dependent, will consider water to be a matter of national security justifying the use of force for its safeguarding [16].

3.2 Shared water systems

Most sub-Saharan Africa freshwater resources are parts of either shared river basins or transboundary watercourse systems. The Senegal River Basin in conjunction with the Organization for the Development of the Senegal River (OMVS) has noted that a strong commitment to regional collaboration must form the basis of the management and protection of these shared systems as the hydrosystem comprised by the Senegal River basin and its tributaries covers a 289,000 km^2 area. The proportions this is shared between Mali, Mauritania, Guinea Conakry and Senegal are 53.5, 26, 11, and 9.5% respectively. The Falémé basin, like the Senegal River, is spread by13,800 km^2 or 47.8% over Mali, 11,500 km^2 or 39.7% over Senegal and 3600 km^2 or 12.5% over Guinea Conakry [17] while Mali and Guinea share the 22,000 km^2 Bafing and the 85,000 km^2 Bakoye basins.

The organization of the Senegal River Basin (OMVS) has as main goal to realize an integrated transnational vision of the Senegal River Basin development where, on the basis of analysis of the basin's water resources/ecosystems, the integration of sectoral objectives will be achieved e.g., hydroelectricity, drinking water and sanitation development, navigation and transport, rural development, mining and industry will be achieved. As seen in **Table 2**, interest in the major components of the OMVS program- energy production, irrigation and navigation-varies according to the riparian country's point of view [18]. Mauritania and Senegal are the two riparian States which exploit nearly 90% of agricultural developments in the basin but whose dependency factor (the total share of water resources produced outside their borders) is the highest in the basin while Mali and Guinea have abundant water resources incommensurably high relative to their agricultural development [19, 20].

Indeed, unlike Mali and Guinea, which border the Niger River basin and have relatively abundant water resources, the Senegal River remains the only source of fresh

Country	Hydroelectric power	Irrigation	Navigation	Flood recession agriculture
Guinea	Potentially high	None	None	None
mali	High	Weak	Very high	Weak
Mauritania	Very high	Very high	High	High
Senegal	Very high	Very high	High	High

Table 2.
Priorities of the riparian States compared to the components of the current OMVS program.

water for Mauritania and Senegal, which exploit nearly 90% agricultural developments in the basin. This situation has led to rivalries for control of the resource, which have resulted in the establishment of a climate of hostility and mistrust between the two neighboring countries for more than two decades now.

In this way the OMVS, in order to manage and develop in a joint way these shared resources, has encompassed a set of instruments specific to this purpose since cooperation is required to achieve sustainable management of the sum total of external resources and as a result e.g., life-improving cross-border hydraulic infrastructures were developed. According to this arrangement, OMVS member countries are obliged to aknowledge its commitments to cooperative management of international waters with its neighbors with goal od promoting the regional interest in terms of peace, security and economic integration.

3.3 Water use in the three countries

While irrigated agriculture is the largest water consumer in these countries water withdrawals vary by country and by sector of activity. Current withdrawals of the irrigation sector in Mali are around 5 billion m^3 in 2006, or 96.4% of the total withdrawal [12], and come almost entirely from mineral resources. Surface water (**Table 3**). In addition, the supply of water to livestock takes about 75 million m^3 industry 4 million and local authorities 107 million (2.06%) in 2006. According to **Table 4** in Senegal (2000), water resources withdrawals were 2221 million m^3 apportioned: for agriculture 2065 million amounting to 93%, for communities 98 million amounting to 4.41% and

Abstraction in millions of m3	**Mali**		**Senegal**		**Mauritania**	
	Value	**%**	**Value**	**%**	**Value**	**%**
Water withdrawal for agriculture (10^9 m3/year)	5075	97.9	2065	93.0	1223	90.6
Water abstraction for industrial uses (10^9 m3/year)	0.004	0.1	58	2.6	31.8	2.4
Water withdrawal for municipalities (10^9 m3/year)	107	2.1	98	4.4	95.4	7.1
Total water withdrawal (sum of sectors) (10^9 m3/year)	5186	100	2221	100	1350.2	100
Progress in water use efficiency	**Mali**		**Senegal**		**Mauritania**	
SDG 6.4.2, Water Stress (%)	8003		11.81		13.25	
SDG 6.4.1, Water use efficiency (US$/m3)	1859		7571		3933	
SDG 6.4.1, Efficiency of water use by irrigated agriculture (US$/m3)	0.047		0.095		0.352	
SDG 6.4.1, Efficiency of water use by industries (US$/m3)	787.1		81.27		47.67	
SDG 6.4.1, Efficiency of water use by services (US$/m3)	58.41		121.5		35.25	

Source: [6].

Table 3.
Water use by main economic sectors in these countries.

Country	Costs assumed	Benefits withdrawn
Mali	35.3%	• 52% of hydroelectric production • opening up thanks to the navigation pane
Mauritania	22.6%	• 15% of hydroelectric production • 33.6% of the 375,000 ha of land made irrigable
Senegal	42.1%	• 33% of hydroelectric production • 64% of the 375,000 ha of land made irrigable

Source: [21].

Table 4.
Cost and benefit distribution key.

for industry 58 million amounting to 2.61%. Similarly, as seen in **Table 3**, in the case of Mauritania (2000), total withdrawals were 1698 million m^3 of which 1500 million went to agriculture amounting to 88.3%, 150 million went for domestic use amounting to 8.83% and 48 million to industry amounting to 2.83%.

Through the UN-Water Initiative for the Integrated Monitoring of Sustainable Development Goal (SDG) 6, the United Nations is committed to supporting countries in monitoring issues related to water and sanitation as part of the 2030 Agenda for Sustainable Development, and in compiling national data to report on global progress towards achieving SDG 6 [19]. Thus the progress made at the scale of the Senegal River basin towards the achievement of SDG 6 is indicated in **Table 3** which shows the values calculated at the national level for the efficiency of water use. While the efficient use of water resources averages just over US$ 15 /m3 worldwide, the countries of the Senegal River Basin record values below this average. Senegal, which records the highest water use efficiency, among the 3 countries in the downstream part of the basin, is at US$ 7571 /m3, against 1933 for Mauritania and 1859 for Mali. However, the situation remains variable depending on the country. Indeed, for the efficiency of water use by industries, Mali is positioned ahead with 787.1 $US /m3, against 81.27 for Senegal and 47.67 for Mauritania. As for the efficiency of the use of water by the services, Senegal stands in front with 121.5 $US /m3, against 58.41 for Mali and 35.25 for Mauritania [20]. At basin scale, the lowest values in this regard are noted in Mauritania.

3.4 Contributing to climate change adaptation from water storage

As the challenges posed by global warming are increasingly understood, it is widely accepted that water is the primary medium through which the societal stresses of climate change will manifest. Although the exact impacts remain uncertain, in many places, even where total precipitation is increasing, climate change will most likely increase precipitation variability. Undoubtedly, those who will be hardest hit are the poor, who are already struggling to cope with the existing variability. It will be increasingly difficult for them to protect their families, their livelihoods and their food supply from the negative effects of seasonal rainfall, droughts and floods, all of which will be exacerbated by climate change.

According to the Intergovernmental Panel on Climate Change [22], climate change is expected to have an impact on the level and variability of rainfall in Africa. Climate change potentially has a significant impact on both water availability and needs in countries bordering the river. Likewise, it alters the hydrological systems and water resources of the Senegal River Basin and reduces water availability. Monitoring the evolution of runoff coefficients at multi-annual, seasonal and monthly scales over the period 1960–2014 clearly shows changes in runoff dynamics and changes in hydrological regimes [23]. Climate change forecast scenarios indicate that climate change will have the effect sometimes of a gradual decrease in runoff from the 2030 horizon to the 2090 horizon, sometimes of a gradual increase in runoff over the basin [24]. What is certain is that climate change will likely lead to more intense and more variable weather events, and will likely lead to more intense and prolonged periods of drought and flooding. The jointly managed water resources of the Senegal River are at the center of climate adaptation strategies, and improved and expanded water storage capacities create buffer zones for periods of water scarcity [25].

Water storage (in all its forms) has a key role to play both in terms of sustainable development and climate change adaptation water storage (in all its forms) plays a key role. Water storage leads to risk reduction risk and, by providing a buffer zone, reduces population vulnerability by offsetting some of the negative impacts induced by climate change. While climate change may impact any water storage option, in general water storage, as seen in **Figure 3**, leads to the improvement of both agricultural productivity and water security.

By altering both water availability and demand, climate change will affect the need, performance and suitability of different water storage options. In some situations, some storage options will be rendered completely impractical while the viability of others may be increased. Storage in ponds, tanks and reservoirs can also be reduced more quickly due to increased evaporation and/or greater sediment inputs. In addition, large and small dams as well as ponds and reservoirs may be at increased risk of eutrophication and flood damage.

In any case, the externalities associated with the different types of storage are also likely to be affected by climate change. Climate change requires fundamental rethinking of how water resources, and in particular water storage options, are planned and managed. In all situations, maximizing the benefits and minimizing the costs of water storage options will, as in the past (but not commonly), require consideration of a wide range of hydrological, social, economic and complex and interdependent environments. However, unlike in the past, planning needs to be much more integrated across a range of levels and scales with greater consideration of the full range of possible options.

The key to planning and managing water storage is determining current and future needs and making appropriate choices from the suite of storage options available. In a given situation, this requires understanding a range of biophysical and socio-economic issues that influence the *need, the effectiveness* and *suitability* of different water storage options, both in isolation and within systems comprising several types. In the past, there has generally been little explicit consideration of these issues, even for the construction of large dams. For the other options, where the planning is generally less formalized, the needs are generally taken for granted and alternative options are only very rarely considered.

3.5 Water storage infrastructure

3.5.1 Investment in water infrastructure

The lack of storage infrastructure has significant negative social and economic impacts, especially in a drought-prone region where water is inadequately stored [13]. Water storage facilitates the supply of water for domestic and industrial use, irrigation for sustainable agriculture, hydropower generation, infrastructure and job creation, improvement of accessibility of regions where large dams are built, the promotion of ecotourism through fishing, canoeing and tourism, and enabling the sale of water to thirsty regions and communities inside and outside of national borders, which constitutes a strong increase in the revenues of local and national governments [26].

Investment in infrastructure for the development and distribution of water resources has shown significant human and macroeconomic benefits. On the other hand, countries that have limited water storage capacity experience damaging shocks from droughts and floods [25]. Investing in improved water storage to even out access to water during and between rainy seasons and being prepared to support flood management are also primal implements in a flood and poverty reduction strategy. A number of declarations has recognized the important role of dams and reservoirs played in sustainable development: The 2002 World Summit on Sustainable Development, The 2004 Beijing Declaration on Hydropower and Sustainable Development, The 2008 Dams and Hydropower for Sustainable Development in Africa and The Ministerial Declarations of the Fifth and Sixth (2009/2012) World Water Forums [27].

As seen in Article 12 of the joint works agreement there is a "key" (**Table 4**) which determines, on the basis of benefits accrued for the individual country, the distribution of investment and operating costs and operating costs. This phase, limited to consultation alone, consists of allocating quotas (water volumes) to the riparian States for the implementation of national development plans for their respective portions of the river [21].

This legal framework testifies to the fact that the political project of the OMVS goes far beyond mere consultation. It aims at regional integration through the development and pursuit of a basin-wide water resources development plan. *A priori,* it therefore surpasses as much the concept of management at the basin scale (just fixed by the doctrine through the Helsinki rules) that of Integrated Water Resources Management (IWRM), which was only formulated in 1992, at the end of the Dublin Ministerial Conference.

3.5.2 Large dams in the Senegal river basin

The OMVS riparian States invested heftily in billions of m^3 storing dams for the dual purpose of overcoming the inequitable water distribution of water and flood/drought management in the basin. As can be seen in **Table 5** an OMVS data-based plan has been set in motion targeted on the construction of water control 'first generation'structures, the Diama and Manantali dams along with their auxiliary structures, which succeded in their target of water availability in guaranteed all year-round sufficient quantity which supports both agricultural development and natural environmental restoration. Case in point, as seen in [28], the Manantali dam, not only accommodates the storage of 11.3 billion m^3 leading, along with Diama Dam to the irrigation capacity of 255,000 ha but also produces 800 GWh/yearand regularizes river flow at Bakel at 300 m^3/s in Bakel along with river navigation.

In **Table 5** can be seen the next generation hydroelectric works planned by OMVS which, as seen in [19], from 2012 onwards targeted the increase the energy supply and full control of basin waters via major hydroelectric dam construction projects. For example, two secong generation structures built along the river, the

Country	Barrage	Watercourse	Storage capacity (Millions of m^3)	Installed power (MW)	Functions
Mali	Manantali	Bafing	11,300	200	Hydroelectricity + Regualtion
	Felou	Senegal	0 (over the water)	70	Hydroelectricity
	Gouina	Senegal	0 (over the water)	140	Hydroelectricity
	Mussala	Faleme	3000	30	Hydroelectricity + Regualtion
	Bindougou	Bafing	2000	49.5	Hydroelectricity + Regualtion
	Boudofora	Bakoye	to be determined	30	Hydroelectricity + Regualtion
	Marela	Bakoye	3000	21	Hydroelectricity + Regualtion
	Badoumbe	Bakoye	10,000	70	Hydroelectricity + Regualtion
Guinea	Koukoutamba	Bafing	3600	280	Hydroelectricity + Regualtion
	Boureya	Bafing	5500	160	Hydroelectricity + Regualtion
	Balassa	Bafing	0 (over the water)	180	Hydroelectricity
Senegal	Gourbassi	Faleme	2100	30	Hydroelectricity + Regualtion

Table 5.
Dams and dam projects in the Senegal River Basin and their storage capacity.

currently operational Félou dam and the under construction Gouina dam, have no water storage capacity. These two dams are said to be second generation structures. Finally, the other planned works (known as third-generation works) in the Senegal River basin consist of two run-of-river hydroelectric works and eight multi-purpose reservoir works. A good part of these structures will therefore be dams with water storage capacity (Badoumbé, Boureya, Moussala, Gourbassi, etc.). Among these planned storage structures, some are in the funding research phase (Koukoutamba, Boureya and Gourbassi) and others are in the study and identification phase (Badoumbé).

Insofar as these developments will create a water reservoir, as is the case for the Gourbassi project (storage volume of 2.1 billion m3 of water), they will contribute to better water control, flowing further downstream and which are not currently controlled (Falémé and Bakoye). They will therefore play an important role in the fight against floods, but require rigorous management and coupled with that of Manantali. On the other hand, run-of-river hydroelectric dams (Félou and Gouina) will not play any role in flood risk management because, being used exclusively for hydroelectricity production, they do not have a sufficient storage volume to rolling floods. In the estuary part, it is planned to extend the embankments of the river upstream from Rosso as far as Dagana, to enable the level of management of the Diama reservoir to be raised to the 250 cm IGN level in the dry season. The fresh water reserve thus created would meet the water needs of 120,000 ha of irrigated crops in the region.

On the technical level, the two major challenges ahead concern the satisfaction of energy and food demands. It is indeed expected that these data will experience an exponential evolution under the effect of the strong demographic growth of the basin: with a rate of 2.7%; the population is expected to double every 25 years. The potential for hydroelectric production and irrigated cultivation is considerable (1200 MW; 375,000 ha) and should make it possible to meet these demands, but this requires developing its potential, which has so far been largely under-exploited: only 25% in the first case, just over 30% in the second. This is the purpose of the medium- and long-term planned projects presented in **Table 5** (the map locating the second-generation projects can be consulted in **Figure 7**). The challenge of the success of these developments does not depend solely on technical factors. The challenge lies in reconciling competing uses, between tradition and modernity, between regional economic integration and sustainable local development. With regard to the first point, the orientation towards multi-use projects responds to the concern of the riparian States to share the benefits according to the interests of each. However, it appears that certain aspects of the OMVS program have suffered from competition between uses for funding. As a result, some States have not obtained satisfaction. One thinks in particular of Mali which had obtained that its opening up be ensured by means of the "navigation" component, from which we have seen that, for lack of funding, nothing or almost nothing has been achieved.

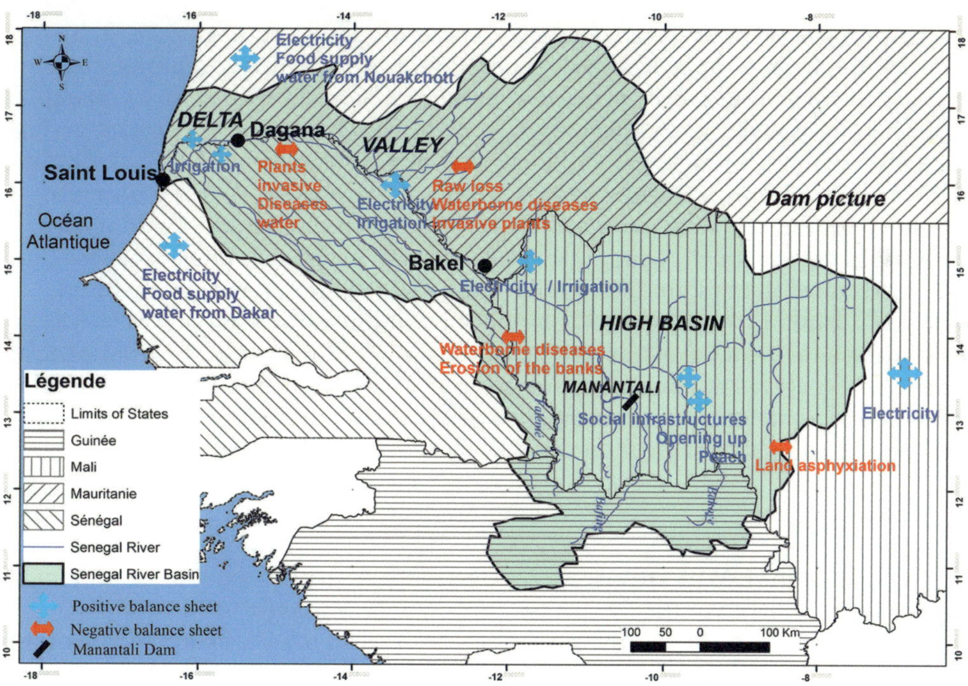

Figure 7.
The Senegal River basin: Perceptions of the results of the program according to the zones of the basin and at the level of the countries bordering the river (Source: [29] modified).

3.6 Water storage capacity

Improved water resource management and water storage capacity makes the economy more resilient to external shocks, such as variability and drought, and thus provides a stable and sustainable basis for productivity and food production and industrial scale [13]. The riparian states, within the framework of the OMVS, continue to develop a large amount of water storage infrastructure, which still needs to be increased in order to improve the water storage capacity in the basin. This storage capacity is 11,300 million m^3 at the level of the Manantali dam and will be around 10,000 million m^3 for Badoumbé, 5500 million for Boureya, 3600 million for Koukoutamba, 3000 million for Moussala and Maréla, 2100 million for Gourbassi and 2000 million for Bindougou. These various facilities should make it possible to store nearly 23 billion m^3 of water, and thus achieve almost total control (more than 97%) of the flows of the Senegal River, by doubling the storage capacities of Manantali and Diama together [19, 30, 31].

At riparian state national level, the largest storage capacity of 13,790 million m^3 is in Mali, the lowest at 250 million m^3 in Senegal while in between lies Mauritania at 500 million m^3. As seen in **Table 1**, in terms of total capacity of dams per inhabitant, Mali is again at the top with 783.5 m^3/inhabitant, Senegal at the bottom with 16.52 m^3/inhabitant and in-between lies Mauritania with 122.9 m^3/inhabitant. Regarding planned storage structures, as seen in **Table 5**, Mali will get the largest one at 29,300 million m^3, Senegal a 2100 million m^3 one and Mauritania none while national rainwater harvesting programs should be supported in all riparian states.

3.7 Water storage for multiple uses

The purpose of water storage includes “water supply for domestic use, industry, livestock and irrigation, hydropower generation, flood and drought protection, fishing and aquaculture, transportation, flood recession cultivation and grazing sites, sinks for pollutants, biodiversity-based tourism, landscape or sporting activities, cultural and religious uses, and biological diversification sites” [32]. According to [13] rural area domestic supply water is used for various samall scale purposes e.g., small businesses, vegetable gardens and stock watering besides cooking and washing. In the case of the Senegal River Basin, water storage is multi-purpose to be distributed in food production irrigation and fisheries, power generation, industrial, drinking, environmental services and mining water supply, as well as in flood management and drought mitigation. In order of importance hydroelectricity production, irrigated agriculture, navigation and other minor activities e.g., fishing and breeding, water supplyand ecological needs are the the main areas of development associated with storage structures.

3.7.1 Hydroelectricity

Throughout the OMVS area, access to electricity is a real obstacle to development. The member countries of the OMVS are, in their entirety, confronted with shortages and growing demands for energy. And yet the watershed has the hydroelectric potential necessary to meet the needs of its populations. Current electricity production represents 16% of the basin’s production capacity. The current demand of the riparian states of the basin is estimated at 4400 GWh/year. If the growth rate is maintained on all the Member States’ electricity networks, energy needs will be around 15,000 GWh in 2040. The Manantali power station, which is currently the only facility operating for energy production can meet 18% of the energy needs of Mali, Mauritania and Senegal [19].

3.7.2 Drinking water supply (DWS)

The member states of the OMVS and more particularly the local populations of the basin mostly use groundwater for their drinking water supply. The weight of abstractions for the DWS is low compared to the volumes of water available on the river. The rate of access to drinking water in the states bordering the river is still very low. Studies carried out by the OMVS in 2008 showed that on average nearly 25% of the population of urban towns and nearly 45% of the population living in rural areas do not have access to drinking water meeting the Objectives of the Millennium Development Goals (MDGs). To date, surface water from the basin is limited to supplying the growing urban population of the cities of Conakry, Bamako, Nouakchott and Dakar. Withdrawals intended for the agglomerations of Dakar from Lake Guiers and Nouakchott from the Aftout Es Sahel, estimated at 27 hm3 per year and water withdrawals intended for the industrial and mining sector, currently estimated at around 40 hm3 per year in the basin, remain negligible compared to withdrawals from irrigated agriculture.

3.7.3 Irrigation

Agriculture is the main activity developed in the basin and irrigation is made possible thanks to the regulation of the hydrological regime since the commissioning of the Manantali and Diama dams. It was the trend deterioration in climatic conditions observed from the 1970s that led the neighboring countries to adopt irrigation as the preferred means of intensifying and securing agricultural production. Irrigated agriculture is thus the sector of activity that consumes the most water in the basin. With 20,000 ha in Guinea, 10,000 ha in Mali, 125,000 ha in Mauritania and 240,000 ha in Senegal, the potential for irrigable land is estimated at 395,000 ha over the entire Senegal River basin [33]. Current surface water withdrawals for irrigated agriculture amount to approximately 1.88 km3 per year, with a peak flow of around 110 m3/s in August [34]. The developed areas are today estimated at 120,000 ha with actual exploitation of 60%. The commissioning of the Manantali dam also made it possible to increase Mali's agricultural potential to more than 10,000 ha/year [35]. Senegal and Mauritania share more than 90% of agricultural development in the basin. The OMVS aims to increase agricultural holdings to 255,000 ha by 2025 in order to improve and secure the productive base of the Senegal River basin.

3.7.4 Navigation

Another major objective of the commissioning of the Manantali dam is to create an uninterrupted waterway (12 months/12), from Saint-Louis to Kayes in Mali over a total distance of 900 km. As Navigation is currently very limited on the basin while there is a desire for expansion, there is currently no concrete plan for its development in the basin. However, as seen in [36], it is foreseen to implement a plan for the development of a 55 m wide navigable channel 55 m wide connecting the towns of Ambidédi (43 km downstream of Kayes in Mali) and Saint-Louis at the mouth of the river. River transport would make it possible to open up priority development areas such as the current and future sites of dams, agricultural and mining production areas, towns isolated from the Atlantic Ocean and between them, and to facilitate the transport of merchandise. Although being a non-consumptive request, navigation still requires a minimum draft of 300 m^3/s at Bakel.

3.7.5 The mines

The mining resources of the river remain very little exploited, being limited to the mines of Mali. These weaknesses are explained by the constraints linked to the permanent availability of water, energy and means of transport. The Falémé mining project relied on the supply of cheap hydroelectric power from the Manantali power station. The reality today is that the energy produced by the dam is intended to satisfy other more pressing needs. The current water needs of the mining sector are estimated at 13 Mm3. They will reach a gross volume of 235 Mm3/year by 2025, 85% of which will be reinjected into the ore processing process. These volumes, although very modest (representing less than 5% of the volumes of water available in the basin) are totally taken from the tributary of the Falémé which, due to a lack of regulation, can restrict access to water during the low water season. The completion of the future river navigation project would propel the mining sector, which could in the long term be one of the basin's development poles. The commissioning of the future development of Gourbassi, although making it possible to reduce the energy deficits of this sector, will not have significant impacts on the reduction of water deficits in the sense that the increase in exploitations is essentially concentrated on the upstream part of the plant.

3.8 Water needs

On the river, irrigation is the sector consuming the most water. The developed perimeters are mainly sown with rice. There are also 10,000 ha of industrial crops [19]. Current irrigation water needs are estimated at 1437 Mm3/year, mainly distributed between Senegal and Mauritania. The combined requirements of Senegal and Mauritania, located on either side of the valley and the delta, represent more than 90% of the total demand of the four countries. The objective of increasing the irrigated area to 255,000 ha will bring irrigation water needs to 5200 Mm3/year by 2025. 90% of agricultural developments in the valley and the delta. The development of irrigation will still remain very low in the upper basin. The areas developed in this part of the basin will represent less than 10% of the total irrigated area. Taking navigation into account in the development objectives of the valley will require having a guaranteed minimum flow of 300 m3/s at Bakel. The combined needs of drinking water supply, livestock and mining are very low and account for less than 10% of the basin's water needs. Demands remain constant throughout the year.

The valley and the delta concentrate more than 90% of low water support requests. The management of resources in this part of the basin is therefore a major challenge for the socio-economic development of the basin. Particular attention will be paid to the Manantali reservoir and later to that of Gourbassi for their locations in relation to the valley and the delta, but also the importance of their mobilizable volumes and which constitute a key variable for low water level support at the level of the valley and the delta and for the overall management, present and future, of the waters of the river.

On the legal level, the Charter specifies the principles which must govern the distribution of water resources (and this in a more detailed manner than the New York Convention since it is a question here of governing a specific situation, and not of establish a framework agreement for the current distribution of the resource by State and by sector (**Table 6**).

3.9 Case study of the large Manantali dam on the Senegal river

The Manantali dam is located on the Bafing River, the main tributary of the Senegal River, 90 km upstream from Bafoulabé (**Figure 7**). Built between 1982 and 1988, the Manantali dam consists of a dam 1460 m long and has a height of 66 m at the foundation. At the IGN filling level of 208 meters, its reservoir has a capacity of 11.3 billion m^3 and covers an area of 477 km^2 [37]. At its minimum operating level (187 m IGN), the reservoir has a volume of 3.4 billion m^3 and covers an area of 275 km^2. The Manantali dam regulates the flow of the Senegal River and makes it possible to irrigate a potential 255,000 ha of land and, in the long term, should allow the navigability of the river over approximately 800 km from the mouth. Added to this is an energy production function [29, 38]. For this, a 200 MW power station was installed and a network made up of 12 transformer stations and approximately 1650 km of high voltage transmission lines of 225,150 and 90 kV, interconnected with the networks of Mali, Mauritania and the Senegal [39].

The Manantali dam is a project multifunction which offers many huge benefits such as flood prevention and surface water availability, power generation, navigation, aquaculture, ecological protection, development-oriented restocking, food self-sufficiency, water transfer and supply in neighboring countries and irrigation [40–42]. Good that the usefulness of the dams is not called into question [43] in intertropical countries whose rivers, while having respectable annual flows, experience wide seasonal and interannual variations, their erection involves a set of irreversible changes. Despite the importance of the advantages, the Manantali dam has its own disadvantages. The impoundment of the Manantali and Diama dams, as well as the resulting developments (embankments, hydro-agricultural developments, etc.) have had negative impacts on the functioning of certain ecosystems in the basin. These impacts are numerous and quite diversified and boil down to the modification of the quantity and quality of water, the pollution of groundwater, the proliferation of aquatic plants, drainage, sedimentation, recurrent diseases), the effect on local culture and the traditional economy [40, 44]. Besides, soil erosion increased, causing collapses of banks and landslides.

Sector/State	Mali	Mauritania	Senegal
Agriculture	1319	1499	1251
Domestic uses	27	101	68
Industry	14	29	41
Total	1360	1630	1380

Table 6.
Current distribution of water resources by State and by sector (in millions of m^3).

Point sight environmental impacts, the dam of Manantali show the proof that there is more advantages than disadvantages. So, he may well be compatible with an ethic of sustainable development and preservation from ecological balances.

4. Conclusions

The variability of precipitation is an important development factor and translates directly into a need for water storage. In Africa, the existing variability and insufficient capacities to manage it are at the root of much of the prevailing poverty and food insecurity. These continents are expected to experience the greatest negative impacts of climate change. By making water available at times when it would not be naturally available, water storage can significantly increase agricultural and economic productivity and improve human well-being. Water storage capacity per person is often cited as an indicator of water security and a measure of large and small scale water infrastructure development [45]. Well-planned and well-managed water storage infrastructure is important to provide a safe and secure water supply for households, agriculture and industry.

In the past, water resources planning has tended to focus on large dams, but dams are only one of many possible water storage options. Other options include natural wetlands, underground aquifers, ponds and small reservoirs. The type of storage to be used in a given location should be suitable for the intended use. Under the right circumstances poverty reduction may benefit by the contribution of each option, while neither is a complete solution since their benefits carry costs and location altering influences poverty reduction in a different way.

There has been very little systematic analysis of alternative storage options in terms of their role in climate change adaptation and poverty reduction. While large dams are the result of central planning and are part of an integral scheme, smaller dams, which are not, result in a piecemeal structure based on local initiatitive and the resulting non-integrated minimal planning based on incomplete data management, erroneous local stakeholder and water resource authorities' interactive communication leading to the expected result of non-optimal investments.

Improved water storage capacity and water security are particularly required in climatic zones characterized by low rainfall and high rainfall variability, such as Mali, Senegal and Mauritania. One of the purposes of water storage in the Senegal River basin is the production of hydroelectricity. Future population growth, combined with climate change, will increase the importance of water storage in many developing countries such as OMVS member countries. However, as water resources are increasingly used and climate variability increases, planning will become even more difficult. Without a better understanding of which types of storage are best used under specific agroecological and social conditions, and without much more systematic planning, there is a risk that many water storage investments will not produce the expected benefits. In some cases, they can even make the most unpleasant impacts of climate change worse.

The need for water storage to support socio-economic development in the countries bordering the Senegal River cannot be overstated. Infrastructure development and management strategy applied in water inbestment is the basic OMVS policy aimed at the support of economic development. Within the OMVS framework these three countries have already invested a lot in the development of the Manantali, Diama and Félou water storage infrastructures in the Senegal river and as there clearly exists more potential for additional infrastructures since these existing dams cannot capture and control all the potentially available water, the OMVS plans for its future exploitment via the programmed dam projects of Koukoutamba, Gourbassi, Bouréya, Balassa and Badoumbé.

Current research aims to better understand water resources and their storage under different social and ecological conditions. This will provide information on the potential impacts of climate change on water supply and demand; the social and environmental impacts of different storage options; the implications of scaling up small-scale interventions; and the reasons for the success/failure of past storage programs. Systematic methods for evaluating the suitability and effectiveness of different storage options are being developed to aid in planning and to facilitate comparison of storage options, individually and within systems.

Conflict of interest

The author declares that there is no conflict of interest.

Author details

Cheikh Faye
Department of Geography, U.F.R. Sciences and Technologies, Assane Seck University of Ziguinchor, Geomatics and Environment Laboratory, Ziguinchor, Senegal

*Address all correspondence to: cheikh.faye@univ-zig.sn

References

[1] Larkin B. The politics and poetics of infrastructure. Annual Review of Anthropology. 2013;**42**:327-343. DOI: 10.1146/annurev-anthro-092412-155522

[2] Crow-Miller B, Webber M, Molle F. The (re)turn to infrastructure for water management? Water Alternatives. 2017;**10**(2):195-207

[3] Commission mondiale des barrages. Barrages et développement: un nouveau cadre pour la prise de décision ; 2000, Earthscan Publications Ltd, 120 Pentonville Road, London N1 9JN, G.B. Dossier 108, Programme Zones Arides, IIED

[4] Molle F, Mollinga PP, Wester P. Hydraulic bureaucracies and the hydraulic mission: Flows of water, flows of power. Water Alternatives. 2009;**2**(3):328-349

[5] Zisopoulou K, Panagoulia D. An in-depth analysis of physical blue and green water scarcity in agriculture in terms of causes and events and perceived amenability to economic interpretation. Water. 2021;**13**(12):1693. DOI: 10.3390/w13121693

[6] Zisopoulou K, Zisopoulos D, Panagoulia D. Water economics: An in-depth analysis of the connection of blue water with some primary level aspects of economic theory I. Water (Switzerland). 2022;**14**(1):103. DOI: 10.3390/w14010103

[7] World Bank. 2022. Available from: https://data.worldbank.org/. [Accessed: April 27, 2022]

[8] Bruun C. The Water Supply of Ancient Rome: A Study of Roman Imperial Administration. Helsinki: The Finnish Society of Sciences and Letters; 1991

[9] OMVS. Projet FEM/Bassin du fleuve Sénégal, 2008. In: Plan d'action stratégique de gestion des problèmes environnementaux prioritaires du bassin du fleuve Sénégal, Version finale. 2008. p. 133. Available from: https://iwlearn.net/resolveuid/81f9d8498bf10c2cdaf710e49b8fd27d

[10] OMVS et HYCOS. Renforcement des capacités nationales et régionales d'observation, transmission et traitement de données pour contribuer au développement durable du bassin du Fleuve Sénégal, Une composante du Système Mondial d'Observation du Cycle Hydrologique (WHYCOS), Document de projet préliminaire ; 2007. p. 53

[11] FAO. FAO global database AQUASTAT. 2022. Available from: https://www.fao.org/aquastat/statistics/query/index.html?lang=fr

[12] MEA. Rapport national sur l'Etat de l'environnement 2009. Ministère de l'Environnement et de l'Assainissement. 2010

[13] Mwendera E, Atyosi Y. A review of water storage for socio-economic development in South Africa. Journal of Water Resource and Protection. 2018;**10**:266-286

[14] CONGAD. Livre bleu « L'eau, la vie, le développement humain » Rapport pays : Sénégal, Document de consultation fourni par SENAGROSOL. CONSULT. 2009:72

[15] Boinet E. La Gestion Intégrée des Ressources en Eau du fleuve Sénégal : bilan et perspectives. Université Paris Sud XI, Mémoire de stage, Faculté Jean Monnet-Promotion. 2011. p. 75

[16] Descroix L, Lasserre F. In: Harmattan L, editor. L'eau dans tous ses états : Chine, Australie, Sénégal, Etats-Unis, Mexique et Moyen-Orient. Paris; 2003. p. 351

[17] Faye C. Evaluation et gestion intégrée des ressources en eau dans un contexte de variabilité hydroclimatique : cas du bassin versant de la Falémé. Thèse de Doctorat: Université Cheikh Anta Diop de Dakar; 2013. p. 309

[18] Niasse M. Prévenir les conflits et promouvoir la coopération dans la gestion des fleuves transfrontaliers en Afrique de l'Ouest. VertigO. 2004;**5**(1):4-16

[19] Thiam NA. Allocation optimale de l'eau dans le bassin versant du fleuve Sénégal. Québec, Canada: Mémoire de Maîtrise en génie des eaux, Université de Laval; 2016. p. 84

[20] FAO. Progrès relatifs à l'efficacité de l'utilisation des ressources en eau—Cadre de référence mondial pour l'indicateur 6.4.1 des ODD. Rome: FAO/ONU-Eau; 2018. p. 56

[21] SOFRECO et al. Etude sur le Programme de développement des Infrastructures en Afrique (PIDA), Rapport de la Phase 1 (provisoire), Secteur GRET (Gestion des Ressources en Eaux Transfrontières), rapport réalisé pour : Nouveau partenariat pour le développement de l'Afrique (NEPAD), Banque Africaine de Développement, Union Africaine, Mars 2011. Available from: http://www.pidafrica.org/phase1/FR/5GRET.pdf

[22] IPCC (Intergovernmental Panel on Climate Change). Climate Change. In: Pachauri RK, Reisinger A, editors. Synthesis Report. Contribution of Working Groups I, II and III to the Fourth Assessment Report of the Intergovernmental Panel on Climate Change. Geneva: IPCC; 2007. p. 104

[23] Faye C, Diop ES, Mbaye I. Impacts des changements de climat et des aménagements sur les ressources en eau du fleuve Sénégal : caractérisation et évolution des régimes hydrologiques de sous-bassins versants naturels et aménagés. Belgeo. 2015;**4**:1-22

[24] Bodian A, Dezetter A, Dacosta H, Ardoin-Bardin S. Impact du changement climatique sur les ressources en eau du haut bassin du fleuve Sénégal. Revue de géographie du laboratoire Leïdi. 2013;**11**:236-251

[25] WWAP. (United Nations World Water Assessment Programme). The United Nations World Water Development Report 2015: Water for a Sustainable World. Paris: UNESCO; 2015. p. 139

[26] Mashinini V. The Lesotho Highlands Water Project and Sustainable Livelihoods. Policy Implications for SADC. AISA Policy Brief Number 22. Pretoria: Africa Institute of South Africa; 2010. p. 12

[27] ICOLD-ICID-IHA-IWRA. Water Storage for Sustainable Development. World Declaration Approved on 5th June 2012 in Kyoto, by The International Commission On Large Dams (ICOLD), The International Commission on Irrigation and Drainage (ICID), The International Hydropower Association (IHA), and the International Water Resources Association (IWRA), Kyoto; 2012. p. 2

[28] Bedredine F. Programme d'Infrastructures Régionales dans le Bassin du fleuve Sénégal. Genève: Haut Commissariat de l'OMVS; 2014. p. 19

[29] Ficatier Y, Niasse M. Volet social et environnemental du barrage de

Manantali. Evaluation rétrospective, Département de la Recherche, Série Evaluation et capitalisation n° 15. Agence Française de Développement. 2008. p. 69. Available from: https://issuu.com/objectif-developpement/docs/15-evaluation-capitalisation

[30] McCartney MP, Smakhtin V. Water Storage in an Era of Climate Change: Addressing the Challenge of Increasing Rainfall Variability. Blue Paper. Colombo, Sri Lanka: International Water Management Institute (IWMI); 2010. p. 14

[31] OMVS. Amélioration de la résilience du bassin du fleuve Sénégal à la variabilité et aux changements climatiques. In: Projet de gestion intégrée des ressources en eau et de développement des usages multiples (PGIRE). Marrakech: COP; 2016. p. 11

[32] Grey D, Sadoff CW. Water for Growth and Development. Thematic Documents of the IV World Water Forum. Mexico City: Comision Nacional del Agua; 2006. p. 55

[33] Programme des Nations Unies pour l'Environnement (PNUE). Travaux de synthèse sur l'avenir de l'Environnement en Afrique; 2006. p. 26

[34] Bader JC, Cauchy S, Duffar L. Saura P. Monographie hydrologique du fleuve Sénégal : De l'origine des mesures jusqu'en 2011 2014; IRD; 2014

[35] FAO. Irrigation potential in Africa. A basin approach. Rome: FAO; 1997 Available from: http://www.fao.org/docrep/w4347e/w4347e10.htm

[36] OMVS. Schéma directeur d'aménagement et de gestion (SDGA) du fleuve Sénégal. Phase 1: Etat des lieux et diagnostic. Organisation pour la Mise en Valeur du Fleuve Sénégal ; 2009. p. 148

[37] Office International de l'Eau (OIEau). Développer les compétences pour mieux gérer l'eau Dialogue autour des grandes infrastructures—Rapport de synthèse—Etape 2. Analyse des processus de décision sur les barrages de Bui (Ghana), Manantali (Sénégal, OMVS) et Kandadji (Niger); 2009. p. 86

[38] Badier JC, Lamagat JP, Guiguen N. Gestion du barrage de Manantali sur le fleuve Sénégal: analyse quantitative d'un conflit d'objectifs. Hydrological Sciences Journal–des Sciences Hydrologiques. 2003;**48**(4):525-537

[39] Dickmann M, Ficatier M, Schmidt M. Évaluation ex post conjointe. Le barrage de Manantali. Coopération financière avec l'Organisation pour la mise en valeur du fleuve Sénégal (OMVS.) 2009. p. 52. Available from: https://www.eib.org/attachments/ev/ev_manantali_rapport_de_synthese_fr.pdf

[40] Faye C. Weight of transformations and major drifts related to major river water projects in Africa: Case of the Manantali dam on the Senegal river basin. Journal of Research in Forestry, Wildlife & Environment. 2018;**10**(3):13-24

[41] Faye C. Examination of water Storage in the Senegal River Basin for a socio-economic development of downstream river countries. International Journal of Current Science and Engineering. 2019;**01**(02):054-062

[42] Faye C, Sow AA, Diop EHS. Caractérisation des effets du barrage de Manantali sur le régime hydrologique du fleuve Sénégal. Revue Roumaine de Géographie/Romanian Journal of Geography. 2019;**63**(1):121-133

[43] Barbier B, Yacouba H, Maïga AH, Mahé G, Paturel J-E. Le retour des grands investissements hydrauliques en Afrique

de l'Ouest : les perspectives et les enjeux. Géocarrefour. 2009;**84**(1-2):31-41

[44] OMVS, Analyse Diagnostique Environnementale Transfrontalière du Bassin du Fleuve Sénégal. Synthèse Régionale, Projet FEM/Bassin du Fleuve Sénégal. Rapport final. 2007:139

[45] SIWI. Making Water a Part of Economic Development: The Economic Benefits of Improved Water Management and Services. Stockholm: Stockholm International Water Institute (SIWI); 2004. p. 48

Chapter 8

Water Availability for the Environmental Flow in Two Rivers of Mexico under Climate Change

Rebeca González-Villela, Alfonso Banderas Tarabay and Marco Mijangos Carro

Abstract

Adaptation to climate change requires, among others, the modification of river flow regimes to account for the change in household, agricultural, industry, and energy water consumption as well as their short/medium/long-term socioeconomic impact. In this study, the comparative analysis of the variation of the precipitation in relation to the availability of water in the Yautepec and Cuautla rivers in Morelos, Mexico, for the previous period and subsequent period is carried out, to determine the change in the availability of water in the ecosystem. In winter (February), an increase in rainfall on the Yautepec and Cuautla River was observed, where annual seasonal agriculture and Pine and Oyamel forest are the characteristic vegetation. In autumn (October), a decrease in precipitation takes place. The flows in some regions do not coincide with the increase in the percentage of precipitation (Oaxtepec and Las Estacas Stations) and point out the synergistic effect of the human use of the water resource and the effects of climate change. On Ticumán Station, the depletion of the flow only can be associated with the use of the resource by human influence. The modifications caused by alteration of a river's flow regime and climatic change must be studied through comparative multidisciplinary studies that give to decision-makers the design of environmental flows.

Keywords: climatic change, environmental flow, vulnerability

1. Introduction

Climate change impacts on the hydrological cycle [1] with particular examples in France [2] and Central Europe [3], both fast and slow [4] where in the case of abrupt changes impacts on the ecosystem [5] and in long-term changes disrupt a pattern of inland moisture advection and convergence zone, increasing cloud base heights and reducing the total column liquid water content over high elevations [6]. Also, this impact has a strong response to global warming [7, 8], influences its extremes [9], and in turn influences via this cycle water resources [10, 11] while, conversely, the hydrological cycle influences climate [5, 12] in general and may, in case of enhancement, moderate transient climate change [13]. Climate changes impact rivers through the hydrological cycle as seen in [14, 15] and directly on river ecosystems as seen in the Danube [16], in the United Kingdom [17, 18], the Narew river [19], and globally [20]. In terms of hydrological cycle "sojourn" river water turnover takes place in 16 days [21]. As a result, river flow is impacted as seen in Europe [22], in the United Kingdom [23], in the Balkans [24], in Ethiopia [25], in India [26], and in West Africa [27]. Precipitation and temperature scenarios of climate change based on atmospheric circulation play an important role [28] and so do diagnostic statistics of daily rainfall variability in an evolving climate [29].

Under local conditions, environmental flows (e-flows) are defined in the 2007 Brisbane conference as "*the quantity, timing, and quality of water flows required to sustain freshwater and estuarine ecosystems and the human livelihoods and wellbeing that depend on these ecosystems.*" [30] and resultant policies showed some moderate success [31] while the general trend of the state of aquatic ecosystems continued to deteriorate [32] due to increased dam building, particularly in ecologically sensitive areas [33]. Subsequently it was refined to "*Environmental flows describe the quantity, timing, and quality of freshwater flows and levels necessary to sustain aquatic ecosystems which, in turn, support human cultures, economies, sustainable livelihoods, and well-being*" in the 20th International Riversymposium and Environmental Flows Conference, held in Brisbane in September 2017, to lend increased support to groundwater-dependent ecosystems (GDEs) [34]. The influence of river flow on environmental flow is seen in [35], that of flow regime type (general regime classification in [36], under a changing climate is seen in [37], in hydroecology context in [38]) on the e-flows releases, and hydropower production is seen in [39], the impact of extreme flow variability on environmental flows is seen in [40], in terms of river basin management in [41, 42], and for natural, hybrid, and novel riverine ecosystems in [43]. A review [44] determined that regarding rivers, at a global level in six world regions encompassing 44 countries, there are applied 207 different environmental flow methodologies focusing onto hydrological (e.g., the 32-parameter range of variability approach (RVA) [45]), hydraulic rating [46], habitat simulation [46], holistic [47], or combinatory approaches [48]. The future is bleak, as a good scenario solution for the year 2050 [49] leads to an 10–20% increase of global virtual water trade so as to retain a semblance of survival in country-level environmental flows.

The adaptation approach is defined as "*Adapting to climate change means taking action to prepare for and adjust to both the current effects of climate change and the predicted impacts in the future*" [50]. The success of an adaptation policy is measured by monitoring, reporting, and evaluation (MRE), where monitoring is "*a continuous process of examining progress made in planning and implementing climate adaptation*" [51], reporting is "*the process by which monitoring and/or evaluation information is formally communicated, often across governance scales*" [52], and evaluation is "*a systematic and objective assessment of the effectiveness of climate adaptation plans, policies and actions, often framed in terms of the impact of reducing vulnerability and increasing resilience*" [51].

Therefore, it is necessary to generate the appropriate analysis models and methodologies to predict trends, capture biophysical impacts and possible variations of climate change [53, 54]. In addition, it is necessary to incorporate socioeconomic elements within the analysis of ecological systems with the purpose of carrying out a sustainable management of the goods and services provided by these ecosystems [55, 56]. The alteration of the river flow regime generally is caused by human activity, an aspect that requires of the studies with a multidisciplinary approach to the analysis of the problem of global change in freshwater systems [57, 58]. In particular, regulation by interbasin transfers, dams, withdrawals, and land cover change are the main human intervention agents [59].

The adaptation to climate change makes necessary the determination of environmental flows in rivers so as to establish the change in water consumption for the population, agricultural activities, and industry and electricity generation, among others. All this is to compensate for variations in annual precipitation by the planning of the water resource security through different actions (transfer of industries to regions of greater humidity, change in the morphology of the cities to compensate for floods, availability of water for irrigation and flood control). These changes have important consequences on economic activities, population health, the ecosystem, and biodiversity [60]. In this sense, it is necessary to generate the tools for ecological, socioeconomic, and political analysis in order to achieve the rational use of aquatic resources in rivers regulated by dams [61]. In Mexico, there is little limnological information available on the country's freshwater systems and the effect that climate change is having on water quality and pollution. In this sense, in the present study, a comparative analysis of the variation of the water availability is applied through percentage of precipitation in the rivers of the Yautepec and Cuautla subbasins for the base period (preimpact) and subsequent period (postimpact) to determine the change in the availability of water in the riparian ecosystem.

2. Methodology

The methods employed are based on the RVA as seen in Richter et al. [45, 62–65], **Figure 1**.

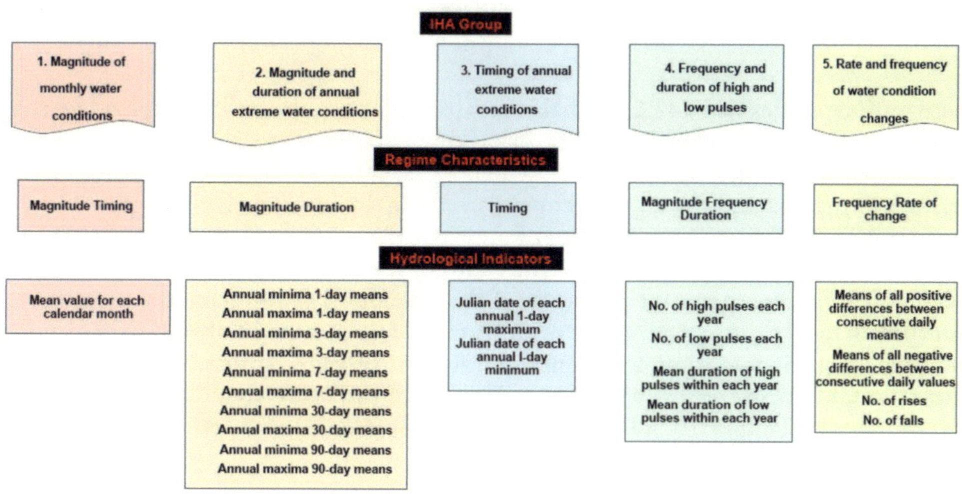

Figure 1.
Range of variability approach (RVA) indicators of hydrological alteration (IHA) [45, 62, 66].

3. Materials and methods

3.1 Study area

Cuautla and Yautepec rivers discharge their waters into the Amacuzac River, main tributary of the Balsas River. Yautepec basin covers an area of 1226 km^2, which represents 25% of the territory of Morelos State. The total population in 2010 in

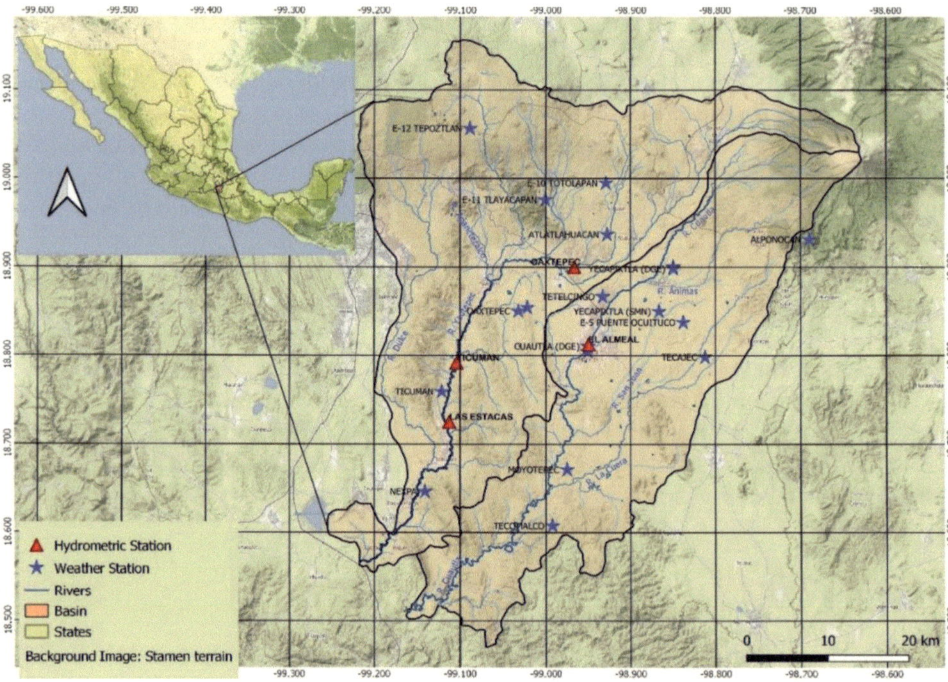

Figure 2.
Location of the subbasins, meteorological and hydrometric stations of the Yautepec and Cuautla River in the Morelos state.

these municipalities was 242,197 inhabitants [67]. The region is characterized by the development of new tourist corridors, urban and industrial areas. There are growing problems of pollution and flood risks, which increase the destruction of historical heritage of bridges and dams, with consequences in the incidence of diseases. The Cuautla River subbasin covers an area of approximately 765 km^2. It is located on the slopes of the Popocatépetl volcano to the south of the Morelos State. In the basin, productive processes generate problems of the extraction of soil from the mountains and the soil loss in the upper parts. In addition, a high extraction of water for human and industrial consumption, and consequently, a strong contamination due to the water input of the users of the irrigation districts of the study area (4500), with an area of irrigated land of 10,500 hectares (**Figure 2**) [68, 69].

3.2 Methods

Precipitation variation percentage in the Yautepec and Cuautla river subbasins in the base period (preimpact) and subsequent period (postimpact) was estimated through the precipitation data of the ERIC III weather stations [70]. The weather stations with data from 1924 to 2010 (**Table 1** and **Figure 1**). Eight of them are located in the Yautepec River subbasin and 11 in the Cuautla River subbasin.

The environmental flow analysis in the Yautepec River included three stations: Oaxtepec (upper part), Ticumán (middle part), and Las Estacas (lower part) of the subbasin. In the Cuautla River, the station El Almeal (high part). The comparative study of the monthly average flows of the hydrometric stations of the Yautepec and Cuautla River subbasins, for the preimpact and postimpact periods, was carried out

Station name	Long.	Latitude	Years of registration	No. years
Atlatlahuacán	−98.90	18.93	1924–2008	84
Cuautla,(SMN)	−98.96	18.81	1926–2006	80
Cuautla, (DGE)	−98.95	18.80	1955–2009	27
Oaxtepec, Yau.	−98.96	18.90	1970–2010	27
Tetelcingo	−98.93	18.86	1942–1973	15
Ticumán, Tlalt.	−99.11	18.76	1955–2008	53
Yautepec, Yau.	−99.08	18.86	1955–2010	55
Yecapixtla	−98.86	18.85	1963–1985	22
Nexpa, Tlalq.	−99.13	18.86	1976–2009	33
Yecapixtla E.T.A. 118	−98.86	18.88	1976–2008	32
Ocuituco E-5	−98.75	18.88	1976–2009	33
Tepoztlán E-12,	−99.11	18.98	1976–2009	33
Tlayacapan	−98.97	19.00	1976–1983	7
Totolalpan	−98.91	18.98	1976–2009	33
Yecapixtla, Yecapixtla	−98.86	18.88	1976–2004	28
Moyotepec, Villa de A.	−98.98	18.71	1978–2009	31
Alpanocan, Tetela de V.	−98.88	18.71	1980–2009	14
Tecajec, Yecapixtla	−98.81	18.78	1981–2009	28
Temoalco, Villa de A.	−98.98	18.63	1981–2009	28

Table 1.
Location of the meteorological stations of the Yautepec and Cuautla river basin.

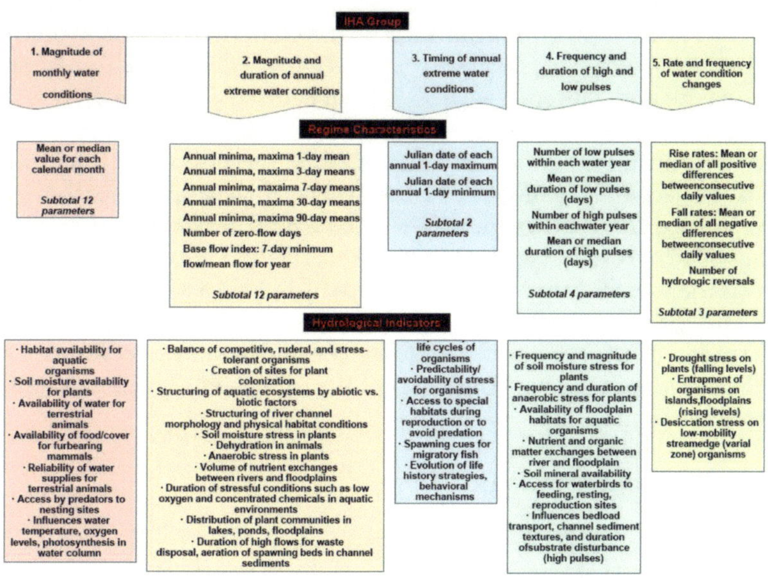

Figure 3.
V7 IHA software 33 IHA parameters and their impact on the environment [73].

Location	Base period (preimpact)	Subsequent period (postimpact)	Long.	Lat.
EL ALMEAL	1948–1978 (31 years)	1979–2011 (32 years)	−98.95	18.81
OAXTEPEC	1949–1979 (31 years)	1980–2011 (26 years)	−98.97	18.90
TICUMAN	1951–1980 (30 years)	1981–2011 (29 years)	−99.10	18.79
ESTACAS	1968–1988 (21 years)	1989–2011 (22 years)	−99.11	18.73

Table 2.
Hydrometric stations location river subbasins Yautepec and Cuautla (Morelos state) and years of registers.

based on the information obtained from the hydrometric stations of CONAGUA (National Water Commission) [71]. Nonparametric graphical and statistical study was analyzed using the software, V7 IHA [72] under the hypothesis that there are no differences between the medians of the preimpact period and the postimpact period (Ho: $\mu1 = \mu2$ and Ha: $\mu1 \neq \mu2$). As can be seen below, the 33 IHA parameter (**Figure 3**), V7 IHA is compatible with **Table 1** and Richter's thesis in terms of [65].

The indicators of hydrological alteration (IHA) provided a quantitative approximation of hydrology through the characterization of intra-annual variation of flow by the use. Also, the comparative studies of hydrological regimes before and after system alteration due to human influence or the effects of climate change [74]. Hydrometric stations' location of the Yautepec and Cuautla rivers are given in **Table 2** and **Figure 2**.

4. Results

4.1 Precipitation

The variation average monthly in the precipitation between the preimpact and postimpact periods in the Yautepec and Cuautla subbasins indicates the largest decreases in February for Tepoztlán (−71.58%) and Nexpa (−66.67%). However, the greatest increases in precipitation were observed in the dry season on the Nexpa station with 62.94% (March); Oaxtepec with 47.24% (February); Totolalpan with 45.49% (January); and Yautepec with 35.50% in February (**Figure 4**).

The months with the highest percentage decreases in precipitation during the year were: February, April, November, and December in the east of the Cuautla subbasin (Alpalocan, Tecajec, and Tecomalco, respectively). However, the northern part (Tetelcino) showed the greatest increases in the winter season (January, February, and March, with 72.76, 47.72, and 98.10%, respectively; **Figure 2**). The month of February was the most favored with respect to the increase in rainfall in the northern part of the Yautepec and Cuautla river subbasins, where annual seasonal agriculture predominates and the dominant vegetation is the pine and oyamel forest. On the contrary, the month of October (high part of the subbasins) was the most affected by the decreases in precipitation.

4.2 Water availability for ecological flows

Flows should be interpreted as below (**Figure 5**).

In the upper basin of the Yautepec River (Oaxtepec Station), the monthly averages of the flows (preimpact period 1949–1979; 31 years) and the postimpact period (1980–2011; 26 years) indicate the significantly decrease in the availability of water

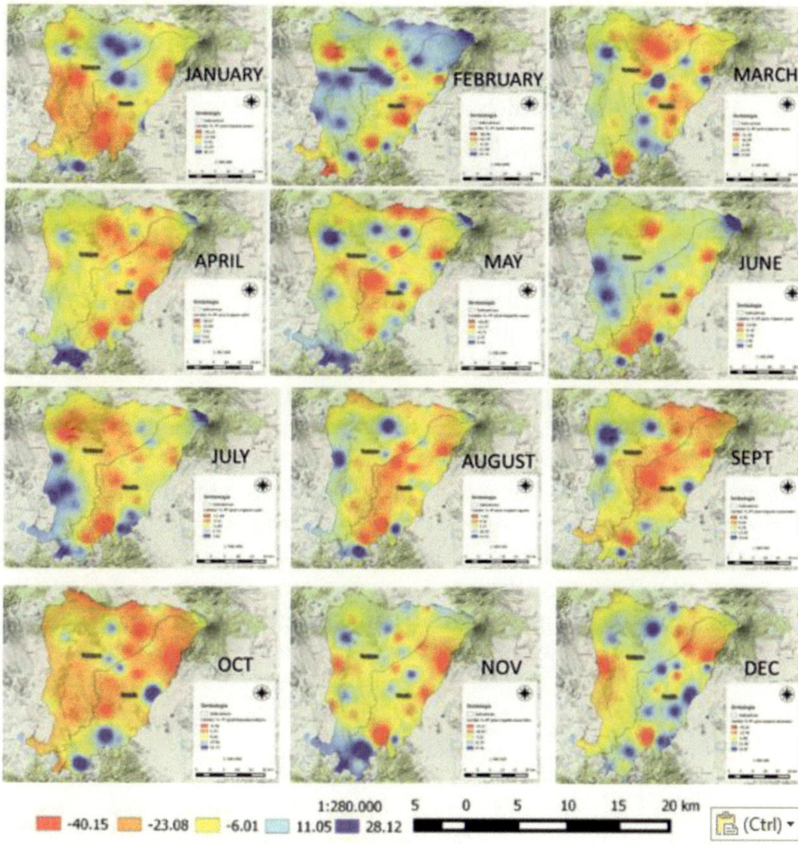

Figure 4.
The percentage variation of the monthly rainfall between the preimpact and postimpact periods for the Yautepec and Cuautla river subbasins.

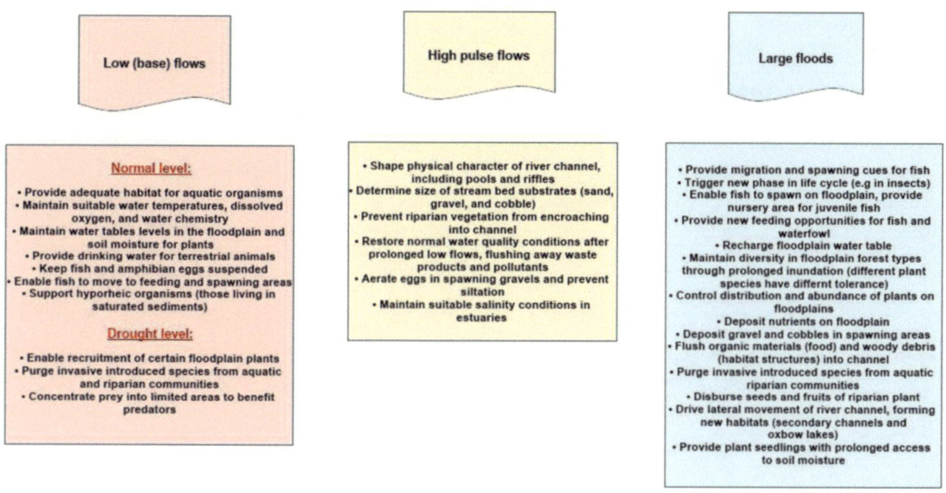

Figure 5.
River flow levels and ecological functions [75].

between the analyzed periods. This can be attributed to the use of the resource for agriculture and by the population after the construction of the little dams. For this hydrometric station, there is a significant decrease in precipitation of −38.57% in March and − 19.84% in April, which coincides with the maximum decrease in flows in this segment in the postimpact period. However, the increase in precipitation was observed for this area in the month of February (47.24%), without the increase in flows as should be expected, which is explained by the use of the resource by the population, nullifying the positive effect of climate change or increase of precipitation percentage on the availability of water for the river (**Figure 6**).

The average monthly flow of the Ticumán Station (middle part of the Yautepec River subbasin), for the preimpact period (1951–1980; 30 years) and postimpact period (1981–2011; 29 years), indicates a significant alteration in the hydrological regime only for the June and November months (**Figure 7**).

For this area, there is a nonsignificant increase in precipitation of 14.04% (September) and 19.63% (November) that coincides with the percentage increases in flows. However, in December, a decrease in the flows percentage of 70% can associated with the use of the resource for activities of anthropic type. Therefore, these variations with a negative tendency in the flow cannot be attributed to the effects of climate change, but to human influence.

On Las Estacas Station, the precipitation percentage variation between the preimpact period and postimpact period was significant for January (−18.8%) and April (−2.9%).

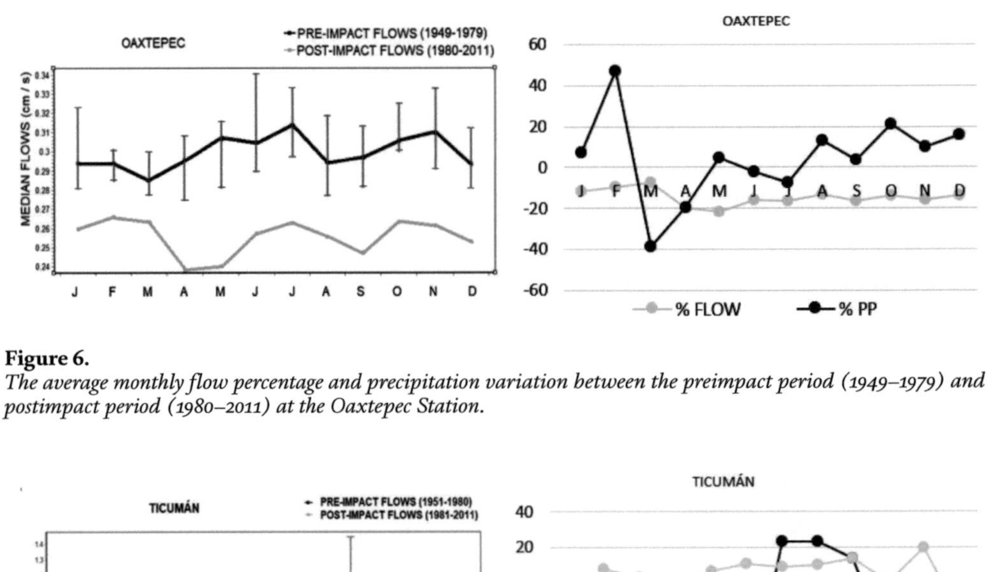

Figure 6.
The average monthly flow percentage and precipitation variation between the preimpact period (1949–1979) and postimpact period (1980–2011) at the Oaxtepec Station.

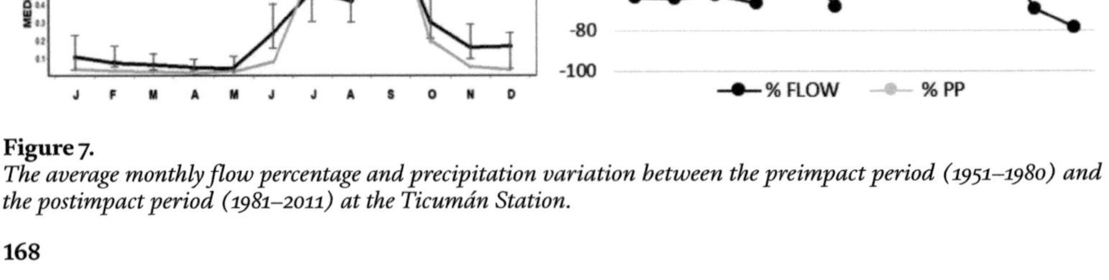

Figure 7.
The average monthly flow percentage and precipitation variation between the preimpact period (1951–1980) and the postimpact period (1981–2011) at the Ticumán Station.

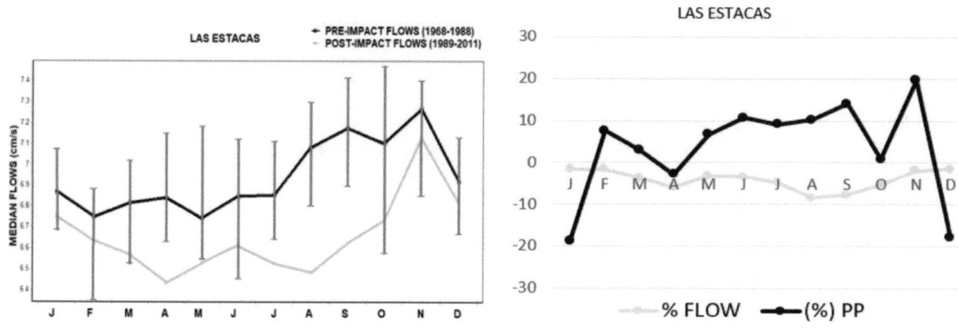

Figure 8.
The flows monthly average percentage and precipitation variation between the preimpact period (1968–1988) and postimpact period (1989–2011) on las Estacas Station.

The flows monthly average percentage for the Las Estacas Station (lower part of the Yautepec River Sub-basin), for the preimpact period (1968–1988; 21 years) and the postimpact period (1989–2011; 22 years), indicates a significant decrement in the flows on April (−5.9%), July (−4.8%), August (−8.5%), and September (−7.7%). This behavior does not coincide with the increase in rainfall from May to September (between 6.8% and 14.04%). Variations in precipitation and flows for this period (July–October) can be attributed to human influence and the effects of climate change (**Figure 8**).

The monthly average flows for the El Almeal Station (upper part) of the Cuautla River subbasin for the preimpact period (1948–1978; 31 years) and postimpact period (1979–2011; 32 years) indicate a significant alteration in the hydrological regime in all months of the year (**Figure 9**).

Coincidentally, in this area there is a significant decrease in precipitation percentages throughout the year (except November), ranging from −29.65% for January to −0.39% in August. Aspect that can be associated with the depletion of flow rates during the year for this season indicates the effects of climate change on the availability of water for the Cuautla River.

4.3 Environmental flows and indicators of hydrological alteration

Increase of low extreme flow rates, decrease of low flow rates, loss of the high flow pulses, small floods, or large floods at the Oaxtepec Hydrometric Station were observed. The hydrological changes occurred in the average monthly flows and the days with

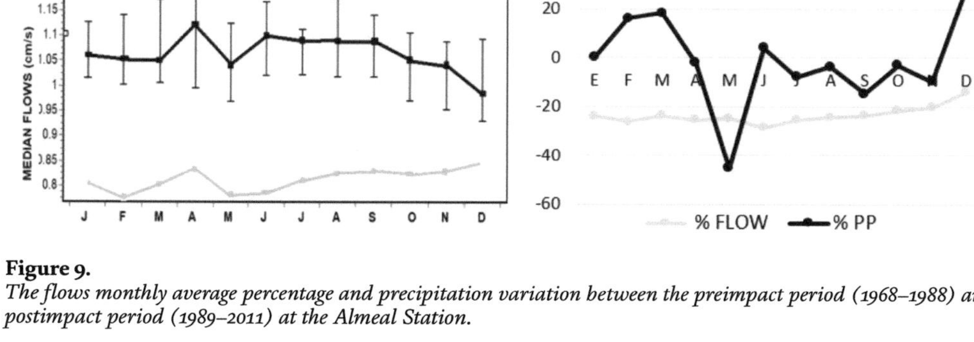

Figure 9.
The flows monthly average percentage and precipitation variation between the preimpact period (1968–1988) and postimpact period (1989–2011) at the Almeal Station.

minimum and maximum flow (**Figure 10**). As well as in the number and duration of low flow pulses, in the increase in high pulses, changes in the rates of variation (to negative), and reversals of the flows. The loss of the frequency, magnitude, and periodicity of the flows during the year indicates the abuse in the use of the resource for agriculture and by the population after the construction of the dams (**Figure 12**).

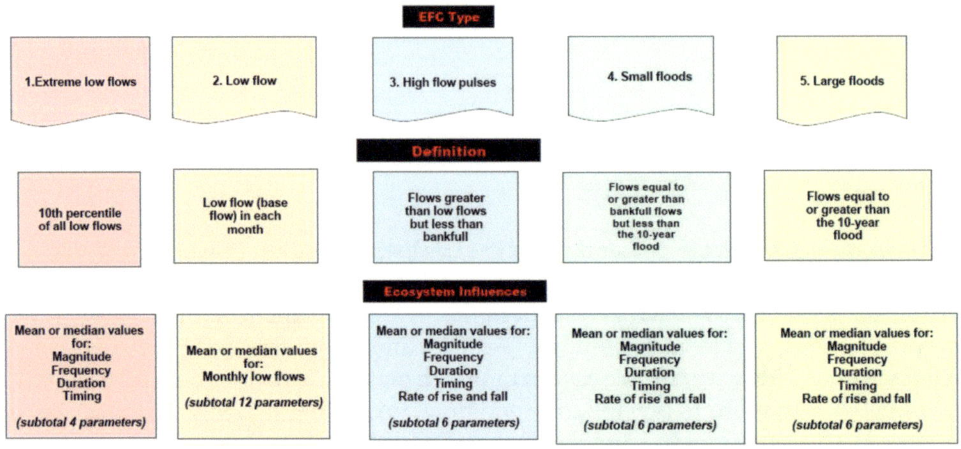

Figure 10.
List of 34-parameter environmental flow components (EFCs) [63], ***Figure 11.***

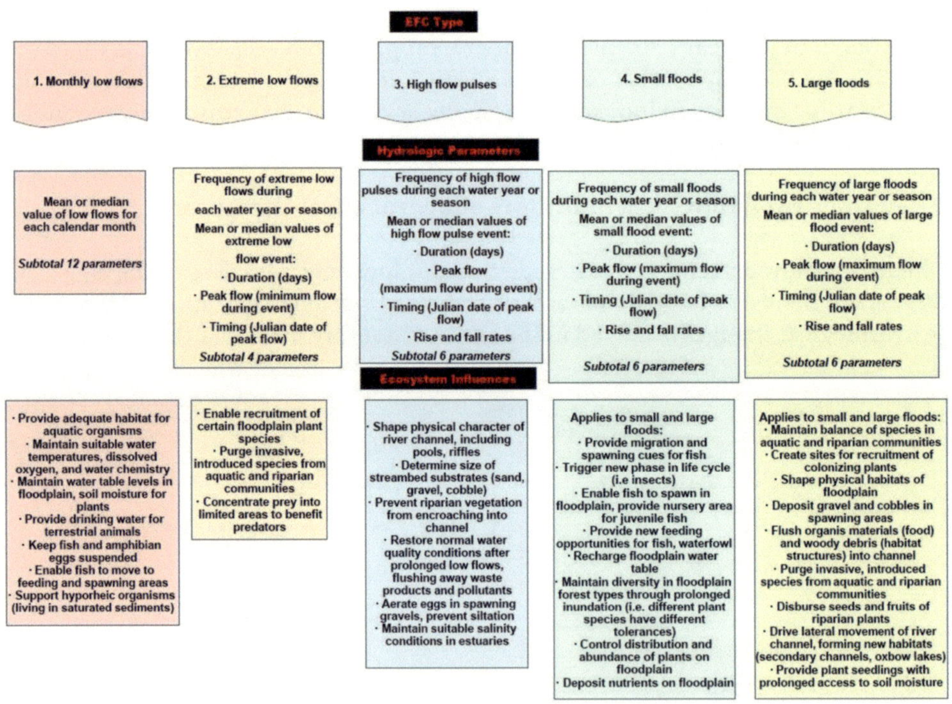

Figure 11.
V7 IHA software 34 IHA parameters and their impact on the environment [73].

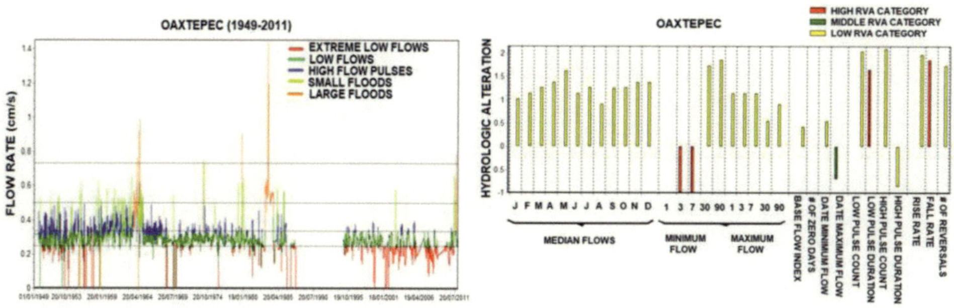

Figure 12.
Ecological flows and hydrological alteration indicators at the Oaxtepec Station.

At the Ticumán hydrometric station, the environmental flows do not observe significant changes. However, the hydrological alteration indexes indicate a decrease in the average flows of January, June, August, and September (rains), as well as in the minimum daily flows with a duration of 3, 7, and 30 days. Also, an increase in the date of the maximum flow, in the duration of the high flow pulses, and a rise in rate flow. The base flow and low flow pulses indicate a tendency toward drought conditions or a tendency to extreme climate and the synergic effect of climate with use of the water by human influence (**Figure 13**).

In Las Estacas hydrometric station (Yautepec River), the environmental flows show alterations in the postimpact period for the small and large floods, in the low flows and high flow pulses. The IHA show changes in the small and large floods for the postimpact period (1980–2000). Also, largest significant decreases in average flows in May, July, October, and November and increases in August and September point out a tendency to extreme weather due to climate change. In addition, decreases in the minimum and maximum daily flows with a duration of 3, 7, 30, and 90 days. The duration of high flow pulses and the rate of increase in flow can be associated with the torrential rains. On the contrary, an increase in the base flow, changes on the date of the maximum flow, and the rate flows decrease indicate shift of the start rainy season. The average flows and the number of investments of the flow can be attributed to human influence and climate change (**Figure 14**).

In El Almeal hydrometric station (Cuautla River), the environmental flows show great changes between the preimpact and the postimpact period. The IHA point out significant differences in all the components of ecological flow, except for the average

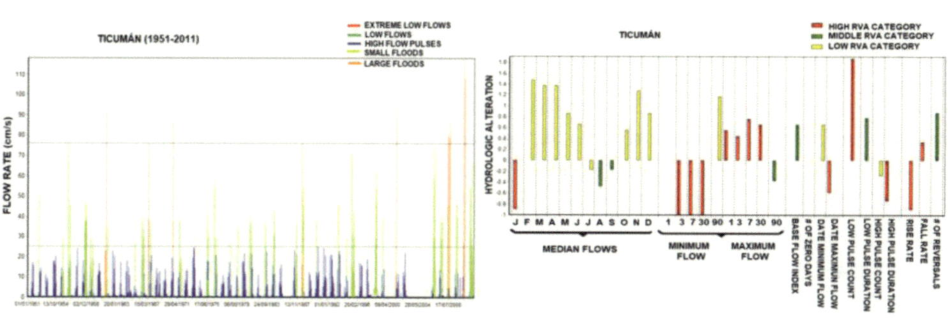

Figure 13.
Ecological flows and hydrological alteration indicators of the Ticumán Station.

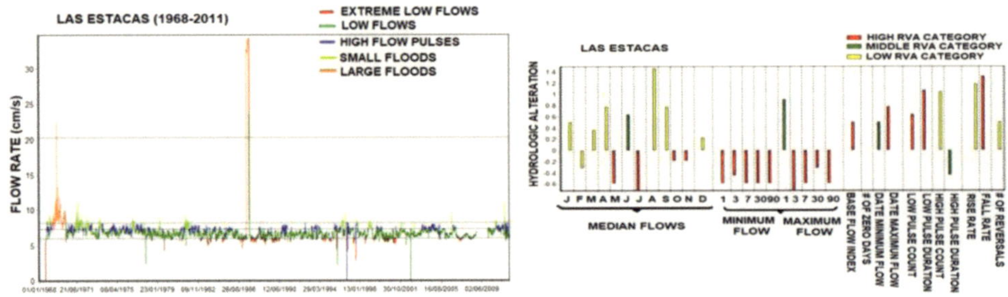

Figure 14.
Ecological flows and hydrological alteration indicators of las Estacas Station.

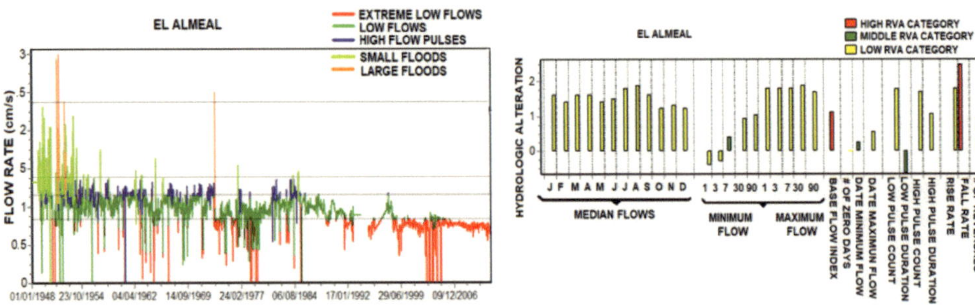

Figure 15.
Ecological flows and hydrological alteration indicators of the El Almeal Station.

flows of March and December, and in the number of investments of the flow. The low extreme flows increased and became more frequent, the low flows and the high flow pulses practically disappeared, as well as small and large floods in the postimpact period, modifications that can be explained by human influence. Changes in the flow averages for the whole year were observed, in the duration of the days with minimum and maximum flow, an increase in the rate of the base flow, as well as in the number of flows pulses (high and low). Also, in the flow rates (increase and decrease in flows), on investments, the shorter duration of low pulses, a situation that reflects an alteration in all components of environmental flows. Therefore, modifications that can be explained by human influence (**Figure 15**). The calculations above lead to the construction of the diagram below (**Figure 16**).

4.4 Regression analysis and R^2 of monthly flows

The regression analysis of monthly flows for Yautepec and Cuautla subbasins in Hidrometric Stations of study period is shown in **Table 3**. All hydrometric stations showed a negative trend throughout the year. Only Las Estacas (January–April, dry season) and Ticumán (May–September, rainy season) showed a positive trend in flows, respectively. These stations are located in the middle and lower part of the Yautepec subbasin where annual, permanent, and semipermanent irrigation agriculture and secondary tree, shrub, and lowland forest vegetation predominate

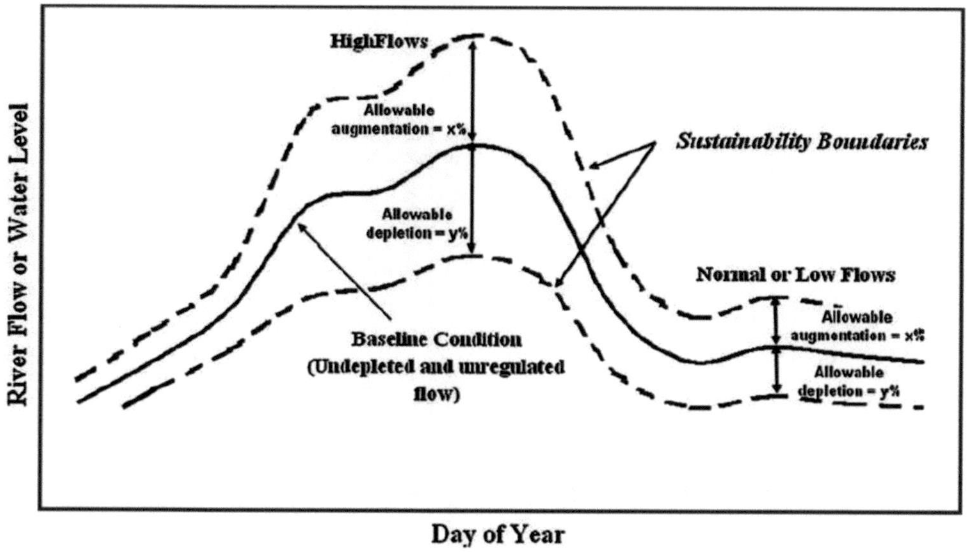

Figure 16.
Use of the sustainability boundary approach (SBA) to set sustainable water management targets [76].

(**Figure 17**). This behavior can be explained by the use of some diversions to irrigate crops in the area during the dry season in Las Estacas and the positive tendency in Ticumán Station on rainy season could be explained for climatic change effect, as can be observed in **Figure 4** during this period, above all in June and July. The Almeal Station in the intermediate part of the Cuautla subbasin with annual, permanent, and semipermanent irrigation agriculture presents the highest values of R^2 (0.23 in October to 0.54 in May) coinciding with the greatest decreases in the percentages of precipitation (**Figure 4**). Aspects that show the synergistic effect of climate change and human influence on the availability of water for rivers.

ARMA (autoregressive moving average) flows analysis of the hydrometric stations.

For the Oaxtepec station, the ARMA analysis of the flow data indicates a homogeneous distribution of the residues with an AR coefficient = −824,201 at $P \leq 0.0517$. However, some events of large floods generate some alterations on residuals and show some cyclicity (**Figure 18**).

For the Ticumán Station, the ARMA analysis of the flow data indicates a homogeneous distribution of the residues with an AR coefficient = −8893 at $P \leq 0.1115$ and shows also some cyclicity (**Figure 19**).

For the Las Estacas Station, the ARMA analysis of the flow data indicates a homogeneous distribution of the residues with an AR coefficient = −79,086 at $P \leq -0.2077$. However, some events of large floods generate some alterations on residuals (**Figure 20**), and the cyclicity of the large floods is not clear.

For the El Almeal Station, the ARMA analysis of the flow data indicates a homogeneous distribution of the residues with an AR coefficient = −8940 at $P \leq -0.1069$. However, some events of large floods generate some alterations on residuals (**Figure 21**).

Month	Stations							
	Oaxtepec		Las Estacas		Ticumán		Almeal	
	Regression	R^2	Regression	R^2	Regression	R^2	Regression	R^2
January	Y = −0.0016X + 3.48	0.12	Y = 0.0098X - 12.8	0.01	Y = −0.0037X + 7.5	0.11	Y = −0.008 + 16.76	0.38
February	Y = −0.0013X + 2.77	0.08	Y = 0.0181X −29.6	0.04	Y = −0.001X + 2.13	0.02	Y = −0.010 + 21.48	0.40
March	Y = −0.0013X + 2.78	0.12	Y = 0.0087X - 10.7	0.01	Y = −0.001X + 2.10	0.06	Y = −0.010X + 21.2	0.43
April	Y = −0.0017X + 3.72	0.18	Y = 0.0009X + 4.77	9E-5	Y = −0.0012X + 2.5	0.13	Y = −0.011X + 21.7	0.38
May	Y = −0.0023X + 4.91	0.22	Y = − 0.0003X + 7.2	1E-5	Y = 0.001X–2.10	1 E-3	Y = −0.010X + 21.3	0.54
June	Y = −0.0018X + 3.95	0.16	Y = −0.0037X + 14.5	1 E-4	Y = 0.006X–10.49	4 E-3	Y = −0.010X + 20.1	0.35
July	Y = −0.0006X + 1.47	0.01	Y = −0.0074X + 21.9	5 E-4	Y = 0.015X–28.08	0.01	Y = −0.009X + 19.4	0.30
August	Y = −0.0005X + 1.27	0.01	Y = −0.0328X + 72.7	0.01	Y = 0.051X–98.50	0.16	Y = −0.012X + 24.2	0.44
September	Y = −0.0002X + 0.04	3 E-4	Y = −0.0245X + 55.8	0.15	Y = 0.032X–59.46	0.04	Y = −0.009X + 19.3	0.32
October	Y = −0.0008X + 1.92	0.22	Y = −0.121X + 31.14	0.04	Y = −0.017X + 35.0	0.05	Y = −0.007X + 14.6	0.23
November	Y = −0.0019X + 4.04	0.18	Y = −0.0172X + 412	0.09	Y = −0.016X + 32.3	0.04	Y = −0.009X + 19.1	0.34
December	Y = −0.0015X + 3.15	0.08	Y = −0.015X + 36.42	0.06	Y = −0.006X + 12.2	0.13	Y = −0.01X + 20.67	0.44

Table 3.
Monthly flows regression analysis of hydrometric stations Yautepec and Cuautla subbasins, and value of R^2 during the all study period.

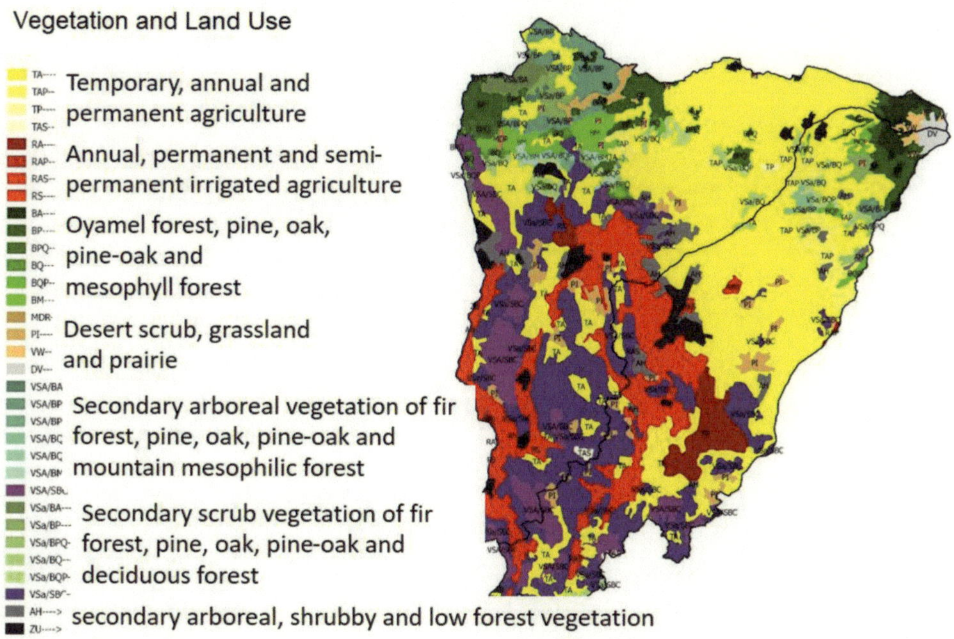

Figure 17.
Vegetation and land use of Yautepec and Cuautla subbasins.

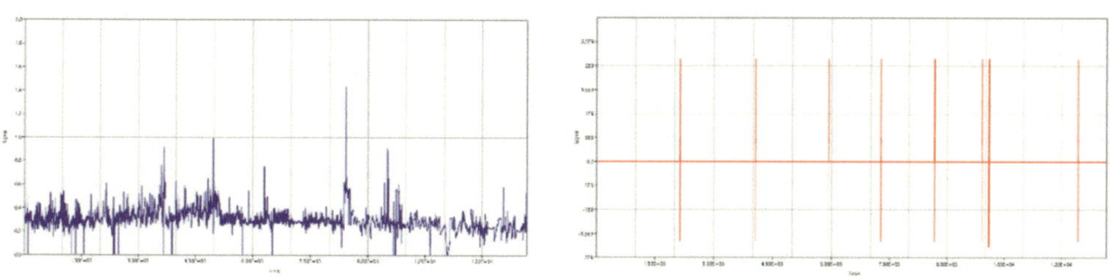

Figure 18.
ARMA analysis of flows of Oaxtepec Station.

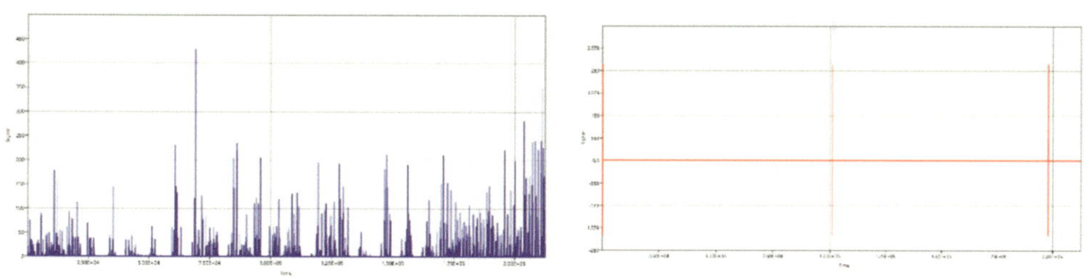

Figure 19.
ARMA analysis of flows of Ticumán Station.

5. Discussion

The impact of climate change on aquatic ecosystems was seen as early as 1999 in eight US regions [77]. In the Fifth Evaluation Report of the Intergovernmental

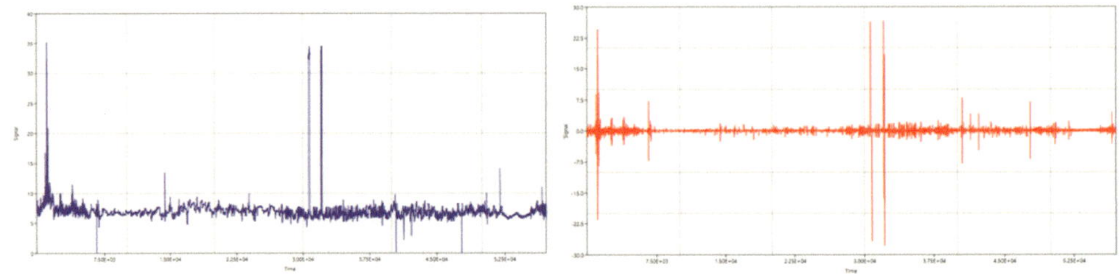

Figure 20.
ARMA analysis of flows of las Estacas Station.

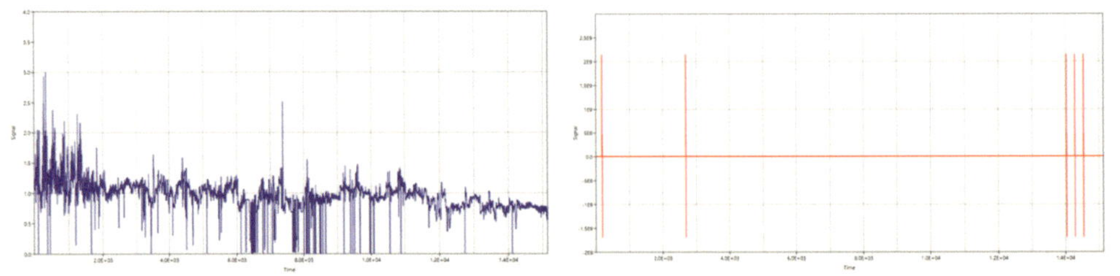

Figure 21.
ARMA analysis of flows of El Almeal Station.

Panel on Climate Change, MacAlister and Subramanyam mentioned that 93% of the impacts associated with climate change will affect aquatic ecosystems [78, 79]. Environmental flows, using the Brisbane definition [30], were incorporated into the "water stress" indicator 6.4.2 [80]. Environmental flows are the source of the "natural" versus "managed" ecosystem support and, as can be seen below (**Figure 22**), due to climate change or direct human intervention, their impact tends to zero as the managed contribution increases to a plateau.

In a study of the Huangqihai River basin in Inner Mongolia, China, it was found that environmental flow requirements (EFRs) contribute to the determination of water scarcity using the QQE indicator that combines the status of quantity, quality, and EFR [82] while if the environmental flow protection is low, 53 countries experience different levels of water shortage, and if it is high, we have an increase to 101 countries [83]. A similar result was found when water withdrawals were replaced by water consumption plus environmental flows where in a global river basin examination for the period 1996–2005, 201 out of 405 river basins examined presented intense water scarcity for at least 1 month per year [84]. Using the environmental water

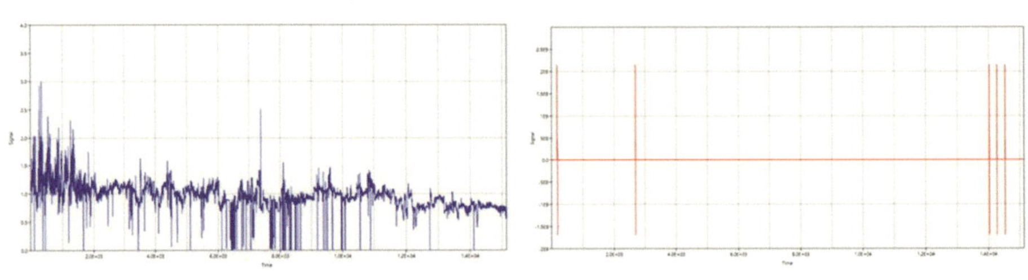

Figure 22.
Benefits from natural (environmental flows) and managed systems [81].

requirement (EWR) as the sum of environmental low-flow requirement (LFR) and environmental high-flow requirement (HFR) shows that if freshwater-dependent ecosystems are to stay in fair condition, 20–50% of the mean annual river flow has to be allocated to them [85]. Hence, it can be said that the constraint of finiteness of water resources imposes a socioeconomic choice regarding water allocation between human use and environmental flow at global, regional, country, and locality levels, which perhaps can be regulated via a scalable framework although the country level plays a decisive role as water is a strategic economic good. Mexico determines the volume of water that is allocated for ecological protection on the basis of the Environmental Flow Mexican Norm (e-flows, NMX-AA-159-SCFI-2012, ratified in 2017) regarding the formation and disposition of environmental water reserves (EWR) 12 of which have a 50-year duration to date (2021), and 75% of them meet up to the theoretical minimum requirement of norm implementation [86].

Annual maximum flood events depend in part on runoff generation and flow routing as seen in [87], while precision moisture estimation [88] may add to the description of the biotic state. Also rainfall and temperature trends analysis [89] plays a determinate role in this description as well. Increases in the winter rainfall in the northern and southern part of the Yautepec and Cuautla River (February) are different as shown by the historic records of the preimpact period. As well as, for the middle part of the subbasin of the Yautepec River where the shrubby secondary vegetation of low deciduous forest predominates and the permanent and semipermanent annual irrigation agriculture. The decreases in rainfall at the end of the rainy season (October) show a climate change in coincidence with other authors and pointed out as one of the most urgent threats to sustainable development worldwide [90]. The significant decrease in the percentage of precipitation for all months of the year at the Oaxtepec and Las Estacas weather stations indicates the synergistic effect of climate change and the use of the resource by the population (mainly agriculture). On the contrary, the effects on the flow depletion can only be associated with the use of the resource by human influence on Ticumán station. The impacts of climate change are exacerbated by rapid population growth, an example of which is seen below (**Figure 23**), rapid urbanization and chaotic economic development, particularly where water demands already exceed limited supplies.

Likewise, climate change is altering precipitation and thawing patterns, affecting the frequency and magnitude of river flows, floods and droughts, and contributing to more extreme weather events and forest fires around the world in a coincident way

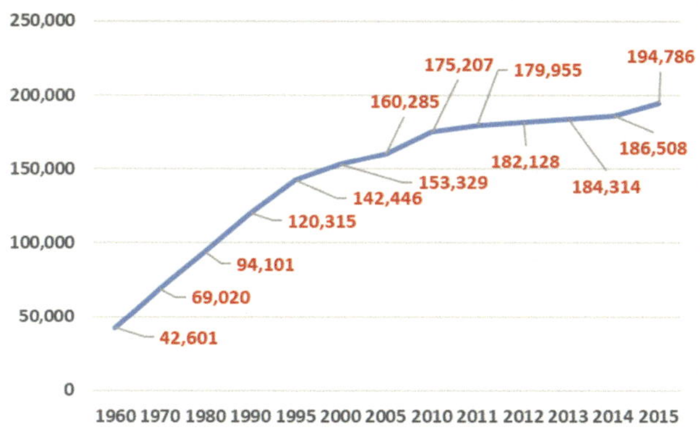

Figure 23.
Population statistics in the Cuautla area [91, 92].

in the subbasins of the Yautepec and Cuautla River [79]. The hydroperiod determines the presence of certain plants and animals in the different strata of the riparian zone and the riverbed, being the dominant factor that makes the difference in the riverbank and riverbeds, and constitutes the most important variable in the corridor structure river [93, 94]. The changes in base flow and low flow pulses on subbasins indicate a tendency toward drought conditions or a tendency to extreme climate and the synergic effect of climate with use of the water by human influence. Moreover, the duration of high flow pulses and the rate of increase in flow can be associated with the torrential rains, as well as the increase in the base flow, changes in the date of the maximum flow, and the rate flows decrease indicate shift of the start rainy season. Variable flow was seen in 52 rivers worldwide whose patterns of flow variability were often correlated with climate [95]. Extreme events, e.g., unusual floods/droughts, may alter the physicochemical conditions under which biotic communities undergo long-term development [96]. Therefore, hydroperiod models are a useful tool in the analysis of the distributions of organisms during the year and the modifications caused by human activity. These models should be studied through comparative multidisciplinary studies to determine the real problems derived from global change in the freshwater systems and to determine the real influence of global warming on the regional climatic conditions of the planet and its influence on river ecosystems [57, 58]. Therefore, the dimensions and processes observed in the development of the watersheds, among them the environmental flows, must be approached in a systemic way, starting from integrative and articulating approaches to generate the actions for the management, conservation, and recovery of the freshwater. As well as, the vulnerability maps and lines of action for climate change adaptation [56, 97].

Acknowledgements

Thanks to Dr. Alejandra Martín Dominguez, Treatment and Water Quality Area Coordinator of the Instituto Mexicano de Tecnología del Agua for the invitation to participate in the project TC1605.1 “Ecological Integrity and Environmental Health Indicators for the Yautepec, Mor., River Basin. 1st., 2nd., and 3rd. Stage” (Indicadores de Integridad Ecológica y Salud Ambiental para la Cuenca del Río Yautepec, Mor. 1ra., 2da., y 3ra. Etapa”), led by Dr. Perla Alonso Elia-Egís.

Author details

Rebeca González-Villela*, Alfonso Banderas Tarabay and Marco Mijangos Carro
Instituto Mexicano de Tecnología del Agua, Morelos, Mexico

*Address all correspondence to: rebeca_gonzalez@tlaloc.imta.mx

References

[1] Kundzewicz ZW. Climate change impacts on the hydrological cycle. Ecohydrology and Hydrobiology. 2008;**8**(2-4):195-203. DOI: 10.2478/v10104-009-0015-y

[2] Dayon G, Boé J, Martin É, Gailhard J. Impacts of climate change on the hydrological cycle over France and associated uncertainties. Comptes Rendus - Geoscience. 2018;**350**(4):141-153. DOI: 10.1016/j.crte.2018.03.001

[3] Stagl J, Mayr E, Koch H, Hattermann FF, Huang S. Effects of climate change on the hydrological cycle in central and Eastern Europe. In: Rannow S, Neubert M, editors. Managing Protected Areas in Central and Eastern Europe under Climate Change. Vol. 58. Dordrecht, Netherlands: Springer; 2014. pp. 31-43. DOI: 10.1007/978-94-007-7960-0

[4] Bala G, Caldeira K, Nemani R. Fast versus slow response in climate change: Implications for the global hydrological cycle. Climate Dynamics. 2010;**35**:423-434. DOI: 10.1007/s00382-009-0583-y

[5] Williams PD, Guilyardi E, Sutton R, Gregory J. A new feedback on climate change from the hydrological cycle. Geophysical Research Letters. 2007;**34**(March):1-5. DOI: 10.1029/2007GL029275

[6] Comarazamy DE, González JE. Regional long - term climate change (1950 – 2000) in the midtropical Atlantic and its impacts on the hydrological cycle of Puerto Rico. Journal of Geophysical Research. 2011;**116**:1-18. DOI: 10.1029/2010JD015414

[7] Held IM, Soden BJ. Robust responses of the hydrological cycle to global warming. Journal of Climate. 2006;**19**:5686-5699

[8] Shiogama H, Hanasaki N, Masutomi Y, Nagashima T, Ogura T, Takahashi K, et al. Emission scenario dependencies in climate change assessments of the hydrological cycle. Climatic Change. 2010;**99**:321-329. DOI: 10.1007/s10584-009-9765-1

[9] Voss R, May W, Roeckner E. Enhanced resolution modelling study on anthropogenic climate change: Changes in extremes of the hydrological cycle. International Journal of Climatology. 2002;**22**:755-777. DOI: 10.1002/joc.757

[10] Yang D, Yang Y, Xia J. Hydrological cycle and water resources in a changing world: A review. Geography and Sustainability. 2021;**2**(2):115-122. DOI: 10.1016/j.geosus.2021.05.003

[11] Hagemann S, Chen C, Clark DB, Folwell S, Gosling SN, Haddeland I, et al. Climate change impact on available water resources obtained using multiple global climate and hydrology models geoscientific. Earth System Dynamics. 2013;**4**:129-144. DOI: 10.5194/esd-4-129-2013

[12] Chahine MT. The hydrological cycle and its ifluence on climate. Nature. 1992;**359**:373-380

[13] Liu M, Vecchi G, Soden B, Yang W, Zhang B. Enhanced hydrological cycle increases ocean heat uptake and moderates transient climate change. Nature Climate Change. 2021;**11**:848-853. DOI: 10.1038/s41558-021-01152-0

[14] Jang CH, Kim HJ, Ahn SR, Kim SJ. Assessment of hydrological changes in a river basin as affected by climate

change and water management practices, by using the cat model. Irrigation and Drainage. 2016;**65**(August):26-35. DOI: 10.1002/ird.2044

[15] Ashmore P, Church M. The impact of climate change on rivers and river processes in Canada. Bulletin / Geological Survey of Canada. Canada: Bulletin; 2001;(555):58

[16] Stagl JC, Hattermann FF. Impacts of climate change on riverine ecosystems: Alterations of ecologically relevant flow dynamics in the Danube River and its major tributaries. Water (Switzerland). 2016;**8**(12):1-20. DOI: 10.3390/w8120566

[17] Visser A, Beevers L, Patidar S. The impact of climate change on hydroecological response in chalk streams. Water (Switzerland). 2019;**11**(3):1-3. DOI: 10.3390/w11030596

[18] Hannaford J. Climate-driven changes in UK river flows: A review of the evidence. Progress in Physical Geography. 2015;**39**(1):29-48. DOI: 10.1177/0309133314536755

[19] Malinowski Ł, Skoczko I. Impacts of climate change on hydrological regimes and water resources management of the Narew River in Poland. Journal of Ecological Engineering. 2018;**19**(4):165-175

[20] Däll P, Zhang J. Impact of climate change on freshwater ecosystems: A global-scale analysis of ecologically relevant river flow alterations. Hydrology and Earth System Sciences. 2010;**14**(5):783-799. DOI: 10.5194/hess-14-783-2010

[21] Shiklomanov IA, Rodda JC. World Water Resources at the Beginning of the Twenty-First Century. Cambridge, U.K.: Cambridge University Press; 2003

[22] Schneider C, Laizé CLR, Acreman MC, Flörke M. How will climate change modify river flow regimes in Europe? Hydrology and Earth System Sciences. 2013;**17**(1):325-339. DOI: 10.5194/hess-17-325-2013

[23] Arnelf NW, Reynard NS. The effects of climate change due to global warming on river flows in Great Britain. Journal of Hydrology. 1996;**183**(3-4):397-424. DOI: 10.1016/0022-1694(95)02950-8

[24] Papadaki C, Dimitriou E. River flow alterations caused by intense anthropogenic uses and future climate variability implications in the Balkans. Hydrology. 2021;**8**(1):1-19. DOI: 10.3390/hydrology8010007

[25] Daba MH, You S. Assessment of climate change impacts on river flow regimes in the upstream of Awash Basin, Ethiopia: Based on IPCC fifth assessment report (AR5) climate change scenarios. Hydrology. 2020;**7**(4):1-22. DOI: 10.3390/hydrology7040098

[26] Abeysingha NS, Islam A, Singh M. Assessment of climate change impact on flow regimes over the Gomti River basin under IPCC AR5 climate change scenarios. Journal of Water and Climate Change. 2020;**11**(1):303-326

[27] Rameshwaran P, Bell VA, Davies HN, Kay AL. How might climate change affect river flows across West Africa? Climatic Change. 2021;**169**(3-4):1-27. DOI: 10.1007/s10584-021-03256-0

[28] Panagoulia D, Bardossy A, Lourmas G. Multivariate stochastic downscaling models for generating precipitation and temperature scenarios of climate change based on atmospheric circulation. Global NEST Journal. 2008;**10**(2):263-272

[29] Panagoulia D, Bárdossy A, Lourmas G. Diagnostic statistics of daily rainfall variability in an evolving climate. Advances in Geosciences. 2006;7:349-354

[30] 10th International Riversymposium and Environmental Flows Conference. The Brisbane declaration: Environmental flows are essential for freshwater ecosystem health and human well-being. In: 10th International River Symposium and International Environmental Flows Conference. Australia: Brisbane, QLD; 2007. pp. 1-7

[31] Harwood A, Johnson S, Richter B, Locke A, Yu X, Tickner D. Listen to the River: Lessons from a Global Review of Environmental Flow Success Stories. Woking, UK: WWF-UK; 2012

[32] Vörösmarty CJ, Pahl-Wostl C, Bunn SE, Lawford R. Global water, the anthropocene and the transformation of a science. Current Opinion in Environment Sustainability. 2013;**5**(6):539-550. DOI: 10.1016/j.cosust.2013.10.005

[33] Winemiller KO, McIntyre PB, Castello L, Fluet-Chouinard E, Giarrizzo T, Nam S, et al. Balancing hydropower and biodiversity in the Amazon, Congo, and Mekong. Science (80-.). 2016;**351**(6269):128-129. DOI: 10.1126/science.aac7082

[34] Arthington AH, Bhaduri A, Bunn SE, Jackson SE, Tharme RE, Tickner D, et al. The Brisbane declaration and global action agenda on environmental flows (2018). Frontiers in Environmental Science. 2018;**6**:1-10. DOI: 10.3389/fenvs.2018.00045

[35] Poff NL, Zimmerman JKH. Ecological responses to altered flow regimes: A literature review to inform the science and management of environmental flows. Freshwater Biology. 2010;**55**(1):194-205. DOI: 10.1111/j.1365-2427.2009.02272.x

[36] Haines AT, Finlayson BL, McMahon TA. A global classification of river regimes. Applied Geography. 1988;**8**(4):255-272. DOI: 10.1016/0143-6228(88)90035-5

[37] Krasovskaia I, Gottschalk L. River flow regimes in a changing climate/Régimes hydrologiques et évolution du climat. Hydrological Sciences Journal. 2002;**47**(4):597-609. DOI: 10.1080/02626660209492962

[38] Harris NM, Gurnell AM, Hannah DM, Petts GE. Classification of river regimes: A context for hydroecology. Hydrological Processes. 2000;**14**(16-17):2831-2848. DOI: 10.1002/1099-1085(200011/12)14:16/17<2831::AID-HYP122>3.0.CO;2-O

[39] Kuriqi A, Pinheiro AN, Sordo-Ward A, Garrote L. Flow regime aspects in determining environmental flows and maximising energy production at run-of-river hydropower plants. Applied Energy. 2019;**256**:17. DOI: 10.1016/j.apenergy.2019.113980

[40] Arthington AH, Balcombe SR. Extreme flow variability and the 'boom and bust' ecology of fish in arid-zone floodplain rivers: A case history with implications for environmental flows, conservation and management. Ecohydrology. 2011;**4**:708-720. DOI: 10.1002/eco

[41] Das Gupta A. Implication of environmental flows in river basin management. Physics and Chemistry of the Earth. 2008;**33**(5):298-303. DOI: 10.1016/j.pce.2008.02.004

[42] Kendy E, Flessa KW, Schlatter KJ, de la Parra CA, Hinojosa Huerta OM, Carrillo-Guerrero YK, et al. Leveraging environmental flows to reform water management policy: Lessons learned from the 2014 Colorado River Delta pulse flow. Ecological Engineering. 2017;**106**:683-694. DOI: 10.1016/j.ecoleng.2017.02.012

[43] Acreman M, Arthington AH, Colloff MJ, Couch C, Crossman ND,

Dyer F, et al. Environmental flows for natural, hybrid, and novel riverine ecosystems in a changing world. Frontiers in Ecology and the Environment. 2014;**12**(8):466-473. DOI: 10.1890/130134

[44] Tharme RE. A global perspective on environmental flow assessment: Emerging trends in the development and application of environmental flow methodologies for rivers. River Research and Applications. 2003;**19**(5-6):397-441. DOI: 10.1002/rra.736

[45] Richter B, Baumgartner J, Wigington R, Braun D. How much water does a river need? Freshwater Biology. 1997;**37**(1):231-249

[46] Arthington AH. Hydraulic rating and habitat simulation methods. In: Environmental Flows: Saving Rivers in the Third Millennium. Berkeley, CA: University of California Press; 2012

[47] Arthington AH, Tharme RE, Brizga SO, Pusey BJ, Kennard MJ. Environment flow assessment with emphasis on holistic methodologies. In: Welcomme RL, Petr T, editors. Proceedings of the Second International Symposium on the Management of Large Rivers for Fisheries: Sustaining Livelihoods and Biodiversity in the New Millennium, 11-14 February 2003. Vol. 2. Phnom Penh, Kingdom of Cambodia. 2004. 199-277. Available from: https://login.proxy.lib.duke.edu/login?url=https://search.proquest.com/docview/20126932?accountid=10598 http://PM6MT7VG3J.search.serialssolutions.com?ctx_ver=Z39.88-2004&ctx_enc=info:ofi/enc:UTF-8&rfr_id=info:sid/Aquatic+Science+%2526+Fisheries+Abstracts+

[48] Książek L, Woś A, Florek J, Wyrębek M, Młyński D, Wałęga A. Combined use of the hydraulic and hydrological methods to calculate the environmental flow: Wisloka river, Poland: Case study. Environmental Monitoring and Assessment. 2019;**191**(4):17. DOI: 10.1007/s10661-019-7402-7

[49] Pastor AV, Palazzo A, Havlik P, Biemans H, Wada Y, Obersteiner M, et al. The global nexus of food–trade–water sustaining environmental flows by 2050. Nature Sustainability. 2019;**2**(6):499-507. DOI: 10.1038/s41893-019-0287-1

[50] European Commission. Adaptation to Climate Change. 2021. European Commission, EU-CLIMATE-ADAPTATION-MISSION@ec.europa.eu. Available from: https://ec.europa.eu/clima/eu-action/adaptation-climate-change_en. (Accessed: March 15, 2022)

[51] European Environmental Agency. National Adaptation Policy Processes across European Countries- 2014 - EEA Report, no. 4. Copenhagen, Denmark: European Environment Agency; 2014

[52] European Environment Agency. National Monitoring, Reporting and Evaluation of Climate Change Adaptation in Europe, no. 20. Copenhagen,Denmark: European Environment Agency; 2015

[53] Cohen SJ. From observer to extension agent-using research experiences to enable proactive response to climate change. Climatic Change. 2010;**100**(1):131-135. DOI: 10.1007/s10584-010-9811-z

[54] Jones PD, Wigley TML. Estimation of global temperature trends: What's important and what isn't. Climatic Change. 2010;**100**(1):59-69. DOI: 10.1007/s10584-010-9836-3

[55] Arnell NW. Adapting to climate change: An evolving research programme. Climatic Change. 2010;**1**:107-111. DOI: 10.1007/s10584-010-9839-0

[56] Yarime M, Takeda Y, Kajikawa Y. Towards institutional analysis of sustainability science: A quantitative examination of the patterns of research collaboration. Sustainability Science. 2010;**5**(1):115-125. DOI: 10.1007/s11625-009-0090-4

[57] Kobashi T, Severinghaus JP, Barnola JM, Kawamura K, Carter T, Nakaegawa T. Persistent multi-decadal Greenland temperature fluctuation through the last millennium. Climatic Change. 2010;**100**(3):733-756. DOI: 10.1007/s10584-009-9689-9

[58] Link PM, Tol RSJ. Estimation of the economic impact of temperature changes induced by a shutdown of the thermohaline circulation: An application of FUND. Climatic Change. 2011;**104**(2):287-304. DOI: 10.1007/s10584-009-9796-7

[59] Caldwell PV, Kennen JG, Sun G, Kiang JE, Butcher JB, Eddy MC, et al. A comparison of hydrologic models for ecological flows and water availability. Ecohydrology. 2015;**8**(8):1525-1546. DOI: 10.1002/eco.1602

[60] Swart R, Bernstein L, Ha-Duong M, Petersen A. Agreeing to disagree: Uncertainty management in assessing climate change, impacts and responses by the IPCC. Climatic Change. 2009;**92**(1-2):1-29. DOI: 10.1007/s10584-008-9444-7

[61] Brown C, King J. Environmental flows: Concepts and methods. In: Davis R, Hirji R, editors. Water Resources and Environment Technical Note E.1. Washington, D.C.: The International Bank for Reconstruction and Development/The World Bank; 2003 Available from: http://www.igemportal.org/Resim/Water Resources Environment Technical Report_Wastewater Reuse-World Bank.pdf

[62] Richter BD, Baumgartner JV, Powell J, Braun DP. A method for assessing hydrologic alteration within ecosystems. Conservation Biology. 1996;**10**(4):1163-1174. DOI: 10.1046/j.1523-1739.1996.10041163.x

[63] Mathews R, Richter BD. Application of the indicators of hydrologic alteration software in environmental flow setting. Journal of the American Water Resources Association. 2007;**43**(6):1400-1413. DOI: 10.1111/j.1752-1688.2007.00099.x

[64] Richter BD, Baumgartner JV, Braun DP, Powell J. A spatial assessment of hydrologic alteration within a river network. River Research and Applications. 1998;**14**(4):329-340. DOI: 10.1002/(sici)1099-1646(199807/08)14:4<329::aid-rrr505>3.0.co;2-e

[65] Richter BD, Warner AT, Meyer JL, Lutz K. A collaborative and adaptive process for developing environmental flow recommendations. River Research and Applications. 2006;**22**(3):297-318. DOI: 10.1002/rra.892

[66] Yu C, Yin X, Yang Z. A revised range of variability approach for the comprehensive assessment of the alteration of flow regime. Ecological Engineering. 2016;**96**:200-207. DOI: 10.1016/j.ecoleng.2015.12.001

[67] Census of population and housing 2010. Census of Population and Housing Units. Census and Intercensal Historical Series 1990-2010. México: Instituto Nacional de Estadística, Geografía e Informática (INEGI); 2010. Available from: https://www.inegi.org.mx/programas/ccpv/2010/

[68] Memorias en Extenso Volumen III - V Congreso Nacional, 1er. Latinoamericano de Manejo de Cuencas 2019. COEPO (Consejo Estatal de Población). Municipio de Cuautla. Región Sierra Occidental. 2014.

Centro Interdisciplinario de Investigaciones y Estudios sobre Medio Ambiente y Desarrollo, IPN, Fundación Gonzalo Río Arronte I.A.P. Red Mexicana de Cuencas, Ciudad de México, México. Available from: www.bing.com/search?q=municipio+de+cuautla&form=

[69] EMDM (Enciclopedia de los Municipios y Delegaciones de México). Estado de Morelos, Cuautla. 2019. INAFED Instituto para el Federalismo y el Desarrollo Municipal. SEGOB (Secretaría de Gobernación). México. 2010, http://www.inafed.gob.mx/work/enciclopedia/EMM17morelos/index.html

[70] ERIC (Extractor Rápido de Información Cimatológica). Eric III, V. 3. 2., SEMARNAT-IMTA. 2019. Secretaría del Medio Ambiente y Recursos Naturales - Instituto Mexicano de Tecnología del Agua, México. Available from: www.imta.gob.mx/productos/software/eric-iii-version-3-2-extractor-rapido-de-informacion-climatolo-detail

[71] CONAGUA (National Water Comsission). Banco Nacional de Aguas Superficiales (BANDAS). Subdirección General Técnica. Gerencia de Aguas Superficiales e Ingeniería de Ríos (GASIR). 2018. Available from: www.conagua.gob.mx/CONAGUA07/Contenido/Documentos/Portada BANDAS.htm

[72] TNC (The Nature Conservancy). Indicators of Hydrologic Alteration. Version 7, User's Manual, Totten Software Design, Smythe Scientific Software, The Nature Conservancy. rPurview LLC - Ted Rybicki. USA. 2006. p. 76. DOI: 10.1002/0470848944. Available from: www.core.ecu.edu/wasklewiczt/GEOG7221/iha_man.pdf

[73] The Nature Conservancy. Indicators of hydrologic alteration version 7.1 User's manual. 2009:76. DOI: 10.1002/0470848944

[74] González-Villela R, Banderas A. Metodogías para el Cálculo de Caudales Ecológicos y Ambientales en Ríos Regulados por Presas. México: Instituto Mexicano de Tecnología del Agua; 2015 Available from: www.imta.gob.mx/biblioteca-digital

[75] Forslund A, et al. Securing Water for Ecosystems and Human Well-Being: The Importance of Environmental Flows. UMEA University, IUCN, WWF, eFlowNet, UNEP-DHI CENTRE, Conservation International, TNC, Deltares, SIWI, UNESCO-IHE, Swedish Water House (SWH). 2009

[76] Richter BD. Re-thinking environmental flows: From allocations and reserves to Sustainabilility boundaries. River Research and Applications. 2010;**26**:1052-1063. DOI: 10.1002/rra

[77] Meyer JL, Sale MJ, Mulholland PJ, Poff NLR. Impacts of climate change on aquatic ecosystem functioning and health. Journal of the American Water Resources Association. 1999;**35**(6):1373-1386. DOI: 10.1111/j.1752-1688.1999.tb04222.x

[78] IPCC (Intergovernmental Panel of climatic Change). Informe de síntesis. Contribución de los Grupos de trabajo I, II y III al Quinto Informe de Evaluación del Grupo Intergubernamental de Expertos sobre el Cambio Climático. Equipo principal de redacción. Ginebra, Suiza: IPCC; 2014

[79] MacAlister C, Subramanyam N. Climate change and adaptive water management: Innovative solutions from the global south. Water International. 2018;**43**(2):133-144. DOI: 10.1080/02508060.2018.1444307

[80] Dickens C, Smakhtin V, Biancalani R, Villholth KG, Eriyagama N, Marinelli M. Incorporating

Environmental Flows into 'Water Stress' indicator 6.4.2. Rome, Italy: FAO; 2019 Available from: https://www.unwater.org/publications/incorporating-environmental-flows-into-water-stress-indicator-6-4-2/

[81] Acreman M. Ethical aspects of water and ecosystems. Water Policy. 2001;**3**(3):257-265. DOI: 10.1016/S1366-7017(01)00009-5

[82] Liu J, Liu Q, Yang H. Assessing water scarcity by simultaneously considering environmental flow requirements, water quantity, and water quality. Ecological Indicators. 2016;**60**:434-441. DOI: 10.1016/j.ecolind.2015.07.019

[83] Vanham D, Alfieri L, Feyen L. National water shortage for low to high environmental flow protection. Scientific Reports. 2022;**12**(1):1-9. DOI: 10.1038/s41598-022-06978-y

[84] Hoekstra AY, Mekonnen MM, Chapagain AK, Mathews RE, Richter BD. Global monthly water scarcity: Blue water footprints versus blue water availability. PLoS One. 2012;**7**(2):9. DOI: 10.1371/journal.pone.0032688

[85] Smakhtin V, Revenga C, Döll P. A pilot global assessment of environmental water requirements and scarcity. Water International. 2004;**29**(3):307-317. DOI: 10.1080/02508060408691785

[86] Salinas-Rodriguez SA, Barba-Macías E, Mata DI, Nava-López M, Neri-Fores I, Varela RD, et al. What do environmental flows mean for long-term freshwater ecosystems ' protection ? Assessment of the Mexican water reserves for the environment program. Sustainability. 2021;**13**:28

[87] Ray RL, Beighley RE, Yoon Y. Integrating runoff generation and flow routing in Susquehanna River basin to characterize key hydrologic processes contributing to maximum annual flood events. Journal of Hydrologic Engineering. 2016;**21**(9):1-3. DOI: 10.1061/(ASCE)HE.1943-5584.0001389

[88] Sobayo R, Hsiang H, Ray RL, Qian L. Integration of Convolutional Neural Network and Thermal Images into Soil Moisture Estimation. 2018. Remote Sensing. 2021;**13**:4964. DOI: 10.1109/ICDIS.2018.00041

[89] Abeysingha NS, Singh M, Sehgal VK, Khanna M, Pathak H. Analysis of rainfall and temperature trends in Gomti river basin. Journal of Agricultural Physics. 2014;**14**(1):56-66

[90] Pambudi AS. Environmental mitigation and adaptation As key factors for increasing water demand. National Academy Science Letters. November 2021;(4069):1-5. DOI: 10.20935/al4069

[91] INEGI. Population Cuautla area. Instituto Nacional de Estadística, Geografía e Informática, México. https://www.inegi.org.mx/ (Accessed: Mar. 16, 2022)

[92] Breuer A, Spring UO. The 2030 agenda as agenda setting event for water governance? Evidence from the Cuautla river basin in Morelos and Mexico. Water (Switzerland). 2020;**12**(2):1-25. DOI: 10.3390/w12020314

[93] Reynolds CS. The ecosystems approach to water management. The main features of the ecosystems concept. Journal of Aquatic Ecosystem Stress and Recovery. 1993;**2**(1):3-8. DOI: 10.1007/BF00043328

[94] Rosenberg DM, McCully P, Pringle CM. Global-scale environmental effects of hydrological alterations: Introduction. Bioscience. 2000;**50**(9):746-751. DOI: 10.1641/

0006-3568(2000)050[0746,GSEEOH]2.
0.CO;2

[95] Puckridge JT, Sheldon F, Walker KF, Boulton AJ. Flow variability and the ecology of large rivers. Marine and Freshwater Research. 1998;**49**:55-72

[96] Naiman RJ, Latterell JJ, Pettit NE, Olden JD. Flow variability and the biophysical vitality of river systems. Comptes Rendus - Geoscience. 2008;**340**(0600):629-643. DOI: 10.1016/j.crte.2008.01.002

[97] De La Maza J, Carabias LJ. Usumacinta bases para una política de sustentabilidad ambiental. A.C., México: Instituto Mexicano de Tecnología del Agua – Natura y Ecosistemas Mexicanos; 2011

Section 4

Water Quality Monitoring and Management

Chapter 9

Monitoring of Rivers and Streams Conditions Using Biological Indices with Emphasis on Algae: A Comprehensive Descriptive Review toward River Management

Ehsan Atazadeh

Abstract

Algal communities are robust indicators of the effect and impact of environmental flows on river-dependent ecosystems as they deflect directly and indirectly those physical chemical and biological changes induced by environmental flows, which alter nutrient concentration, salinity, and alkalinity. Algal periphyton communities are the deterministic indicators of many aspects of ecological disturbance and its response, providing valuable evidential data at intertemporal scale of riverine status in terms of both health and quality, and their collection is comparatively simple, inexpensive, and environmental friendly.

Keywords: biological indices, algae, rivers, biological monitoring, ecological assessment

1. Introduction

River health is defined as follows:

"(i) the absence of distress defined by measured characteristics or indicators;

(ii) the ability of an ecosystem to handle stress, or bounce back its resilience [1];

(iii) the identification of risk factors such as industrial or sewage effluents." [2].

River health consists of both ecological and human values, as shown in **Figure 1**, and is mostly dependent on river condition, which is measured mostly by a large variety of qualitative indices (poor-to-excellent scaling system) as seen in Ladson et al. [5], Hill et al. [6], Gordon et al. [7], Acreman and Ferguson [8], and Atazadeh et al. [3]. Physicochemical indices are the most common indices for lotic water, for example, in the U.S. as seen in Toxic and Priority Pollutants Under the Clean Water Act [9], which are extended to Chemical, Microbiological, Whole Effluent Toxicity, Radiochemical,

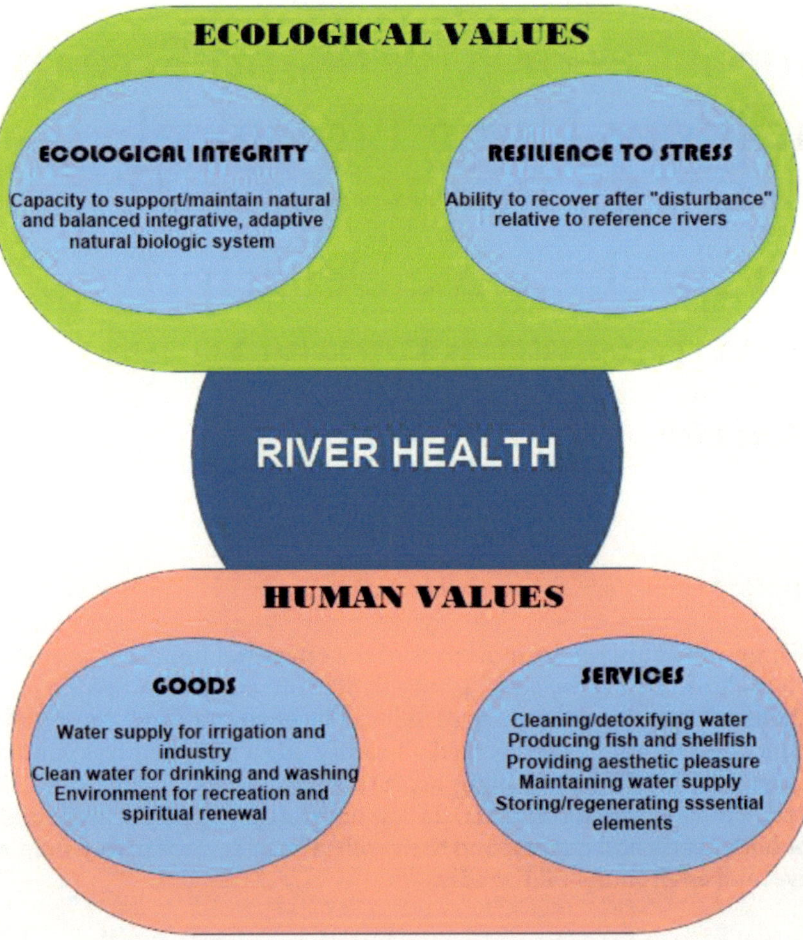

Figure 1.
Ecological and human value contribution to river health as modified from [3, 4].

Industry-Specific and Biosolids [10]. However, these proved inadequate in achieving full spectrum protection [11, 12].

The first to point out that organic matter concentration areas in streams attract invertebrates was Hynes in [13] based on the previous works [14, 15] on the bottom fauna distribution and quantity. This was followed by the River Continuum Concept where "*the structural and functional characteristics of stream communities are adapted to conform to the most probable position or mean sate of the physical system*" [16], the framework for a spatiotemporal hierarchical classification system "*among and within stream systems*" [17] and the collection of articles in Boon and Raven [18]. These are some of the cornerstones that led to the employment of bioindicators, biomonitoring, and bioassessment. Biological indicators (bioindicators) are defined as "*an organism (or part of an organism or a community of organisms) that contains information on the quality of the environment (or a part of the environment)*" [19]. Biomonitoring is "the systematic use of living organisms or their responses to determine the condition or changes of the environment" [20]. Bioassessment is defined as "*an evaluation of the condition of a waterbody using biological surveys and other direct measurements of the resident biota in surface water*" [21].

Consequently, biological indicators fill in the gaps left by physiochemical indices as being more integrative [22] and range from lower trophic-level organisms (e.g., algae or benthic macroinvertebrates) all the way to upper trophic-level species (e.g., fish and mussels). These, combined with river geomorphological and hydrological indices [7], form a more dependable framework upon which river health assessment can rely [23]. Algae entered the phase of scientific study well over a hundred years ago and its connection with riverine environmental condition started in 1908 [24–30].

2. Methodology

The PRISMA method [31] was applied in the usual way with a multitude of pertinent keywords and exclusion of (sea, ocean, and lake) where appropriate. The goal of this review is to present a cogent well-sourced picture of the state of monitoring of rivers and streams condition using biological indices with a particular focus on algae. To this end, methods, frameworks, and diverse indices were descriptively enumerated and the general role of algae was expanded upon.

3. Results

3.1 Aggregate indices and monitoring frameworks

Indices may be simple or aggregate, that is, an index comprised of subindices. An example of the latter is the Aggregate Water Quality Index (AWQI). This was shown to have the capacity of being formulated without the problems of ambiguity (subindices show use-targeted acceptable water quality but the aggregated index fails to do so), eclipsing (aggregated index does not reflect sufficiently poor water quality shown by water quality variables), and rigidity (more variables have to be included targeted particular water quality aspects) as seen in Swamee and Tyagi [32]. To set up an AWQI, the process is as follows [33]:

- selecting the perceived to be significant water quality parameters
- forming subindices
- establishing weights for relative parameters
- selecting an aggregation process of the subindices

The aggregation process ranges from the simple weighted additive method to the modified additive method [34] and more complicated methods [35].

An example of river condition index is the aggregate-type River Condition Index in New South Wales, which is comprised of the following subindices:

- "River Styles® (River [36]) Geomorphic Condition assessment – surrogate input under FARWH "Physical Form" category (RSGC)" [37].

- "Riparian vegetation cover assessment (native woody vegetation) – surrogate input under FARWH "Fringing Zone" category (RVC)" [37]

- "Macro water planning: hydrologic stress or risk rating – surrogate input under FARWH "Hydrological Change" category" [37].

- "River biodiversity condition data – surrogate input under FARWH "Aquatic Biota" category (RBCI)" [37]

- "Catchment Disturbance Index – surrogate input under FARWH "Catchment Disturbance" category (CDI)" [37].

The River Condition Index (RCI) score is computed by employing Euclidean distance for the subindices [38].

$$RCI = 1 - \frac{\sqrt{(1-RSC)2+(1-CDI)2+(1-HS)2+(1-RBCI)2+(1-RVC)2}}{\sqrt{5}}$$

To interpret the results, **Table 1** converts metric to qualitative results using the Framework for Assessing River and Wetland Health (FARWH) [39].

The South African Government's River Health Program in the process of assessing both river health and stream condition developed an Index of Stream Geomorphology based on the measurement of geomorphic variables in view of the fact that they are the main constituents of the channel morphology impacting river aquatic biota [7, 40, 41]. This is comparable to the underlying logic in Chessman et al. [42] associating downward geomorphology changes with downward changes in assemblages of macrophytes and macroinvertebrates, while the latter are seen to display new sensitivities [43], a reaction seen also in freshwater mussels [44]. Also, positive result of fluvial geomorphology is associated with maintaining river health framework structural elements [45].

The River Habitat Quality survey framework as seen in Fox et al. [46] is used extensively worldwide, especially in Europe and the U.S., to assess both river health and condition. Its field survey protocols converge with those of SERCON (System for Evaluating Rivers for Conservation) [47] and was compared to the Systeme d'Evaluation de la Qualite du Milieu Physique (SEQ-MP) from France, and the field survey method of the Landerarbeitsgemeinschaft Wasser (LAWA-vor-Ort) from Germany [48]. A subset of its features leads to distinguishing between lowland, Alpine, and southern European rivers in terms of hydromorphological character [49]

Score	Condition Bands
0.81–1	Very Good (equivalent to FARWH "Largely Unmodified")
0.61–0.8	Good (equivalent to FARWH "Slightly Modified")
0.41–0.6	Moderate (equivalent to FARWH "Moderately Modified")
0.21–0.4	Poor (equivalent to FARWH "Substantially Modified")
0–0.2	Very Poor (equivalent to FARWH "Severely modified")

Table 1.
Conversion of metric to qualitative results [37].

and is amenable to prediction techniques [50]. In the U.K., it was used for the period 1994–1997 for quality assessment [51], for the determination of characteristics and controls of Gravel-Bed Riffles [52], for the classification of urban rivers [53], and for the aims of the WFD hydro-morphological assessment (STAR Project overview) [54]. Also, in exploring the interactions between flood defense maintenance works and river habitats [55], for environmental assessment and catchment planning [56] and for the evaluation of the effects of riparian restoration [57]. In Serbia, it was used in the Golijska Moravica and Jerma basins [58], in Austria for the identification of rivers with high-and-good habitat quality [59], in Germany for river habitat monitoring and assessment [60], in the U.S. for the measurement of Little Tallahatchie River in northern Mississippi [61], and in Portugal for fluvial hydromorphological assessment [62]. Also, in Poland regarding its seasonal diversification [63] and in Southern Europe [64].

River geomorphology is a characteristic often used for river health evaluation as seen [7] and is deemed to be quite important [65]. Geomorphology impacts water quality [66], and its assessment is used in place of "command-and-control" practices that cause environmental damaging biodiversity reduction and lessening provision of ecosystem services [67], and its change is associated with river rehabilitation [68] while *via* the River Styles framework links policy with action-on-the-ground. Also, along with ecology and river channel, habitats constitute a mesoscale approach to basin-scale challenges [69], assist in determining the ecological health of wadable streams [70], and affect riparian habitat within alluvial channel-floodplain river systems [71], and its spatial variability impacts the disturbance temporal patterns influencing ecosystem structure/dynamics [72], acting as a framework for the analysis of microplastics in riverine sediments [73].

The Index of Stream Condition (ISC) was developed, tested, and applied in Australian regions [5], for example, in Victoria [74] as seen in **Figure 2**.

The basis of the ISC system lies in a subjective ranking system whereby the current condition of a river is compared to a known/modeled "pristine" condition across the index/subindex groups seen in **Figure 1** which are scaled to (0–10) values and added as seen in **Figure 3**.

It should be noted that macro-benthos data are included [5, 7].

Hydrology	Physical Form	Streamside zone	Water Quality	Aquatic Life
Hydrology refers to the amount of water that is within the river channel at a particular point in time at a particular location. A minimum of 15 years of monthly flow data is used.	Physical form takes into account the river bank condition as well as instream habitat (logs or 'snags') and major barriers to fish migration, such as dams and artificial weirs.	Streamside zone measures characteristics of the woody vegetation within 40 metres of the river's edge.	Water quality is the quality of water in the river.	Aquatic life is based on the number and type of aquatic macroinvertebrates found within the river.
• Low flows • High flows • Zero flows • Seasonality • Variability	• Bank condition • Artificial barriers • Instream woody habitat	• Width • Fragmentation • Overhang • Cover of trees and shrubs • Structure • Large Trees • Weeds	• Total Phosphorus • Turbidity • Salinity (EC) • pH	• AUSRIVAS • SIGNAL • EPT • Number of Families

Figure 2.
Third benchmark index of stream condition subindices and metrics of hydrology [75].

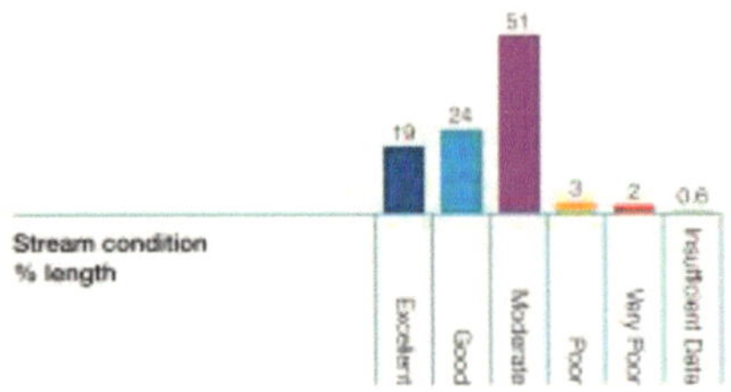

(a)

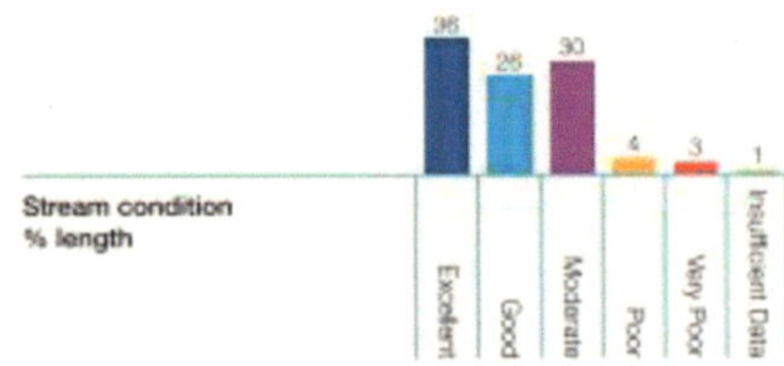

(b)

Basin	Reach	Reach Length (Km)	River	Hydrology	Physical Form	Streamside Zone	Water Quality	Aquatic Life	ISC Score	Condition
1	1	43.3	Mitta Mitta River	2	7	4	9	4	20	Poor
1	2	42.9	Mitta Mitta River	2	5	4		7	19	VPoor
1	3	17.4	Mitta Mitta River	2	5	7	9	6	24	Moderate
1	4	38.4	Little Snowy Creek	7	7	8		10	38	Good
1	5	23	Snowy Creek	9	7	9		9	42	Excellent

(c)

Figure 3.
*North east region: (a) index of stream condition, (b) upper Murray index of stream condition, (c) part of the upper Murray calculation * [76]. *insufficient data are not added.*

3.2 Biological monitoring systems

Biological monitoring systems are usually based on fish, benthic macro-invertebrates as well as macrophytes, riparian vegetation, and algae [7, 77–91].

In determining water quality from the assessment of river ecosystem health, macro-invertebrates are considered to be both important, as they are a critical part of the aquatic food framework [92] and preferable to other targets for the following reasons [93]:

a. They are both differentially sensitive to various pollutant types and rapidly reactive to them while being capable of responding in a graded way to a variety and different levels of stress.

b. They are abundant as well as ubiquitous and easily collectible and identifiable [79, 94–99], while identification and enumeration are easy in comparison with microorganisms and plankton.

c. If they are benthic the fact that they are sedentary makes them more representational of local conditions.

d. Their lifespans are long enough to provide adequate environmental quality records, though temporal/seasonal variability in index calibration should be introduced [100–102].

e. Their communities include a number of phyla which makes them heterogeneous.

Generally, in-stream biomonitoring can be employed using some or all of the aforesaid macro- and micro-organisms as biological indicators/indices. However, from the point of view of spatial and temporal invariance this leads to problems.

Benthic macro-invertebrates are used by a large number of scientists because they are sensitive to water degradation and river health and depend on sediment quality [103]. A major issue is eutrophication, "*excessive plant and algal growth due to the increased availability of one or more limiting growth factors needed for photosynthesis*" [104], which usually occurs in rivers that are passing through urbanized/agricultural areas [105] and seems to depend strongly on the local stream and its surroundings characteristics [106] in case of which both reaction and response become limited [107, 108]. Another issue is that in order to assess human-influenced events separately from the naturally caused ones, that is, natural seasonal or successional variation [109], a stressor-specific multimetric approach is needed [110].

In a 2008 statistical analysis [111], the findings were as below:

- Benthic macro-invertebrates (19 sources) 68% reported sampling difficulties, 58% reported not enough taxonomic keys available, 42% reduced sensitivity to disturbance, and 21% drifting (not a good local indicator).

- In the case of algae (nine sources), 78% reported insufficient metrics/indices, 67% reported not enough taxonomic keys available, and 33% reported that the group is a poor accumulator.

- For fish (14 sources), 64% reported drifting (not a good local indicator), 36% reported sampling difficulties, and 21% reported insufficient metrics/indices.

- For zooplankton (nine sources), 67% reported sampling difficulties, 67% reported insufficient metrics/indices, 50% reported heavy affectation by non-anthropogenic conditions, 33% reported that the group is a poor accumulator, and 33% reported not regular occurrence in habitat under study.

In rivers where macrophytes are not abundant, bottom-lying biofilm is the main agent of nutrient uptake, a stratum that consists [112] of algae, bacteria, and fungi ensconced in a polysaccharide matrix. In the case of nutrient change, algae react directly but invertebrates generally respond indirectly depending on the water quality intensity of influence on the habitat. The mechanism explaining this [113, 114] is that an initial subsidy effect consisting of increasing nutrients leads to the direct stimulation of algal productivity and this, in the role of a mediator, causes, through increased trophic resources, the macroinvertebrate's response stimulation. For this reason, some approaches to understanding river conditions have been based on As algal biofilm/diatom communities are sensitive and responsive to river physical, chemical, and biological changes [24, 29, 115–118] and there are a lot of approaches for river condition assessment based on them.

Biofilms are a major element of river food webs [119] and important for stream biogeochemical and nutrient processes [120–126]. Microalgae are the main food source for fauna in freshwater ecosystems. Algae-based processes lead to the production and synthesis of organic matter (carbon) and allow its entry into the food web *via* which is available to higher trophic consumers such as fish and waterbirds [127–129], and consequently, algae, in terms of freshwater ecosystems [126], are considered to be the most essential part of food webs and biogeochemical cycling, for example, carbon cycling [130]. Epiphytic algae are one of the appropriate food sources for stream invertebrates in an interactive way [131] since freshwater algae carry high concentrations of polyunsaturated fatty acids and in stream food webs, high-quality algae enhances the food value of low-quality riparian leaf litter [129, 132, 133]. In terms of algal groups, diatoms and cryptophytes supply aquatic invertebrate food of higher quality due to long-chain omega-3 polyunsaturated fatty acids [132, 134–136], while omega-3 (n-3) long-chain essential fatty acids (EFA) are higher in running water than brackish [136] and are projected to decrease as world temperature rises [137]. As these species are the important indicators of river health, their primary production is equally important and may be decreased by turbidity and shading, due to light blocking [138, 139], while shear stress and low nutrients are important inhibitors particularly for algal primary production along with temperature and grazing [140].

Primary producer community structure (PPCS) in rivers and streams is influenced by the general state of hydrodynamics [141, 142] as flow velocity increase is correlated positively with increased nutrient delivery by increasing PPCS productivity through thinning the diffusive boundary layer up to a point where it becomes negatively correlated due to dislodgement [143–145]. As seen in Gurnell [146], PPCS influences ecosystem structure *via* their hydrodynamics and morphology by flow-vegetation-sediment feedbacks.

Algae, having increased sensitivity, often signal changes in environmental conditions by responding well before effects on higher organisms manifest themselves [29, 78]. The reduction of river flow affects biofilm structure [125], for example, in terms of causing increased algae bloom [147], which prevents sunlight from penetrating the water surface [148] as well as ecosystem processes in general [149–151]. In the biofilm structure, diatom assemblages are shown to be highly responsive to water quality variation [152] since their assemblages are used to measure water quality [153].

3.3 Biological indices

There are many biological indices for water quality assessment, which may depend either on many parameters or on a particular one. The algal periphyton system, in terms of similarity, diversity, evenness, structure, and dominance, has been employed in the construction of a variety of biological indices of both kinds [154–156]. There has been criticism regarding the processes, which reduces the indices to a single quantitative or qualitative result regarding its representational effect [157] due to seasonal variability [158–161] or regionality [162, 163]. Despite the objections, this type of indices has been employed in many countries including the U.K., the U.S.A, Spain, and Canada. Most countries have passed legislation according to which government entities controlling rivers and water bodies in general are obliged to use biological indices, for example, Italy [164], in order to assess stream condition in terms of water quality and water abstraction impact [7]. The European Water Framework Directive [165] included biological monitoring as a stream health assessment tool, and in the U.S.A., macroinvertebrate community assessment is used under the Clean Water Act [166].

Biological indices are used by conjunctive employment with multivariate statistical analysis, which leads to a good understanding of aquatic biota-sensitivity and to the determination of the driver-response relationship [7, 91, 167–169]. Biological indices resulting from multivariate analysis techniques are as follows:

- The U.K. derived RIVPACS (River Invertebrate Prediction And Classification Scheme) [170–174].
- The Canadian BEAST (BEnthic Assessment of SedimenT) [99, 175].
- The Australian AusRivAS (Australian River Assessment System) [176, 177] but according to Chessman [178], its applications in terms of biological health assessment do not have consistent or quantified status of any nature rendering them virtually meaningless.
- The Australian SIGNAL (Stream Invertebrate Grade Number Average Level) [43, 179].
- The South African Scoring System (SASS) [180, 181].

Multi-metric techniques (biotic integrity indices) [182] are employed as an approach where an integrated balance is maintained in adaptive biological systems between elements and processes such as species, genus, assemblage and biotic interaction, nutrient and energy dynamic, meta-population process respectively in natural habitats [183, 184]. The initial concept of biotic integrity, the Index of Biotic Integrity (IBI) has been developed for fish in shallow rivers [185] in the USA measuring trophic composition, species composition, and abundance and health of fish [183, 186], and Karr's work is re-evaluated in Capmourteres et al. [187]. According to Gordon et al. [7], in the case of the biotic integrity index small disturbance to the system has negligible effect on the biological integrity of the system, which was one of the presuppositions of its design [185]. However, a unified conclusion regarding the regularity of different group-based IBI evaluation results has not been reached [188–190] as seen in Huang et al. [191]. Various biotic integrity-based biotic indices besides IBI exist:

- BIBI (Benthic Index for Biotic Integrity) employing macro-invertebrates [192] uses 10 metrics of stream macroinvertebrate communities integrated into one value that has numerical/qualitative range 10 (poor) to 50 (excellent) [193]. Also, it is related to environmental factors [194], upon occasion may show opposite results to those obtained using the Organism – Sediment Index (OSI) [195], and in a Korean study [196], the BIBI Korean variant by [197] was found to be negatively correlated with the Korean Saprobic Index (KSI), which is based on the saprobic valency concept as seen in Zelinka and Marvan [198].
- PIBI (Periphyton Index for Biotic Integrity) where algal periphyton is the main element [6, 24, 199, 200].
- DSIAR (Diatom Species Index for Australian Rivers) based on diatoms is correlated at a significant level to ARC_E (Assessment of River Condition, Environment) [116, 201].
- BII (Biotic Integrity Index) that employs diatom community structural metrics [202].
- MBII (Macroinvertebrate Biotic Integrity Index) where data collection is performed by employing a probability design, evaluating five characteristics

(precision, range, responsiveness to disturbance, relationship to catchment area, and redundancy with respect to other metrics), while a continuous scale is employed for scoring [203].

3.4 Riverine ecological assessment: The role of algae

Algae are in general one of the primary producers in aquatic ecosystems [204, 205] taking into consideration that Water N:P molar ratios could result in being restrictive for river algal communities' population dynamics and species coexistence [206].

Algae react to riverine ecosystem disturbances and show sensitivity to changes in environmental conditions [29, 126, 150, 207–214]. However, riverine algae are not as sensitive to changes in environmental conditions as periphytic algae, which grow by substrate attachment, and in case of negative environmental changes move away [215].

In effect, they have the characteristics necessary to become prime environmental conditions monitors in aquatic ecosystems at a global applicative level [28, 29, 90, 108, 126, 216–221]. In particular, their properties [120, 220, 222, 223] are seen below:

- high sensitivity to environmental changes
- easy to sample
- the majority of species are both cosmopolitan and with well-known autecology
- possess a wide spectrum of structural (biomass, composition) and functional (metabolism) attributes are valuable for their use in monitoring ecosystems.

Community structure, biomass standing crop, and species composition have been employed in the assessment of riverine ecological condition both directly and indirectly [224–227]. Riverine biofilm structure [125, 207, 228–232] and ecosystem processes [151] have been shown to be affected by flow variation. Within the biofilm, diatom algal assemblages within the biofilm are highly responsive to water chemistry variations [152], and consequently, their composition can give away any ecological responses to flow-driven changes occurring in stream water quality. Using algae as an ecological state assessment tool leads to the detection of harmful riverine ecosystem human activity [126, 233, 234], providing thus the evidence necessary for carrying out water resource managerial decisions.

Flow current exerts influence over algal immigration [235], reproduction by varying nutrient supply rates [236], and community physiognomy by decreasing attachment strength [237].

High stream discharge velocities may affect benthic algae in different ways, which depend on both frequency and intensity [238, 239] and change both physiological and structural properties of the community [240, 241].

In lotic and lentic freshwater ecosystems, algae are the main primary producers as, for example, trophic status *via* the trophic state index (TSI) is determined by algal levels [242], seen in Round [243], Stevenson et al. [244], and Allan and Castillo [245] and being the main source of energy for first-order consumers such as small herbivores places them in an important role in the food web. Algae growth is dependent on riverine nutrient concentration, mainly on phosphorus and nitrogen, and also on benthos-concentrated ones [244, 246], while other factors, such as predation and hydrology, have a significant contribution [141, 244, 247–250].

As seen in Steinman [251], there are various forms of benthic algae assembly, for example, stalked (colonial) aggregates, unicellular states [252], and filamentous [253]. Benthic algal biomass constitutes an excellent water quality indicator [254–257] and, through that, of river condition and therefore health [258]. Algal biomass analyses are often used for river health evaluation [259] as well as of riverine ecosystems anthropogenic modifications analysis, for example, of dry mass [260], of chlorophyll-*a* concentration [261–263], bio-volume and peak biomass [244, 260, 264], and ash-free dry mass.

The flow regime, stream velocity in effect, is in negative correlation negatively with chlorophyll-*a* concentration [228, 265, 266]. While chlorophyll-*a* concentration tends to increase downstream in a state of constant flow, there are also upstream-caused downstream effects [267]. The flow-related disturbance effect on biomass is also [228, 265, 266, 268–270] as well as in rainforest streams [271] and in the creation of gradient of metacommunity types within stream networks [272]. Algal biomass is seen to decrease due to suspended solids and grazers, that is, fish and invertebrates, substratum instability, flow disturbance, that is, velocity where stream algal biomass responds to nutrient enrichment depending on the velocity [273]. Conversely, light, temperature, and nutrients are seen to be the main promoting resources of algal biomass [141, 274].

Nutrients as well as grazing pressure and light influence algal growth and community structure [275, 276]. In terms of algal biomass control, the main top-down controllers are nutrients and light are top-down and grazers, mainly fish and snails, are the main bottom-up [207, 251, 277], while under certain conditions there is feedback between the two processes [278]. A controller of algal community structure is lotic system flow disturbance [245, 279–281].

Shifts in water quality and flow variation affect algal colonization and structure [125, 228, 245, 282–285]. Flow regime impacts on both water quality, that is, temperature, suspended solids, oxygen level, organic matters, and other nutrients in general, and the metabolism of rivers or streams and biotic structure and function [7, 149, 286, 287]. Climate change impacts water quality [288], and as seen in Baron et al. [289], environmental factors impact the structure and function of aquatic ecosystems and flow regimes, sediment and organic materials, water quality, nutrients and other chemicals elements, light, temperature, for example, "brownification" where, as seen in De Wit et al. [290], a 10% increase in precipitation will result in increasing by 30% the soil transfer of OC to freshwaters.

The list of environmental factors affecting the structure and function of benthic algae in riverine ecosystems was compiled and analyzed, in particular grazers, temperature, pH, light, hydraulics, and nutrients (N, P, Si). While, under fast flow and low nutrients algal community structure, species composition, biomass, and standing crop decrease slow flow and a high concentration of nutrients increase algal biomass and community structure.

Climate variability/change is seen to affect algae directly [291–294]. As seen in Sinha and Michalak [295], precipitation, which is climate induced, is a preeminent factor in the variability of riverine nitrogen. Moreover, precipitation plays a role in algae affecting stream flow velocity [296, 297] as seen in the Heavy Precipitation Index [298], which is a part of the Streamflow Indicator as defined in [299]. This makes the climate/precipitation mechanisms in [300–303], and flood events [304] lead to effects, which influence algae in an important and multifaceted way.

3.5 Algae in the role of indicators in the assessment of stream condition

Algae are considered to be primary producers in aquatic ecosystems powering both food webs and biogeochemical cycling [126], even rare metals [305]. Algae are present in almost every aquatic environment including fresh, brackish, marine, and hypersaline water [306, 307]. Algae communities in rivers are usually diverse and inhomogeneous [308, 309], and their types are as in **Table 2**.

The floristic composition of algae in the benthos could be employed in water quality, stream condition, and eutrophication monitoring [24, 90, 108, 311–315]. A number of studies show a preference for diatoms since diatom-based methods in bio-monitoring approaches demonstrating a tendency for higher success rate to be the most successful [29, 316]. In practice, other algal groups present bigger difficulties in sampling and quantitative estimation in comparison with diatoms. Moreover, common river algae, in particular the green algae [78], show a demonstrable lack of identification keys although partial country-wide lists exist, for example, [317] where 321 out of 500 genera are identified. However, these other groups may provide information that diatom-based measures cannot provide easily; for example, eutrophication can be monitored by cyanobacterial and green algae biomass and diversity could be used to monitor eutrophication [108, 313, 318–320].

Diatoms also play important roles in biotechnology, engineering, biology, and material science [321] but their main role in the general riverine environmental condition [29, 322, 323], water quality ecological assessment of aquatic systems [107, 152, 324–328], eutrophication [78, 314, 329], pollution [330, 331], bioassessment [107, 116, 332–334], and urbanization [335–338].

3.6 Biological index-based water body classification

Water body classification is now a function of water chemistry, biological, and hydrological characteristics as it is necessary to include pollutant effects on biota since the nature of the receiving waters influences the effect on water quality as seen in **Figure 4**. Also, river water is classifiable on the basis of biology, hydrology, and quality, into different condition ecological categories in the qualitative scale of bad, poor, moderate, good, or high [8].

Epilithon	on rock
epidendron or epixylon	on woody debris
epiphyton	on plants
episammon	on sand
epipelon	on mud
epizoon	on animals

Table 2.
Algae community types [308, 310].

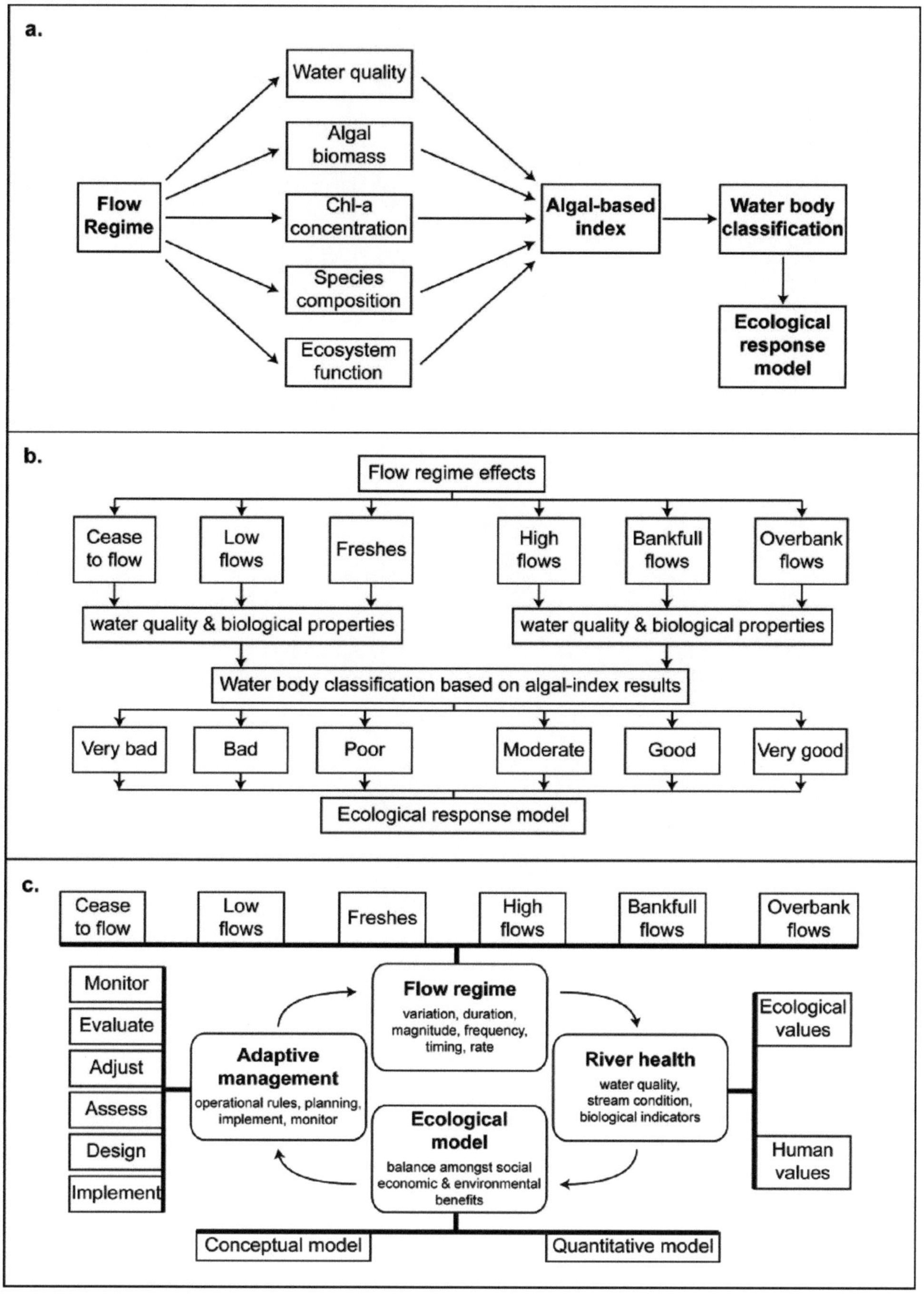

Figure 4.
Water body classification employing ecological models biological indices, and adaptive management.

Author details

Ehsan Atazadeh
Faculty of Natural Sciences, Department of Plant, Cell and Molecular Biology, University of Tabriz, Tabriz, Iran

*Address all correspondence to: ehsan.atazadeh@gmail.com; atazadeh@tabrizu.ac.ir

References

[1] Holling CS. Resilience and stability of ecological systems. Annual Review of Ecology and Systematics. 1973;**4**:1-23

[2] Norris RH, Thoms MC. What is river health? Freshwater Biology. 1999;**41**(2):197-209. DOI: 10.1046/j.1365-2427.1999.00425.x

[3] Atazadeh E, Barton A, Shirinpour M, Zarghami M, Rajabifard A. River management and environmental water allocation in regulated ecosystems of arid and semiarid regions–A review. Fundamental and Applied Limnology. 2020a;**193**(4):327-345

[4] Boulton AJ. An overview of river health assessment: Philosophies, practice, problems and prognosis. Freshwater Biology. 1999;**41**(2):469-479. DOI: 10.1046/j.1365-2427.1999.00443.x

[5] Ladson AR, White LJ, Doolan JA, Finlayson BL, Hart BT, Lake PS, et al. Development and testing of an index of stream condition for waterway management in Australia. Freshwater Biology. 1999;**41**:453-468

[6] Hill BH, Herlihy AT, Kaufmann PR, DeCelles SJ, Vander Borgh MA. Assessment of streams of the eastern United States using a periphyton index of biotic integrity. Ecological Indicators. 2003;**2**:325-338

[7] Gordon ND, McMahon TA, Finlayson BL, Gippel CJ, Nathan RJ. Stream Hydrology: An Introduction for Ecologists. 2nd ed. John Wiley & Sons Ltd: Sussex; 2004. p. 429

[8] Acreman MC, Ferguson AJD. Environmental flows and the European Water Framework Directive. Freshwater Biology. 2010;**55**:32-48

[9] EPA. 2022b. Toxic and Priority Pollutants Under the Clean Water Act. U.S. Government. 2022. Available from: https://www.epa.gov/eg/toxic-and-priority-pollutants-under-clean-water-act#priority

[10] EPA. 2022a. Clean Water Act Analytical Methods. U.S. Government. 2022. Available from: https://www.epa.gov/cwa-methods

[11] Kazancı N, Dügel M, Girgin S. Determination of indicator genera of benthic macroinvertebrate communities in running waters in western Turkey. Review of Hydrobiology. 2008;**1**:1-16

[12] United States Environmental Protection Agency (USEPA). Water Quality Standards Academy: Basic Course. Washington District of Columbia, U.S.A: Office of Water; 2005. p. 152

[13] Hynes HBN. The stream and its valley. Internationale Vereinigung Für Theoretische Und Angewandte Limnologie: Verhandlungen. 1975;**19**(1):1-15. DOI: 10.1080/03680770.1974.11896033

[14] Egglishaw HJ. The distributional relationship between the bottom fauna and plant detritus in streams. The Journal of Animal Ecology. 1964;**33**:463-476

[15] Egglishaw HJ. The quantitative relationship between bottom fauna and plant detritus in streams of different calcium concentrations. Journal of Applied Ecology. 1968;**5**:731-740

[16] Vannote RL, Wayne Minshall G, Cummins KW, Sedell JR, Cushing CE. The river continuum concept (Vannote et Al. 1980). Canadian Journal of Fisheries and Aquatic Sciences. 1980;**37**:130-137

[17] Frissell CA, Liss WJ, Warren CE, Hurley MD. A hierarchical framework for stream habitat classification: Viewing streams in a watershed context. Environmental Management. 1986;**10**(2):199-214. DOI: 10.1007/BF01867358

[18] Boon PJ, Raven PJ, editors. River Conservation and Management. Chichester, U.K.: Wiley-Blackwell; 1992

[19] Markert BA, Breure AM, Zechmeister HG, editors. Bioindicators & Biomonitors: Principles, Concepts and Applications. Amsterdam, The Netherlands: Elsevier; 2003. DOI: 10.1016/b978-0-444-51554-4.00030-4

[20] Oertel N, Salanki J. Biomonitoring and bioindicators in aquatic ecosystems. In: Ambasht RS, Ambasht NK, editors. Modern Trends in Applied Aquatic Ecology II Biomonitoring and Bioindicators in Aquatic Ecosystems. Boston, MA, U.S.A: Springer; 2003. pp. 219-246

[21] Barbour MT, Gerritsen J, Snyder BD, Stribling JB. Rapid Bioassessment Protocols for Use in Streams and Wadeable Rivers:Periphyton, Benthic Macroinvertebrates and Fish. 2nd ed. Washington, District of Columbia, U.S.A.: EPA 841-B-99-002. U.S. Environmental Protection Agency. Office of Water; 1999

[22] Karr JR. Defining and measuring river health. Freshwater Biology. 1999;**41**:221-234

[23] Moog O, Schmutz S, Schwarzinger I. Biomonitoring and bioassessment Otto. In: Schmutz S, Sendzimir J, editors. Riverine Ecosystem Management: Science for Governing Towards a Sustainable Future. Cham, Switzerland: Springer Open; 2018. pp. 371-390

[24] Hill BH, Herlihy AT, Kaufmann PR, Stevenson RJ, McCormick FH, Johnson CB. Use of periphyton assemblage data as an index of biotic integrity. Journal of the North American Benthological Society. 2000;**19**:50-67

[25] Hustedt F. Systematische und ökologische Untersuchungen über die Diatomeen-Flora von Java, Bali und Sumatra nach dem Material der Deutschen Limnologischen Sunda-Expedition. Archiv für Hydrobiologie. 1937;**15**(SUPPL):187-295

[26] Kolkwitz R, Marson M. Ökologie der pflanzlichen Saprobien. Berichte der Deutschen Botanischen Gesellschaft. Ber Dtsch Bot Ges. 1908;**26**:505-519

[27] Lowe RL. Environmental Requirements and Pollution Tolerance of Freshwater Diatoms. Cincinnati, Ohio: Environmental Protection Agency; 1974

[28] Sládecek V. System of Water Quality from the Biological Point of View. Schweizerbart'sche: Verlagsbuchhandlung; 1973

[29] Stevenson JR, Pan Y, Van Dam H. Assessing environmental conditions in rivers and streams with diatoms. In: Smol J, Stoermer E, editors. The Diatoms: Application for the Environmental and Earth Sciences. Cambridge: Cambridge University Press; 2010. pp. 57-85

[30] Watanabe T, Asai K, Houki A. Numerical estimation to organic pollution of flowing water by using the epilithic diatom assemblage-----diatom assemblage index (DAIpo). Science of the Total Environment. 1986;**55**:209-218

[31] PRISMA. PRISMA (Preferred Reporting Items for Systematic Reviews and Meta-Analyses) 2020. Checklist. The BMJ. San Francisco: PRISMA; 2021. Available from: http://www.prisma-statement.org/

[32] Swamee PK, Tyagi A. Improved method for aggregation of water quality subindices. Journal of Environmental Engineering. 2007;**133**(2):220-225. DOI: 10.1061/(asce)0733-9372(2007)133:2(220)

[33] Sutadian AD, Muttil N, Yilmaz AG, Perera B. Development of river water quality indices - A review. Environmental Monitoring and Assessment. 2016;**188**:58-90

[34] Bordalo AA, Teixeira R, Wiebe WJ. A water quality index applied to an international shared river basin: The case of the Douro River. Environmental Management. 2006;**38**:910-920

[35] Banda TD, Kumarasamy MV. Development of water quality indices (WQIs): A review. Polish Journal of Environmental Studies. 2020;**29**:2011-2021

[36] River Styles. The River Styles Framework. 2022. Available from: https://riverstyles.com/river-styles-framework/

[37] Healey M, Raine A, Parsons L, Cook N. River Condition Index in New South Wales: Method Development and Application. Sydney, Australia: NSW Office of Water; 2012

[38] Norris R, Dyer F, Hairsine P, Kennard M, Linke S, Merrin L, et al. Australian Water Resources 2005: A Baseline Assessment of Water Resources for the National Water Initiative, Level 2 Assessment, River and Wetland Health Theme: Assessment of River and Wetland Health: Potential Comparative Indices, May 2007. National Water Commission: Canberra; 2007a

[39] Norris R, Dyer F, Hairsine P, Kennard M, Linke S, Merrin L, et al. Australian Water Resources 2005: A Baseline Assessment of Water Resources for the National Water Initiative, Level 2 Assessment, River and Wetland Health Theme: Assessment of River and Wetland Health: A Framework for Comparative Assessment of the Ecological Condition of Australian Rivers and Wetlands, May 2007. Canberra: National Water Commission; 2007b

[40] Rowntree K, Wadeson R. A geomorphological framework for the assessment of instream flow requirements. Aquatic Ecosystem Health & Management. 1998;**1**:125-141

[41] Rowntree K, Ziervogel G. Development of an Index of Stream Geomorphology for the Assessment of River Health Index of Stream Condition NAEBP Report Series No 7. Pretoria, South Africa: Department of Water Affairs and Forestry; 1999

[42] Chessman BC, Fryirs KA, Brierley GJ. Linking geomorphic character, behaviour and condition to fluvial biodiversity: Implications for river management. Aquatic Conservation: Marine and Freshwater Ecosystems. 2006;**16**:267-288

[43] Chessman BC. New sensitivity grades for Australian river macroinvertebrates. Marine and Freshwater Research. 2003;**54**:95-103

[44] Jones HA, Byrne M. The impact of catastrophic channel change on freshwater mussels in the Hunter River system, Australia: A conservation assessment. Aquatic Conservation: Marine and Freshwater Ecosystems. 2010;**20**:18-30

[45] Petts GE, Gurnell AM. Dams and geomorphology: Research progress and future directions. Geomorphology. 2005;**71**:27-47

[46] Fox PJA, Naura M, Scarlett P. An account of the derivation and testing of a standard field method, river

habitat survey. Aquatic Conservation: Marine and Freshwater Ecosystems. 1998;**8**(4):455-475. DOI: 10.1002/(SICI)1099-0755(199807/08)8:4<455::AID-AQC284>3.0.CO;2-7

[47] Wilkinson J, Martin J, Boon PJ, Holmes NTH. Convergence of field survey protocols for SERCON (System for Evaluating rivers for Conservation) and RHS (River Habitat Survey). Aquatic Conservation: Marine and Freshwater Ecosystems. 1998;**8**:579-596

[48] Raven PJ, Holmes NTH, Charrier P, Dawson FH, Naura M, Boon PJ. Towards a harmonized approach for hydromorphological assessment of rivers in Europe: A qualitative comparison of three survey methods. Aquatic Conservation: Marine and Freshwater Ecosystems. 2002;**12**(4):405-424. DOI: 10.1002/aqc.536

[49] Szoszkiewicz K, Buffagni A, Davy-Bowker J, Lesny J, Chojnicki BH, Zbierska J, et al. Occurrence and variability of river habitat survey features across Europe and the consequences for data collection and evaluation. Hydrobiologia. 2006;**566**(1):267-280. DOI: 10.1007/s10750-006-0090-7

[50] Jeffers JNR. Characterization of river habitats and prediction of habitat features using ordination techniques. Aquatic Conservation: Marine and Freshwater Ecosystems. 1998;**8**(4):529-540. DOI: 10.1002/(SICI)1099-0755(199807/08)8:4<529::AID-AQC301>3.0.CO;2-9

[51] Raven PJ. River Habitat Quality: The Physical Character of Rivers and Streams in the UK and Isle of Man. Bristol: Environment Agency; 1998

[52] Emery JC, Gurnell AM, Clifford NJ, Petts GE. Characteristics and controls of gravel - bed riffles : An analysis of data from the river- habitat survey. Water and Environment Journal. 2004;**18**(4):210-216

[53] Davenport AJ, Gurnell AM, Armitage PD. Habitat survey and classification of urban rivers. River Research and Applications. 2004;**20**(6):687-704. DOI: 10.1002/rra.785

[54] Erba S, Buffagni A, Holmes N, O'Hare M, Scarlett P, Stenico A. Preliminary testing of river habitat survey features for the aims of the WFD hydro-morphological assessment: An overview from the STAR project. Hydrobiologia. 2006;**566**(1):281-296. DOI: 10.1007/s10750-006-0089-0

[55] Harvey GL, Wallerstein NP. Exploring the interactions between flood defence maintenance works and river habitats: The use of river habitat survey data. Aquatic Conservation: Marine and Freshwater Ecosystems. 2009;**19**:689-702. DOI: 10.1002/aqc

[56] Raven PJ, Holmes NTH, Vaughan IP, Hugh Dawson F, Scarlett P. Benchmarking habitat quality: Observations using river habitat survey on near-natural streams and Rivers in northern and Western Europe. Aquatic Conservation: Marine and Freshwater Ecosystems. 2010;**20**(SUPPL. 1):13-30. DOI: 10.1002/aqc.1103

[57] Clews E, Vaughan IP, Ormerod SJ. Evaluating the effects of riparian restoration on a Temperate River-system using standardized habitat survey. Aquatic Conservation: Marine and Freshwater Ecosystems. 2010;**20**(SUPPL. 1):96-104. DOI: 10.1002/aqc.1096

[58] Urosev M, Milanovic A, Milijasevic D. Assessment of the river habitat quality in undeveloped areas of Serbia applying the RHS (River Habitat Survey) method. Journal of

the Geographical Institute Jovan Cviji, SASA. 2009;**59**(2):37-58. DOI: 10.2298/ijgi0902037u

[59] Muhar S, Schwarz M, Schmutz S, Jungwirth M. Identification of rivers with high and good habitat quality: Methodological approach and applications in Austria. Hydrobiologia. 2000;**422-423**:343-358. DOI: 10.1007/978-94-011-4164-2_28

[60] Kamp U, Binder W, Hölzl K. River habitat monitoring and assessment in Germany. Environmental Monitoring and Assessment. 2007;**127**(1-3):209-226. DOI: 10.1007/s10661-006-9274-x

[61] Shields FD, Rigby JR. River habitat quality from river velocities measured using acoustic Doppler current profiler. Environmental Management. 2005;**36**(4):565-575. DOI: 10.1007/s00267-004-0292-6

[62] Ferreira J, Pádua J, Hughes SJ, Cortes RM, Varandas S, Holmes N, et al. Adapting and adopting river habitat survey: Problems and solutions for fluvial hydromorphological assessment in Portugal. Limnetica. 2011b;**30**(2):263-272. DOI: 10.23818/limn.30.20

[63] Kiraga M. The diversification of river habitat survey output during the four seasons: Case studies of three lowland Rivers in Poland. Journal of Ecological Engineering. 2020;**21**(6):116-126. DOI: 10.12911/22998993/123248

[64] Buffagni A, Kemp JL. Looking beyond the shores of the United Kingdom: Addenda for the application of river habitat survey in southern European Rivers. Journal of Limnology. 2003;**61**(2):199-214. DOI: 10.4081/jlimnol.2002.199

[65] Hazbavi Z. Importance of geology and geomorphology in watershed health assessment. Agriculture and Forestry. 2018;**64**(4):277-287. DOI: 10.17707/agricultforest.64.4.27

[66] Kamarudin MK, Amri AM, Nalado ME, Toriman HJ, Umar R, Ismail A, et al. Evolution of river geomorphology to water quality impact using remotesensing and GIS technique. Desalination and Water Treatment. 2019;**149**:258-273. DOI: 10.5004/dwt.2019.23838

[67] Brierley G. Geomorphology and river management. Kemanusiaan. 2008;**15**:13-26. DOI: 10.1672/0277-5212(2006)26[884:garm]2.0.co;2

[68] Wolfert HP. Geomorphological Change and River Rehabilitation: Case Studies on Lowland Fluvial Systems in the Netherlands. Wageningen, Netherlands: Alterra; 2001

[69] Newson MD, Newson CL. Geomorphology, ecology and river channel habitat: Mesoscale approaches to basin-scale challenges. Progress in Physical Geography. 2000;**24**(2):195-217. DOI: 10.1177/030913330002400203

[70] Lepper TA. Assessing the Value of a Geomorphic Toolbox to Assist with Determining Ecological Health of Wadable Streams within the Waikato Region. New Zealand: Massey University; 2020

[71] Steiger J, Tabacchi E, Dufour S, Corenblit D, Peiry JL. Hydrogeomorphic processes affecting riparian habitat within Alluvial Channel-Floodplain River systems: A review for the temperate zone. River Research and Applications. 2005;**21**(7):719-737. DOI: 10.1002/rra.879

[72] Montgomery DR. Process domains and river continuum. Journal of the American Water Resources Association. 1999;**35**(2):397-410

[73] Miller J, Orbock-Miller SM. A geomorphic framework for the analysis of microplastics in riverine sediments. E3S Web of Conferences. 2020;**202**:1-10. DOI: 10.1051/e3sconf/202020201002

[74] Ladson T. An Index of Stream Condition: Reference Manual. Melbourne, Victoria, Australia: Government of Victoria; 1999

[75] DEPI. Index of Stream Condition: The Third Benchmark of Victorian River Condition ISC3. Melbourne, Victoria, Australia: State Government of Victoria; 2010a

[76] DEPI. North East Region - Index of Stream Condition. Melbourne, Victoria, Australia: State Government of Victoria; 2010b

[77] Faith DP. Benthic macroinvertebrates in biological surveillance: Monte Carlo significance tests on functional groups' responses to environmental gradients. Environmental Monitoring and Assessment. 1990;**14**(2-3):247-264. DOI: 10.1007/BF00677920

[78] Kelly MG, Whitton BA. The trophic diatom index: A new index for monitoring eutrophication in rivers. Journal of Applied Phycology. 1995;7:433-444

[79] Rosenberg DM, Resh VH. Freshwater Biomonitoring and Benthic Macroinvertebrates. New York: Chapman & Hall; 1993

[80] Aguiar FC, Fernandes MR, Teresa Ferreira M. Descripteurs de La VAcgActation Rivulaire Comme Guide de La Restauration Accologique Au Sein Des Paysages de Cours Daeau. Knowledge and Management of Aquatic Ecosystems. 2011;**402**:1-12. DOI: 10.1051/kmae/2011074

[81] Asadi-Sharif E, Yahyavi B, Bayrami A, Rahim-Pouran S, Atazadeh E, Singh R, et al. Physicochemical and biological status of Aghlagan river, Iran: Effects of seasonal changes and point source pollution. Environmental Science and Pollution Research. 2021;**28**(12):15339-15349

[82] Bytyqi P, Czikkely M, Shala-Abazi A, Fetoshi O, Ismaili M, Hyseni-Spahiu M, et al. Macrophytes as biological indicators of organic pollution in the Lepenci River Basin in Kosovo. Journal of Freshwater Ecology. 2020;**35**(1):105-121. DOI: 10.1080/02705060.2020.1745913

[83] Harris J, Silveira R. Large-scale assessments of river health using an index of biotic integrity with low-diversity fish communities. Freshwater Biology. 1999;**41**:235-252

[84] Hawkes H. Origin and development of the biological monitoring working party score system. Water Research. 1997;**32**(3):964-968

[85] Ibelings B, Admiraal W, Bijkerk R, Ietswaart T, Prins H. Monitoring of algae in Dutch Rivers: Does it meet its goals? Journal of Applied Phycology. 1998;**10**(2):171-181. DOI: 10.1023/A:1008049000764

[86] Marchand L, Nsanganwimana F, Cook BJ, Vystavna Y, Huneau F, Le Coustumer P, et al. Trace element transfer from soil to leaves of macrophytes along the Jalle d'Eysines River, France and their potential use as contamination biomonitors. Ecological Indicators. 2014;**46**:425-437. DOI: 10.1016/j.ecolind.2014.07.011

[87] Munné A, Prat N, Solà C, Bonada N, Rieradevall M. A simple field method for assessing the ecological quality of riparian habitat in rivers and streams: QBR index. Aquatic Conservation: Marine and Freshwater Ecosystems. 2003;**13**:147-163

[88] Onaindia M, Amezaga I, Garbisu C, García-Bikuña B. Aquatic Macrophytes as biological indicators of environmental conditions of rivers in north-eastern Spain. Annales de Limnologie. 2005;**41**(3):175-182. DOI: 10.1051/limn:20054130175

[89] Thorne RSJ, Williams WP, Cao Y. The influence of data transformations on biological monitoring studies using macroinvertebrates. Water Research. 1999;**33**(2):343-350. DOI: 10.1016/S0043-1354(98)00247-4

[90] Whitton B, Kelly M. Use of algae and other plants for monitoring rivers. Australian Journal of Ecology. 1995;**20**:45-56

[91] Norris RH, Norris KH. The need for biological assessment of water quality: Australian perspective. Australian Journal of Ecology. 1995;**20**:1-6

[92] Edelstein K. Pond and Stream Safari: A Guide to the Ecology of Aquatic Invertebrates. Ithaca, NY, U.S.A: Cornell Cooperative Extension; 1993

[93] Metcalfe JL. Biological water quality assessment of running waters based on macroinvertebrate communities: History and present status in Europe. Environmental Pollution. 1989;**60**(1-2):101-139. DOI: 10.1016/0269-7491(89)90223-6

[94] Hayslip G, editor. Methods for the Collection and Analysis of Benthic Macroinvertebrate Assemblages in Wadeable Streams of the Pacific Northwest. Cook, Washington, U.S.A: Pacific Northwest Aquatic Monitoring Partnership; 2007

[95] Hirst H, Jüttner I, Ormerod SJ. Comparing the responses of diatoms and macroinvertebrates to metals in upland streams of Wales and Cornwall. Freshwater Biology. 2002;**47**(9):1752-1765. DOI: 10.1046/j.1365-2427.2002.00904.x

[96] King JM. Abundance, biomass and diversity of benthic macro-invertebrates in a Western Cape River, South Africa. Transactions of the Royal Society of South Africa. 1983;**45**(1):11-34. DOI: 10.1080/00359198309520092

[97] Metzeling L, Perriss S, Robinson D. Can the detection of salinity and habitat simplification gradients using rapid bioassessment of benthic invertebrates be improved through finer taxonomic resolution or alternative indices? Hydrobiologia. 2006;**572**:235-252

[98] Ojija F, Laizer H. Macro invertebrates as bio indicators of water quality in Nzovwe Stream, in Mbeya, Tanzania. International Journal of Scientific & Technology Research. 2016;**5**(06):211-222 Available from: www.ijstr.org

[99] Reynoldson T, Norris R, Resh V, Day K, Rosenberg D. The reference condition: A comparison of multimetric and multivariate approaches to assess water-quality impairment using benthic macroinvertebrates. Journal of the North American Benthological Society. 1997;**16**:833-852

[100] Atazadeh E, Barton A, Razeghi R. Importance of environmental flows in the Wimmera catchment, Southeast Australia. Limnological Review. 2020b;**20**(4):185-198

[101] Dallas HF. Seasonal variability of macroinvertebrate assemblages in two regions of South Africa: Implications for aquatic bioassessment. African Journal of Aquatic Science. 2004;**29**(2):173-184. DOI: 10.2989/16085910409503808

[102] Mazor RD, Purcell AH, Resh VH. Long-term variability in bioassessments:

A twenty-year study from two northern California streams. Environmental Management. 2009;**43**(6):1269-1286. DOI: 10.1007/s00267-009-9294-8

[103] Townsend KR, Pettigrove VJ, Carew ME, Hoffmann AA. The effects of sediment quality on benthic macroinvertebrates in the river Murray, Australia. Marine and Freshwater Research. 2009;**60**(1):70-82. DOI: 10.1071/MF08121

[104] Schindler DW. Recent advances in the understanding and management of eutrophication. Limnology and Oceanography. 2006;**51**:356-363

[105] Omernik JM. Nonpoint Source–Stream Nutrient Level Relationships: A Nationwide Survey. EPA-600/3-77-105, Ecological Research Series. Washington, DC: United States Environmental Protection Agency; 1977

[106] Corkum LD. Responses of chlorophyll-a, organic matter, and macroinvertebrates to nutrient additions in rivers flowing through agricultural and forested land Archiv für Hydrobiologie. 1996;136(3):391-411

[107] Atazadeh I, Sharifi M, Kelly MG. Evaluation of the trophic diatom index for assessing water quality in River Gharasou, western Iran. Hydrobiologia. 2007;**589**:165-173

[108] Kelly MG, Whitton BA. Biological monitoring of eutrophication in rivers. Hydrobiologia. 1998;**384**:55-67

[109] Karr JR, Chu EW. Restoring Life in Running Waters: Better Biological Monitoring. Washington D.C.: Island Press; 1999

[110] Ofenböck T, Moog O, Gerritsen J, Barbour M. A stressor specific multimetric approach for monitoring running waters in Austria using benthic macro-invertebrates. Hydrobiologia. 2004;**516**(1-3):251-268. DOI: 10.1023/B:HYDR.0000025269.74061.f9

[111] Resh VH. Which group is best? Attributes of different biological assemblages used in freshwater biomonitoring programs. Environmental Monitoring and Assessment. 2008;**138**(1-3):131-138. DOI: 10.1007/s10661-007-9749-4

[112] Lock MA. Attached microbial communities in rivers. In: Ford TE, editor. Aquatic Microbiology—An Ecological Approach. Oxford: Blackwell; 1993. pp. 113-138

[113] Feminella JW, Hawkins CP. Interactions between stream herbivores and periphyton: A quantitative analysis of past experiments. Journal of the North American Benthological Society. 1995;**14**:465-509

[114] Hillebrand H. Top-down versus bottom-up control of autotrophic biomass - a meta-analysis on experiments with periphyton. Journal of the North American Benthological Society. 2002;**21**:349-369

[115] Acs E, Szabo K, Toth B, Kiss KT. Ivestigation of benthic algal communities, especially diatoms, of some Hungarian streams in connection with reference conditions of the water framework directives. Acta Botanica Hungarica. 2004;**46**(3-4):255-277

[116] Chessman BC, Bate N, Gell PA, Newall P. A diatom species index for bioassessment of Australian rivers. Marine and Freshwater Research. 2007;**58**:542-557

[117] Hayashi S, Eun J, Kazuhito J. Construction of river model biofilm for assessing pesticide effects. Archives

of Environmental Contamination and Toxicology. 2011;**60**(1):44-56. DOI: 10.1007/s00244-010-9531-4

[118] Makk J, Ács É, Márialigeti K, Kovács G. Investigations on diatom-associated bacterial communities colonizing an artificial substratum in the river Danube. Biologia, Bratislava. 2003;**58**(4):729-742

[119] Weitere M, Erken M, Majdi N, Arndt H, Norf H, Reinshagen M, et al. The food web perspective on aquatic biofilms. Ecological Monographs. 2018;**88:1-17**. DOI: 10.1002/ecm.1315

[120] Burns A, Ryder DS. Potential for biofilms as biological indicators in Australian riverine systems. Ecological Management & Restoration. 2001;**2**:53-64

[121] Graba M, Sauvage S, Majdi N, Mialet B, Moulin FY, Urrea G, et al. Modelling epilithic biofilms combining hydrodynamics , invertebrate grazing and algal traits. Freshwater Biology. 2014;**2014**(59):1213-1228. DOI: 10.1111/fwb.12341

[122] Metting FB Jr. Biodiversity and application of microalgae. Journal of Industrial Microbiology. 1996;**17**: 477-489

[123] Neu TR, Lawrence JR. Development and structure of microbial biofilms in river water studied by confocal laser scanning microscopy. FEMS Microbial Ecology. 1997;**24**:11-25

[124] Prieto DM, Rosa D-R, Rubinos DA, Díaz-Fierros F, Barral MT. Biofilm formation on river sediments under different light intensities and nutrient inputs: A flume Mesocosm study. Environmental Engineering Science. 2016;**33**(4):12. DOI: 10.1089/ees.2015.0427

[125] Ryder DS, Watts RJ, Nye E, Burns A. Can flow velocity regulate epixylic biofilm structure in a regulated floodplain river? Marine and Freshwater Research. 2006;**57**:29-36

[126] Stevenson J. Ecological assessments with algae: A review and synthesis. Journal of Phycology. 2014;**50**:437-461

[127] Bunn SE, Balcombe SR, Davies PM, Fellows CS, McKenzie-Smith FJ. Aquatic productivity and food webs of desert river ecosystems. In: Ecology of Desert Rivers. Queensland, Australia: Griffth University; 2006. pp. 76-99

[128] Fogg GE. The role of algae in organic production in aquatic environments. British Phycological Bulletin. 1963;**2**(4):195-205. DOI: 10.1080/00071616300650021

[129] Guo F, Kainz MJ, Sheldon F, Bunn SE. The importance of high-quality algal food sources in stream food webs– current status and future perspectives. Freshwater Biology. 2016a;**61**:815-831

[130] Fabian J. Effects of Algae on Microbial Carbon Cycling in Freshwaters. Potsdam: Universität Potsdam und IGB Leibniz-Institut für Gewässerökologie und Binnenfischerei; 2018

[131] Fenoglio S, Tierno M, Doretto A, Falasco E, Bona F. Aquatic insects and benthic diatoms : A history of biotic relationships in freshwater ecosystems. Water (Switzerland). 2020;**12**(10):2934

[132] Guo F, Kainz MJ, Valdez D, Sheldon F, Bunn SE. High-quality algae attached to leaf litter boost invertebrate shredder growth. Freshwater Science. 2016b;**35**:000-000

[133] Torres-Ruiz M, Wehr JD, Perrone AA. Trophic relations in a stream food web: Importance of fatty acids for

macroinvertebrate consumers. Journal of the North American Benthological Society. 2007;**26**:509-522

[134] Brett M, Muller-Navarra D. The role of highly unsaturated fatty acids in aquatic foodweb processes. Freshwater Biology. 1997;**38**:483-499

[135] Guo F, Bunn SE, Brett MT, Fry B, Hager H, Ouyang X, et al. Feeding strategies for the acquisition of high-quality food sources in stream macroinvertebrates : Collecting , integrating, and mixed feeding. Limnology and Oceanography. 2018;**63**:1964-1978. DOI: 10.1002/lno.10818

[136] Peltomaa E, Johnson MD, Taipale SJ. Marine Cryptophytes are great sources of EPA and DHA. Marine Drugs. 2018;**16**(1):1-11. DOI: 10.3390/md16010003

[137] Hixson SM, Arts MT. Climate warming is predicted to reduce Omega-3, long-chain, polyunsaturated fatty acid production in phytoplankton. Global Change Biologyhange Biology. 2016;**22**(8):2744-2755. DOI: 10.1111/gcb.13295

[138] Bunn S, Davies P, Mosisch T. Ecosystem measures of river health and their response to riparian and catchment degradation. Freshwater Biology. 1999;**41**:333-345

[139] Davies PM, Bunn SE, Hamilton SK, Dudgeon D. Primary production in tropical streams and rivers. In: Tropical Stream Ecology. Canberra: Academic Press; 2008. pp. 23-42

[140] Allan JD, Castillo MM, Capps KA. Primary producers. In: Stream Ecology. Cham: Springer; 2021. pp. 141-176. DOI: 10.1007/978-3-030-61286-3_6

[141] Biggs BJF. Patterns in benthic algae of streams. In: Stevenson RJ, Bothwell ML, Lowe RL, editors. Algal Ecology: Freshwater Benthic Ecosystems. San Diego, CA, USA: Academic Press; 1996. pp. 31-56

[142] Biggs BJ, Stokseth S. Hydraulic habitat suitability for periphyton in rivers. River Research and Applications. 1996;**12**(2-3):251-261

[143] Larned ST, Nikora VI, Biggs BJ. Mass-transfer-limited nitrogen and phosphorus uptake by stream periphyton: A conceptual model and experimental evidence. Limnology and Oceanography. 2004;**49**(6):1992-2000

[144] Nepf HM. Flow and transport in regions with aquatic vegetation. Annual Review of Fluid Mechanics. 2012a;**44**(1):123-142

[145] Nepf HM. Hydrodynamics of vegetated channels. Journal of Hydraulic Research. 2012b;**50**(3):262-279

[146] Gurnell A. Plants as river system engineers. Earth Surface Processes and Landforms. 2014;**39**(1):4-25

[147] Kim S, Chung S, Park H, Cho Y, Lee H. Analysis of environmental factors associated with cyanobacterial dominance after river weir installation. Water (Switzerland). 2019;**11**:1163. DOI: 10.3390/w11061163

[148] Liu C, Chen Y, Zou L, Cheng B, Huang T. Time-lag effect : River algal blooms on multiple driving factors. Frontiers in Earth Science. 2022;**9**:1-12. DOI: 10.3389/feart.2021.813287

[149] Palmer M, Ruhi A. Linkages between flow regime, biota, and ecosystem processes: Implications for river restoration. Science. 2019;**365**(365):1264. DOI: 10.1126/science.aaw2087

[150] Ryder DS. Response of epixylic biofilm metabolism to water level

variability in a regulated floodplain river. Journal of the North American Benthological Society. 2004;**23**:214-223

[151] Ryder DS, Miller W. Setting goals and measuring success: Linking patterns and processes in stream restoration. Hydrobiologia. 2005;**552**:147-158

[152] Reid MA, Tibby JC, Penny D, Gell PA. The use of diatoms to assess past and present water quality. Australian Journal of Ecology. 1995;**20**:57-64

[153] Gevrey M, Rimet F, Park YS, Giaraudel J-L, Ector L, Lek S. Water quality assessment using diatom assemblages and advanced modelling techniques. Freshwater Biology. 2004;**49**:208-220

[154] Singh S, James A, Bharose R. Biological assessment of water pollution using periphyton productivity : A review. Nature Environment and Pollution Technology. 2017;**16**(2):559-567

[155] Xuwang Y, Xiaodong Q, Qingnan L, Ying L, Yuan Z, Wei M. Using periphyton assemblages to assess stream conditions of Taizi River Basin, China. Acta Ecologica Sinica. 2012;**6**:1677-1691

[156] Ziglio G, Flaim G, Siligardi M. Biological Monitoring of Rivers. London, UK: John Wiley & Sons; 2006

[157] Suter GW. A critique of ecosystem health concepts and indexes. Environmental Toxicology and Chemistry. 1993;**12**:1533-1539

[158] Leunda PM, Oscoz J, Miranda R, Arino AH. Longitudinal and seasonal variation of the benthic macroinvertebrate community and biotic indices in an undisturbed Pyrenean River. Ecological Indicators. 2009;**9**:52-63. DOI: 10.1016/j.ecolind.2008.01.009

[159] Munoz CZ, Sainz-Cantero CE, Sanchez-Ortega A, Alba-Tercedor J. Are biological indices BMPW'and ASPT'and their significance regarding water quality seasonally dependent? Factors explaining their variations. Water Research. 1995;**29**(I):285-290

[160] Murphy PM. The temporal variability in biotic indices. Environmental Pollution. 1978;**17**:227-236

[161] Stark JD, Phillips N. Seasonal variability in the macroinvertebrate community index: Are seasonal correction factors required? Seasonal variability in the macroinvertebrate community index: Are seasonal correction factors required? New Zealand Journal of Marine and Freshwater Research. 2009;**43**(4):867-882. DOI: 10.1080/00288330909510045

[162] Dallas HF. Rapid bioassessment protocols using aquatic macroinvertebrates in Africa – Considerations for regional adaptation of existing biotic indices. Frontiers in Water. 2021;**3**:1-12. DOI: 10.3389/frwa.2021.628227

[163] Tang T, Jan Stevenson R, Infante DM. Accounting for regional variation in both natural environment and human disturbance to improve performance of multimetric indices of lotic benthic diatoms. The Science of the Total Environment. 2016;**568**:1124-1134. DOI: 10.1016/j.scitotenv.2016.03.060

[164] Government of Italy. Decreto Legislativo 11 Maggio 1999, n. 152 (Modificato Dal Decreto Legislativo 18 Agosto 2000, n. 258). Disposizione Sulla Tutela Delle Acque Dall'inquinamento e Recepimento Della Direttiva 91/271/CEE Concernente Il Trattamento Delle Acque Reflue Urbane E. Italy. 1999

[165] EU. The EU Water Framework Directive - Integrated River Basin Management for Europe. 2000. Available

from: https://ec.europa.eu/environment/water/water-framework/index_en.html

[166] Govenor B, Merckaert K, Vanderborght B, Nicotra MM. Invariant set distributed explicit reference governors for provably safe on-board control of nano-quadrotor swarms. Frontiers in Robotics and AI. 2017;**8**:663809. DOI: 10.3389/frobt.2021.663809

[167] Downes B, Barmuta L, Fairweather P, Faith D, Keough M, Lake P, et al. 2002. Monitoring Ecological Impacts: Concepts and Practice in Flowing Waters. New York, USA: Cambridge University Press; 2002

[168] Singh G, Patel N, Jindal T, Srivastava P, Bhowmik A. Assessment of spatial and temporal variations in water quality by the application of multivariate statistical methods in the Kali River, Uttar Pradesh, India. Environmental Monitoring and Assessment. 2020;**192**(6):26. DOI: 10.1007/s10661-020-08307-0

[169] Unda-Calvo J, Ruiz-Romera E, Martínez-Santos M, Vidal M, Antigüedad I. Multivariate statistical analyses for water and sediment quality index development: A study of susceptibility in an Urban River. Science of the Total Environment. 2020;**711**:135026. DOI: 10.1016/j.scitotenv.2019.135026

[170] Clarke RT, Wright JF, Furse MT. RIVPACS models for predicting the expected macroinvertebrate fauna and assessing the ecological quality of Rivers. Ecological Modelling. 2003;**160**(3):219-233

[171] Clarke RT, Murphy JF. Effects of locally rare taxa on the precision and sensitivity of RIVPACS bioassessment of freshwaters. Freshwater Biology. 2006;**51**(10):1924-1940

[172] Hargett EG, ZumBerge JR, Hawkins CP. Development of a Rivpacs Model for Wadeable Streams of Wyoming Wyoming Department of Enviroznmental Quality. Cheyenne,WY, U.S.A: Wyoming Department of Environmental Quality; 2005

[173] Wright J, Furse M, Armitage P. RIVPACS-a technique for evaluating the biological quality of rivers in the UK. European Water Pollution Control. 1993;**3**:15-15

[174] Wright J, Furse M, Moss D. River classification using invertebrates: RIVPACS applications. Aquatic Conservation: Marine and Freshwater Ecosystems. 1998;**8**:617-631

[175] Reynoldson T, Day K, Pascoe T, Wright J, Sutcliffe D, Furse M. The development of the BEAST: A predictive approach for assessing sediment quality in the north American Great Lakes. In: Assessing the Biological Quality of Fresh Waters: RIVPACS and Other Techniques. Proceedings of an International Workshop Held in Oxford, UK, on 16-18 September 1997. Oxford, UK: Freshwater Biological Association (FBA); 2000. pp. 165-180

[176] Australian Government. Australian River Assessment System (AusRivAS) Bioassessment: Macroinvertebrate. 2022. Available from: https://wetlandinfo.des.qld.gov.au/wetlands/resources/tools/assessment-search-tool/australian-river-assessment-system-ausrivas-bioassessment-macroinvertebrate/

[177] Coysh J, Nichols S, Ransom G, Simpson J, Norris R, Barmuta L, et al. AUSRIVAS Macroinvertebrate Bioassessment Predictive Modelling Manual. Canberra: Cooperative Research

Centre for Freshwater Ecology; 2000. pp. 15-30

[178] Chessman BC. What's wrong with the Australian River Assessment System (AUSRIVAS)? Marine and Freshwater Research. 2021;**72**(8):1110-1117. DOI: 10.1071/MF20361

[179] Chessman BC, Growns JE, Kotlash AR. Objective derivation of macro invertebrate family sensitivity grade numbers for the SIGNAL biotic index: Application to the Hunter River system, New South Wales. Marine and Freshwater Research. 1997;**48**:159-172

[180] Bowd R, Kotze DC, Morris CD, Quinn NW. Testing the applicability of the SASS5 scoring procedure for assessing wetland health: A case study in the KwaZulu-Natal Midlands, South Africa. African Journal of Aquatic Science. 2006;**31**(2):229-246

[181] Dickens CWS, Graham PM. South African scoring system (SASS 5), rapid bioassessment method for rivers. African Journal of Aquatic Science. 2002;**27**:1-10. DOI: 10.2989/16085914.2002.9626569

[182] Davis WS, Simon TP, editors. Biological Assessment and Criteria: Tools for Water Resource Planning and Decision Making. Boca Raton, Florida, USA: Lewis Publishers; 1995

[183] Karr JR. Assessing Biological Integrity in Running Waters : A Method and its Rationale Special Publication No. 05. Urbana-Champaign, IL, U.S.A: Illinois Natural History Survey; 1986

[184] Karr JR. Ecological integrity and ecological health are not the same. Engineering within Ecological Constraints. 1996;**97**:109

[185] Leonard PM, Orth DJ. Application and testing of an index of biotic integrity in small, coolwater streams. Transactions of the American Fisheries Society. 1986;**115**(3):401-414. DOI: 10.1577/1548-8659(1986)115<401:aatoai>2.0.co;2

[186] Karr JR. Assessment of biotic integrity using fish communities. Fisheries. 1981;**6**:21-27

[187] Capmourteres V, Rooney N, Anand M. Assessing the causal relationships of ecological integrity: A re-evaluation of Karr's iconic index of biotic integrity. Ecosphere. 2018;**9**(3):e02168. DOI: 10.1002/ecs2.2168

[188] Dolph CL, Huff DD, Chizinski CJ, Vondracek B. Implications of community concordance for assessing stream integrity at three nested spatial scales in Minnesota, U.S.a. Freshwater Biology. 2011;**56**:1652-1669

[189] Johnson RK, Daniel H. Spatial congruency of benthic diatom, invertebrate, macrophyte, and fish assemblages in European streams. Ecological Applications. 2010;**20**:978-992

[190] Paavola R, Muotka T, Virtanen R, Heino J, Jackson D, Mäki-Petäys A. Spatial scale affects community concordance among fishes, benthic macroinvertebrates, and bryophytes in streams. Ecological Applications. 2006;**16**:368-379

[191] Huang X, Jing X, Liu B, Guan X, Li J. Assessment of aquatic ecosystem health with indices of biotic integrity (IBIs) in the Ganjiang River system, China. Water (Switzerland). 2022;**14**:278 https://doi.org/doi:10.3390/w14030278

[192] Kerans BL, Karr JR. A benthic index of biotic integrity (B-IBI) for rivers of the Tennessee Valley. Ecological Applications. 1994;**4**:768-785

[193] Fore LS, Karr JR, Wisseman RW. Assessing invertebrate responses to human activities: Evaluating

alternative approaches. Journal of the North American Benthological Society. 1996;**15**(2):212-231. DOI: 10.2307/1467949

[194] Li F, Cai Q, Ye L. Developing a benthic index of biological integrity and some relationships to environmental factors in the subtropical Xiangxi River, China. International Review of Hydrobiology. 2010;**95**(2):171-189. DOI: 10.1002/iroh.200911212

[195] Diaz RJ, Randy Cutter G, Dauer DM. A comparison of two methods for estimating the status of benthic habitat quality in the Virginia Chesapeake Bay. Journal of Experimental Marine Biology and Ecology. 2003;**285-286**:371-381. DOI: 10.1016/S0022-0981(02)00538-5

[196] Jun YC, Kim NY, Kwon SJ, Han SC, Hwang IC, Park JH, et al. Effects of land use on benthic macroinvertebrate communities: Comparison of two mountain streams in Korea. Annales de Limnologie. 2011;**47**:S35-S49. DOI: 10.1051/limn/2011018

[197] MOE/NIER. Study on Development of Methods for Synthetic Assessment of Water Environment (III) – Survey and Evaluation of Aquatic Ecosystem Health. Seoul, Korea: The Ministry of Environment/National Institute of Environmental Research MOE/NIER; 2006

[198] Zelinka M, Marvan P. Zur pra¨ zisierung der biologische klassifikation der reinheit fliessender gewa¨ sser. Archiv für Hydrobiologie. 1961;**57**:389-407

[199] Hamsher SE, Vis ML. Utility of the periphyton index of biotic integrity (PIBI) as an indicator of acid mine drainage impacts in Southeastern Ohio. Journal of Phycology. 2003;**39**(1):20-21

[200] Limbeck R, Smith G. Pilot Study: Implementation of a Periphyton Monitoring Network for the Non-Tidal Delaware River Delaware River Biomonitoring Program Delaware River Basin Commission. West Trenton, NJ, U.S.A.: Delaware River Biomonitoring Program Delaware River Basin Commission; 2007

[201] Chessman BC, Townsend SA. Differing effects of catchment land use on water chemistry explain contrasting behaviour of a diatom index in tropical northern and temperate southern Australia. Ecological Indicators. 2010;**10**:620-626. DOI: 10.1016/j.ecolind.2009.10.006

[202] Wang Y-k, Jan Stevenson R, Meitzmeir L. Development and evaluation of a diatom-based index of biotic integrity for the interior plateau ecoregion, USA. Journal of the North American Benthological Society. 2005;**24**(4):990-1008. DOI: 10.1899/03-028.1

[203] Klemm DJ, Karen A, et al. Development and evaluation of a macroinvertebrate biotic integrity index (MBII) for regionally assessing mid-Atlantic highlands streams. Environmental Management. 2003;**31**(5):656-669. DOI: 10.1007/s00267-002-2945-7

[204] Kvíderová J, Elster J. Standardized algal growth potential and/or algal primary production rates of maritime Antarctic stream waters (King George Island, south Shetlands). Polar Research. 2013;**32**(SUPPL):1-17. DOI: 10.3402/polar.v32i0.11191

[205] Webster IT, Rea N, Padovan AV, Dostine P, Townsend SA, Cook S. An analysis of primary production in the Daly River, a relatively unimpacted tropical river in Northern Australia.

Marine and Freshwater Research. 2005;**56**(3):303-316. DOI: 10.1071/MF04083

[206] Sabater S, Artigas J, Corcoll N, Proia L, Timoner X, Tornés E. Ecophysiology of river algae. In: Jr ON, editor. River Algae. Berlin, Germany: Springer; 2016. pp. 197-218. DOI: 10.1007/978-3-319-31984-1

[207] Biggs BJ, Goring DG, Nikora VI. Subsidy and stress responses of stream periphyton to gradients in water velocity as a function of community growth form. Journal of Phycology. 1998;**34**:598-607

[208] Dixit SS, Smol JP, Kingston JC, Charles DF. Diatoms: Powerful indicators of environmental change. Environmental Science & Technology. 1992;**26**:22-33

[209] Lacoursière S, Lavoie I, Rodríguez MA, Campeau S. Modeling the response time of diatom assemblages to simulated water quality improvement and degradation in running waters. Canadian Journal of Fisheries and Aquatic Sciences. 2011;**68**:487-497

[210] McCormick PV, Cairns J Jr. Algae as indicators of environmental change. Journal of Applied Phycology. 1994;**6**:509-526

[211] Patrick R. A proposed biological measure of stream conditions based on a survey of the Conestoga Basin, Lancaster County, Pennsylvania. Proceedings of the Academy of Natural Sciences of Philadelphia. 1949;**101**:277-341

[212] Potapova MG, Charles DF. Benthic diatoms in USA rivers: Distributions along spatial and environmental gradients. Journal of Biogeography. 2002;**29**:167-187

[213] Smol JP, Cumming BF. Tracking long-term changes in climate using algal indicators in lake sediments. Journal of Phycology. 2000;**36**:986-1011

[214] Stevenson RJ, Bahls LL. Periphyton protocols. In: Barbour MT, Gerritsen J, Snyder BD, Stribling JB, editors. Rapid Bioassessment Protocols for Use in Streams and Wadeable Rivers: Periphyton, Benthic Macroinvertebrates, and Fish. 2nd ed. Washington, DC, U.S.A: Office of Water US Environmental Protection Agency; 1999. pp. 1-22

[215] Puche E, Rojo C, Ramos-Jiliberto R, Rodrigo MA. Structure and vulnerability of the multi-interaction network in macrophytedominated lakes. Oikos. 2019;**129**:35-48

[216] Bartleson RD, Kemp WM, Stevenson JC. Use of a simulation model to examine effects of nutrient loading and grazing on *Potamogeton perfoliatus* L. communities in microcosms. Ecological Modelling. 2005;**185**:483-512

[217] Dong XY, Li B, He FZ, Gu Y, Sun MQ, Zhang HM, et al. Flow directionality, mountain barriers and functional traits determine diatom metacommunity structuring of high mountain streams. Scientific Reports. 2016;**6**:24711

[218] Lange K, Townsend CR, Matthaei CD. A trait-based framework for stream algal communities. Ecology and Evolution. 2016;**6**:23-36

[219] Lowe RL, Pan Y. Benthic algal communities as biological indicators. In: Stevenson RJ, Bothwell ML, Lowe RL, editors. Algal Ecology. San Diego: Freshwater Benthic Ecosystems. Academic Press; 1996. pp. 705-739

[220] Van Dam H, Mertens A, Sinkeldam J. A coded checklist and ecological indicator values of freshwater diatoms from the Netherlands.

Netherland Journal of Aquatic Ecology. 1994;**28**:117-133

[221] Wu NC, Schmalz B, Fohrer N. Development and testing of a phytoplankton index of biotic integrity (P-IBI) for a German Lowland River. Ecological Indicators. 2012;**13**:158-167

[222] Porter SD. Algal Attributes: An Autecological Classification of Algal Taxa Collected by the National Water-Quality Assessment Program (Data Series 329). Reston, VA: US Geological Survey; 2008. Available from: http://pubs.usgs.gov/ds/ds329/

[223] Environmental Protection Authority Victoria. Rapid Bioassessment Methodology for Rivers and Streams. Melbourne: EPA; 2003

[224] Bothwell ML. Phosphorus-limited growth dynamics of lotic periphytic diatom communities: Areal biomass and cellular growth rate responses. Canadian Journal of Fisheries and Aquatic Sciences. 1989;**46**:1293-1301

[225] Breuer F, Janz P, Farrelly E, Ebke K-p. Environmental and structural factors influencing algal communities in small streams and ditches in Central Germany. Journal of Freshwater Ecology. 2017;**32**(1):65-83. DOI: 10.1080/02705060.2016.1241954

[226] McCormick PV, Stevenson RJ. Periphyton as a tool for ecological assessment and management in the Florida Everglades. Journal of Phycology. 1998;**34**:726-733

[227] Zhang J, Shu X, Zhang Y, Tan X, Zhang Q. The responses of epilithic algal community structure and function to light and nutrients and their linkages in subtropical rivers. Hydrobiologia. 2020;**847**(3):841-855. DOI: 10.1007/s10750-019-04146-4

[228] Biggs BJ, Hickey CW. Periphyton responses to a hydraulic gradient in a regulated river in New Zealand. Freshwater Biology. 1994;**32**:49-59

[229] Osorio V, Proia L, Ricart M, Pérez S, Ginebreda A, Luís J, et al. Hydrological variation modulates pharmaceutical levels and bio fi lm responses in a Mediterranean River. The Science of the Total Environment. 2014;**472**(2014):1052-1061. DOI: 10.1016/j.scitotenv.2013.11.069

[230] Ponsatí L, Acuña V, Aristi I, Arroita M, Schiller D, Elosegi A, et al. Biofilm responses to flow regulation by dams in Mediterranean Rivers. River Research and Applications. 2015;**31**(1003-1016):1003-1016. DOI: 10.1002/rra

[231] Ponsati L, Corcoll N, Petrovic M, Pico Y, Ginrebreda A, Tornes E, et al. Multiple-stressor effects on river biofilms under different hydrological conditions. Freshwater Biology. 2016;**2016**(61):2102-2115. DOI: 10.1111/fwb.12764

[232] Tsai Y-p. Impact of flow velocity on the dynamic behaviour of biofilm bacteria. Biofouling. 2005;**21**(5/6):267-277. DOI: 10.1080/08927010500398633

[233] Poikane S, Kelly M, Cantonati M. Benthic algal assessment of ecological status in European lakes and rivers : Challenges and opportunities. The Science of the Total Environment. 2016;**568**:603-613. DOI: 10.1016/j.scitotenv.2016.02.027

[234] Schneider SC, Kahlert M, Kelly MG. Interactions between PH and nutrients on benthic algae in streams and consequences for ecological status assessment and species richness patterns. The Science of the Total Environment. 2013;**444**:73-84. DOI: 10.1016/j.scitotenv.2012.11.034

[235] Stevenson RJ, Peterson CG. Variation in benthic diatom (Bacillariophyceae) immigration with habitat characteristics and cell morphology. Journal of Phycology. 1989;**25**:120-129

[236] Horner RR, Welch EB, Seeley MR, Jacoby JM. Responses of periphyton to changes in current velocity, suspended sediment and phosphorus concentration. Freshwater Biology. 1990;**24**:215-232

[237] Poff NL, Voelz NJ, Ward JV. Algal colonization under four experimentally-controlled current regimes in a high mountain stream. Journal of the North American Benthological Society. 1990;**9**:303-318

[238] Biggs BJF, Close ME. Periphyton biomass dynamics in gravel bed Rivers: The relative effects of flow and nutrients. Freshwater Biology. 1989;**22**:209-231

[239] Grimm NB, Fisher SG. Stability of periphyton and macroinvertebrates to disturbance by flash floods in a desert stream. Journal of the North American Benthological Society. 1989;**8**:293-307

[240] Peterson CG, Stevenson RJ. Post-spate development of epilithic algal communities in different current environments. Canadian Journal of Botany. 1990;**68**:2092-2102

[241] Steinman AD, McIntire CD. Effects of current velocity and light energy on the structure of periphyton assemblages in laboratory streams. Journal of Phycology. 1986;**22**:352-361

[242] Tas B. Diversity of phytoplankton and trophic status in the Gaga Lake, Turkey. Energy Education Science and Technology Part A: Energy Science and Research. 2012;**30**(1):33-44

[243] Round FE. Biology of the Algae. London, UK: Arnold; 1970

[244] Stevenson RJ, Bothwell ML, Lowe RL, Thorp JH. Algal Ecology: Freshwater Benthic Ecosystem. Michigan, USA: Academic Press; 1996

[245] Allan JD, Castillo MM. Stream Ecology: Structure and Function of Running Waters. New York, USA: Springer Science & Business Media; 2007

[246] Rober AR, Wyatt KH, Stevenson RJ. Regulation of algal structure and function by nutrients and grazing in a boreal wetland. Regulation. 2011;**30**:787-796

[247] Law RJ. A review of the function and uses of, and factors affecting, stream phytobenthos. Freshwater Reviews. 2011;**4**:135-166

[248] Shen L, Dou M, Xia R, Li G, Yang B. Effects of hydrological change on the risk of riverine algal blooms : Case study in the mid-downstream of the Han River in China. Environmental Science and Pollution Research. 2021;**28**:19851-19865

[249] Xia R, Zhang Y, Wang G, Zhang Y, Dou M. Multi-factor identifi cation and modelling analyses for managing large river algal blooms. Environmental Pollution. 2019;**254**:113056. DOI: 10.1016/j.envpol.2019.113056

[250] Long T, Funk A, Pletterbauer F, et al. Management challenges related to long-term ecological impacts, complex stressor interactions, and different assessment approaches in the Danube River Basin. River Research and Applications. 2019;**35**:500-509. DOI: 10.1002/rra.3243

[251] Steinman AD. Effects of grazers on freshwater benthic algae. In: Stevenson RJ, Bothwell ML, Lowe RL, editors. Algal Ecology; Freshwater Benthic Ecosystems. San Diego, CA: Academic Press; 1996. pp. 341-373

[252] Richardson K, Beardall J, Raven JA. Adaptation of unicellular algae to irradiance: An analysis of strategies. New Phytologist. 1983;**93**(2):157-191. DOI: 10.1111/j.1469-8137.1983.tb03422.x

[253] Dean CA, Selckmann GM. Ecology and Management of Filamentous Green Algae. Rockville, MD, U.S.A: Interstate Commission on the Potomac River Basin (ICPRB); 2015

[254] Gökçe D. Algae as an indicator of water quality. In: Thajuddin N, Dhanasekaran D, editors. Algae - Organisms for Imminent Biotechnology. London: IntechOpen; 2016. DOI: 10.5772/62916

[255] Berkman JAH, Canova MG. Chapter 7.4, Algai Biomass Indicators: U.S. Geological Survey Techniques of Water-Resources Investigations, Book 9. Colorado: U.S. Geological Survey; 2007. DOI: 10.3133/twri09A7.4

[256] Paul MJ, Walsh B, Oliver J. Algal Indicators in Streams: A Review of their Application in Water Quality Management of Nutrient Pollution. EPA 822-B-17-002. Ohio, USA: Environmental Protection Agency; 2017. Available from: https://www.epa.gov/nutrient-policy-data/algal-indicators-streams-review-their-application-water-quality-management

[257] Vis C, Hudon C, Cattaneo A, Pinel-Alloul B. Periphyton as an indicator of water quality in the St Lawrence River (Quebec, Canada). Environmental Pollution. 1998;**101**(1):13-24. DOI: 10.1016/S0269-7491(98)00042-6

[258] Raschke RL, Schultz DA. The use of the algal growth potential test for data assessment. Water Pollution Control Federation. 1987;**59**:222-227

[259] Hodson R, De Silva N. Assessing the State of Periphyton in Southland Streams and Rivers. Invercargill, New Zealand: Environment Southland Division; 2018

[260] Lund EM, Drake DC. Species-specific wet-dry mass calibrations for common submersed macrophytes in the upper Mississippi River. Aquatic Botany. 2021;**169**(February 2021):103344

[261] Maslukah L, Zainuri M, Wirasatriya A, Salma U. Spatial distribution of chlorophyll-a and its relationship with dissolved inorganic phosphate influenced by rivers in the north coast of Java. Journal of Ecological Engineering. 2019;**20**(7):18-25. DOI: 10.12911/22998993/108700

[262] Pinder LCV, Marker AFH, Pinder AC, Ingram JKG, Leach DV, Collett GD. Concentrations of suspended chlorophyll a in the Humber Rivers. Science of the Total Environment. 1997;**194-195**:373-378. DOI: 10.1016/S0048-9697(96)05376-4

[263] Turner RE, Milan CS, Swenson EM, Lee JM. Peak chlorophyll a concentrations in the lower Mississippi River from 1997 to 2018. Limnology and Oceanography. 2022;**67**(3):703-712. DOI: 10.1002/lno.12030

[264] Gyeom H, Hong S, Jeong K-s, Kim D-k, Joo G-j. Determination of sensitive variables regardless of hydrological alteration in artificial neural network model of chlorophyll a : Case study of Nakdong River. Ecological Modelling. 2019;**398**(March):67-76. DOI: 10.1016/j.ecolmodel.2019.02.003

[265] Biggs BJ. Hydraulic habitat of plants in streams. Regulated Rivers: Research & Management. 1996b;**12**(2-3):131-144

[266] Biggs BJ, Smith RA, Duncan MJ. Velocity and sediment disturbance of periphyton in headwater streams: Biomass and metabolism. Journal of the

North American Benthological Society. 1999;**18**:222-241

[267] Saul BC, Hudgens MG, Mallin MA. Downstream effects of upstream causes. Journal of the American Statistical Association. 2019;**114**(528):1493-1504. DOI: 10.1080/01621459.2019.1574226

[268] Leland HV. The influence of water depth and flow regime on phytoplankton biomass and community structure in a shallow, lowland river. Hydrobiologia. 2003;**506**:247-255

[269] Riseng C, Wiley M, Stevenson R. Hydrologic disturbance and nutrient effects on benthic community structure in midwestern US streams: A covariance structure analysis. Journal of the North American Benthological Society. 2004;**23**:309-326

[270] Taylor SL, Roberts SC, Walsh CJ, Hatt BE. Catchment urbanisation and increased benthic algal biomass in streams: Linking mechanisms to management. Freshwater Biology. 2004;**49**:835-851

[271] Mosisch TD, Bunn SE. Temporal patterns of rainforest stream Epilithic algae in relation to flow-related disturbance. Aquatic Botany. 1997;**58**(2):181-193. DOI: 10.1016/S0304-3770(97)00001-6

[272] Campbell RE, Winterbourn MJ, Cochrane TA, McIntosh AR. Flow-related disturbance creates a gradient of metacommunity types within stream networks. Landscape Ecology. 2015;**30**(4):667-680. DOI: 10.1007/s10980-015-0164-x

[273] Stevenson RJ, Rier ST, Riseng CM, Schultz RE, Wiley MJ. Comparing effects of nutrients on algal biomass in streams in two regions with different disturbance regimes and with applications for developing nutrient criteria. Hydrobiologia. 2006;**561**(1):149-165. DOI: 10.1007/s10750-005-1611-5

[274] Juneja A, Ceballos RM, Murthy GS. Effects of environmental factors and nutrient availability on the biochemical composition of algae for biofuels production: A review. Energies. 2013;**6**(9):4607-4638. DOI: 10.3390/en6094607

[275] Biggs BJ, Lowe RL. Responses of two trophic levels to patch enrichment along a New Zealand stream continuum. New Zealand Journal of Marine and Freshwater Research. 1994;**28**:119-134

[276] Rosemond AD, Mulholland PJ, Elwood JW. Top-down and bottom-up control of stream periphyton: Effects of nutrients and herbivores. Ecology. 1993;**74**:1264-1280

[277] McCormick PV. Resource competition and species coexistence in freshwater benthic algal assemblages. In: Stevenson RJ, Bothwell ML, Lowe RL, editors. Algal Ecology, Freshwater Benthic Ecosystems. London: Academic Press; 1996. pp. 229-252

[278] Iannino A, Fink P, Weitere M. Feedback between bottom-up and top-down control of stream biofilm mediated through eutrophication effects on grazer growth. Scientific Reports. 2021;**11**(1):1-10. DOI: 10.1038/s41598-021-00856-9

[279] Biggs BJ, Smith RA. Taxonomic richness of stream benthic algae: Effects of flood disturbance and nutrients. Limnology and Oceanography. 2002;**47**:1175-1186

[280] Cullis JD, Crimaldi JP, McKnight DM. Hydrodynamic shear removal of the nuisance stalk-forming diatom Didymosphenia geminata. Limnology and Oceanography: Fluids and Environments. 2013;**3**:256-268

[281] Ledger ME, Harris RML, Armitage PD, Milner AM. Disturbance frequency influences patch dynamics in stream benthic algal communities. Oecologia. 2008;**155**(4):809-819. DOI: 10.1007/s00442-007-0950-5

[282] Chester ET, Robson BJ. Do recolonisation processes in intermittent streams have sustained effects on benthic algal density and assemblage composition? Marine and Freshwater Research. 2014;**65**:784-790

[283] Lukács Á, Kókai Z, Török P, Bácsi I, Borics G, Várbíró G, et al. Colonisation processes in benthic algal communities are well reflected by functional groups. Hydrobiologia. 2018;**823**(1):231-245. DOI: 10.1007/s10750-018-3711-z

[284] Robson BJ. Role of residual biofilm in the recolonization of rocky intermittent streams by benthic algae. Marine and Freshwater Research. 2000;**51**:725-732

[285] Robson BJ, Matthews TG, Lind PR, Thomas NA. Pathways for algal recolonization in seasonally-flowing streams. Freshwater Biology. 2008;**53**:2385-2401

[286] Bunn SE, Arthington AH. Basic principles and ecological consequences of altered flow regimes for aquatic biodiversity. Environmental Management. 2002;**30**:492-507

[287] Goshtasbi H, Atazadeh E, Fathi M, Movafeghi A. Using physicochemical and biological parameters for the evaluation of water quality and environmental conditions in international wetlands on the southern part of Lake Urmia, Iran. Environmental Science and Pollution Research. 2022;**29**(13):18805-18819

[288] Woodward G, Perkins DM, Brown LE. Climate change and freshwater ecosystems: Impacts across multiple levels of organization. Philosophical Transactions of the Royal Society B: Biological Sciences. 2010;**365**(1549):2093-2106. DOI: 10.1098/rstb.2010.0055

[289] Baron JS, Poff NL, Angermeier PL, Dahm CN, Gleick PH, Hairston NG, et al. Meeting ecological and societal needs for freshwater. Ecological Applications. 2002;**12**:1247-1260

[290] De Wit E, van Doremalen N, Falzarano D, Munster VJ. SARS and MERS: Recent insights into emerging coronaviruses. Nature Reviews Microbiology. 2016;**14**(8):523-534. DOI: 10.1038/nrmicro.2016.81

[291] Viney NR, Bates BC, Charles SP, Webster IT, Bormans M. Modelling adaptive management strategies for coping with the impacts of climate variability and change on riverine algal blooms. Global Change Biology. 2007;**13**:2453-2465. DOI: 10.1111/j.1365-2486.2007.01443.x

[292] Quiel K, Becker A, Kirchesch V, Scho A. Influence of global change on phytoplankton and nutrient cycling in the Elbe River. Regional Environmental Change. 2011;**11**:**405**-**421**. DOI: 10.1007/s10113-010-0152-2

[293] Zhou S, Naicheng W, Zhang M, Peng W, He F, Guo K. Local environmental, geo-climatic and spatial factors interact to drive community distributions and diversity patterns of stream benthic algae, macroinvertebrates and fishes in a large basin, Northeast China. Ecological Indicators. 2020;**117**(June):106673. DOI: 10.1016/j.ecolind.2020.106673

[294] Johnson BT, Hennessy EA. Systematic reviews and meta-analyses in the health sciences: Best practice

methods for research syntheses. Social Science & Medicine. 2019;**233**:237-251. DOI: 10.1016/j.socscimed.2019.05.035

[295] Sinha E, Michalak A. Precipitation dominates interannual variability of riverine nitrogen loading across the continental United States. Environmental Science & Technology. 2016;**50**(23):12874-12884. DOI: 10.1021/acs.est.6b04455

[296] Pourfallah K, Hassan RM, Khaledian MR. Impact of precipitation and flow rate changes on the water quality of a Coastal River. Shock and Vibration. 2021;**2021**:13. DOI: 10.1155/2021/6557689

[297] Shang H, Menenti M, Jia L. Modelling stream flow with a discrete rainfall and 37 GHz PDBT microwave observations: The Xiangjiang River basin case study. Hydrology and Earth System Sciences Discussions. 2016;**no. June**:1-34. DOI: 10.5194/hess-2016-129

[298] Environmental Protection Agency. Climate Change Indicators: Heavy Precipitation. 2021. Available from: https://www.epa.gov/climate-indicators/climate-change-indicators-heavy-precipitation

[299] Environmental Protection Agency (EPA). Climate Change Indicators in the United States: Streamflow. Washigton, U.S.A: Environmental Protection Agency (EPA); 2016

[300] Panagoulia D. Catchment hydrological responses to climate changes calculated from incomplete climatological data. IAHS Publication. 1993;**212**:461-468

[301] Panagoulia D. Assessment of daily catchment precipitation in mountainous regions for climate change interpretation. Hydrological Sciences Journal. 1995;**40**(3):331-350

[302] Panagoulia D, Bardossy A, Lourmas G. Multivariate stochastic downscaling models for generating precipitation and temperature scenarios of climate change based on atmospheric circulation. Global NEST Journal. 2008;**10**(2):263-272

[303] Panagoulia D, Tsekouras GJ, Kousiouris G. A multi-stage methodology for selecting input variables in ANN forecasting of river flows. Global NEST Journal. 2017;**19**:49-57

[304] Ray RL, Beighley RE, Yoon Y. Integrating runoff generation and flow routing in Susquehanna River basin to characterize key hydrologic processes contributing to maximum annual flood events. Journal of Hydrologic Engineering. 2016;**21**(9):117-138

[305] Mann H, Fyfe WS. Biogeochemical cycling of the elements in some fresh water algae from gold and uranium mining districts. Biorecovery. 1988;**1**:3-26

[306] Bold CH, Wynne MJ. Introduction to the Algae: Structure and Reproduction. Michigan: Prentice Hall; 1984

[307] Shetty P, Gitau MM, Maróti G. Salinity stress responses and adaptation mechanisms in eukaryotic green microalgae. Cell. 2019;**8**(12):1657. DOI: 10.3390/cells8121657

[308] Kelly MG, Adams C, Graves AC, Jamieson J, Krokowski J, Lycett EB, et al. The Trophic Diatom Index: A user's Manual. Environment Agency: Bristol; 2001. p. 135

[309] Kelly MG, Cazaubon A, Coring E, Dell'Uomo A, Ector L, Goldsmith B, et al. Recommendations for the routine sampling of diatoms for water quality assessments in Europe. Journal of Applied Phycology. 1998;**10**:215-224

[310] Lowe RL, Laliberte GD. Benthic Stream Algae: Distribution and Structure. San Diego, CA: Academic Press; 1996. pp. 269-293

[311] Huh MK. River ecosystem and floristic characterization of riparian zones at the Youngjeong River, Sacheon-Ci, Korea. Journal of Life Science. 2017;**27**(3):301-309. DOI: 10.5352/jls.2017.27.3.301

[312] Lebkuecher JG, Tuttle EN, Johnson JL, Willis NK. Use of algae to assess the trophic state of a stream in Middle Tennessee. Journal of Freshwater Ecology. 2015;**30**:349-376

[313] Perona E, Bonilla I, Mateo P. Epilithic cyanobacterial communities and water quality: An alternative tool for monitoring eutrophication in the Alberche River (Spain). Journal of Applied Phycology. 1998;**10**:183-191

[314] Potapova M, Charles DF. Diatom metrics for monitoring eutrophication in rivers of the United States. Ecological Indicators. 2007;**7**:48-70

[315] Whaeeb ST, Alsadoon Z, Fadhil AH. Echinococcosis-review. International Journal of Advanced Research in Biological Sciences. 2021;**8**(9):1-3. DOI: 10.22192/ijarbs.2021.08.9.001

[316] Kelly MG. Use of the trophic diatom index to monitor eutrophication in rivers. Water Research. 1998;**32**:236-242

[317] Greeson, Phillip E. Annotated Key To the Identification of Commonly Occurring and Dominant Genera of Algae Observed in the Phytoplankton of the United States. US Geological Survey Water Supply Paper 279. 1982

[318] Codd GA. Cyanobacterial toxins, the perception of water quality, and the prioritisation of eutrophication control. Ecological Engineering. 2000;**16**:51-60

[319] Ferreira JG, Andersen JH, Borja A, Bricker SB, Camp J, Da Silva MC, et al. Overview of eutrophication indicators to assess environmental status within the European Marine Strategy Framework Directive. Estuarine, Coastal and Shelf Science. 2011a;**93**:117-131

[320] Vasconcelos V. Eutrophicatton, toxic cyanobacteria and cyanotoxins: When ecosystems cry for help. Limnetica. 2006;**25**(1-2):425-432. DOI: 10.23818/limn.25.30

[321] Gordon R, Losic D, Tiffany MA, Nagy SS, Sterrenburg FAS. The glass menagerie: Diatoms for novel applications in nanotechnology. Trends in Biotechnology. 2009;**27**:116-127

[322] Blimm WD, Herbst DB. Use of Diatoms and Soft Algae as Indicators of Environmental Determinants in the Lahontan Basin, USA. Sacramento, CA, U.S.A.: CALIFORNIA STATE WATER RESOURCES BOARD; 2003

[323] Fore LS, Grafe C. Using diatoms to assess the biological condition of large rivers in Idaho (USA). Freshwater Biology. 2002;**47**:2015-2037

[324] Kelly MG, Juggins S, Bennion H, Burgess A, Yallop M, Hirst H, et al. Use of Diatoms for Evaluating Ecological Status in UK Freshwaters. Bristol, U.K.: Environment Agency; 2008

[325] Kwandrans J, Eloranta P, Kawecka B, Wojtan K. Use of benthic diatom communities to evaluate water quality in rivers of southern Poland. Journal of Applied Phycology. 1998;**10**:193-201

[326] Patrick R. Use of algae, especially diatoms, in the assessment of water quality. In: Biological Methods for the Assessment of Water Quality. Virginia, USA: ASTM International; 1973

[327] Tan X, Ma P, Xia X, Zhang Q. Spatial pattern of benthic diatoms and water quality assessment using diatom indices in a subtropical river, China. CLEAN–soil. Air, Water. 2014;**42**:20-28

[328] Masouras A, Karaouzas I, Dimitriou E, Tsirtsis G, Smeti E. Benthic diatoms in river biomonitoring—present and future perspectives within the water framework directive. Water. 2021;**13**:478. DOI: 10.3390/w13040478

[329] Cox EJ, Reid G. River Nar Eutrophication Studies-Diatoms. London, U.K.: The Natural History Museum; 1994

[330] Rimet F. Recent views on river pollution and diatoms. Hydrobiologia. 2012;**683**(1):1-24. DOI: 10.1007/s10750-011-0949-0

[331] Wu J-T, Kow L-T. Applicability of a generic index for diatom assemblages to monitor pollution in the tropical River Tsanwun, Taiwan. Journal of Applied Phycology. 2002;**14**:63-69

[332] Atazadeh E, Gell P, Mills K, Barton A, Newall P. Community structure and ecological responses to hydrological changes in benthic algal assemblages in a regulated river: Application of algal metrics and multivariate techniques in river management. Environmental Science and Pollution Research. 2021;**28**(29):39805-39825

[333] Çelekli A, Lekesiz Ö, Yavuzatmaca M. Bioassessment of water quality of surface waters using diatom metrics. Turkish Journal of Botany. 2021;**45**(5):379-396. DOI: 10.3906/bot-2101-16

[334] John J. Bioassessment of health of aquatic systems by the use of diatoms. In: Modern Trends in Applied Aquatic Ecology. Netherlands: Springer; 2003. pp. 1-20

[335] John J. A Diatom Prediction Model and Classification for Urban Streams from Perth, Western Australia. Germany: Koeltz Scientific Books; 2012

[336] Newall P, Walsh CJ. Response of epilithic diatom assemblages to urbanization influences. Hydrobiologia. 2005;**532**:53-67

[337] Sonneman JA, Walsh CJ, Breen PF, Sharpe AK. Effects of urbanization on streams of the Melbourne region, Victoria, Australia. II. Benthic Diatom Communities. Freshwater Biology. 2001;**46**:553-565

[338] Szczepocka E, Nowicka-Krawczyk P, Kruk A. Deceptive ecological status of urban streams and Rivers—Evidence from diatom indices. Ecosphere. 2018;**9**(7):e02310. DOI: 10.1002/ecs2.2310

Chapter 10

Pollution Evaluation of Industrial Effluents from Consolidated Breweries: A Case Study from Benue State, Nigeria

Eucheria N. Nweke, Victor U. Okechukwu, Daniel O. Omokpariola, Theresa C. Umeh and Nwanneamaka R. Oze

Abstract

Industrial effluent discharged into surface water is an environmental concern, as it affects the esthetics, water quality as well as microbial and aquatic flora. Brewery effluents were analyzed for physicochemical parameters (pH, temperature, conductivity, turbidity, total dissolved solids (TDS), dissolved oxygen (DO), biological oxygen demand (BOD), chemical oxygen demand (COD), nitrate, and sulfate, chloride) and heavy metals (As, Cd, Co, Cr, Fe, Mn, Ni, Pb, and Zn). Atomic absorption spectrophotometer was used to characterize heavy metals using standard analytical methods and compared with WHO standards. The result showed that pH (6.2–6.98), conductivity (137–273 μS/cm), chloride (31–53 mg/l), nitrate (7.53–10.72 mg/l), BOD, and DO were within the WHO limit. However, turbidity, sulfate, and phosphate were above the WHO limit. Heavy metal concentrations Cr, Ni, Pb, Mn, As, and Cd were higher than the WHO limit and vice versa for Fe, Zn, and Co. Ecological risk assessment revealed that effluent samples pose low to moderate ecological risk, for As, Pb, and Ni. Therefore, there is a need for proper treatment and continual monitoring before discharge into the environment.

Keywords: brewery effluents, heavy metal, ecological risk, contamination factor, pollution load index

1. Introduction

Due to an increase in industrial activities, environmental pollution is one of the most critical problems in developing countries. More challenging is the unsafe disposal of solid wastes/industrial wastes into the ambient environment. Industries that use large amounts of water in their processes include chemical manufacturing, steel plants, metal processors, etc. Effluents and most products from industries create serious pollution to water bodies and soils. Water bodies especially freshwater reservoirs, and rivers are the most affected. This has rendered underground and surface waters unsafe for human, recreational, and agricultural use. Biotic life is destroyed and natural ecosystems are infected. Human life is at risk and the principle of sustainable development is compromised [1].

Moderately or untreated industrial effluents may contain high levels of pollutants which in water body systems cause an increase in BOD, COD, Total Dissolved Solids (TDS), Total Suspended Solids (TSS), toxic metals such as Cd, As, Cr, Ni, and Pb, and fecal coliform. Hence make such water unsuitable for drinking, irrigation, and aquatic life support. Industrial wastewater impacts include high BOD from biodegradable wastes such as those from human sewage, pulp and paper industries, slaughterhouses, tanneries, and chemical industries [2–4].

Brewery wastewater effluent is highly variable in quality and composition. The products of the brewery operations include large volumes of wastewater from liquors pressed from grains and yeast recovery, from the Clean-in-place system located in the brewing house, cellar house, and bottling house, which is discharged into the nearby River. These industrial wastewaters are the main source of heavy metals since nearly all industrial by-products consist of some level of heavy metals [5].

Wastewater shows different degrees of environmental nuisance and contamination hazards due to its chemical and microbiological characteristics. Excessive nutrients (primarily, nitrogen and phosphorus) in wastewater, sludge, and excreta may contaminate surface waters and cause eutrophication, which affects the esthetics of water bodies (lakes, rivers), and results in odor and appearance problems, which was evident in the physiochemical evaluation of brewery effluents in Enugu State [6] and Edo State [7] both in Nigeria.

Previous research has shown that the release of untreated effluents has the potential to negatively impact aquatic organisms, by decreasing pH to acid level, increasing conductivity, temperature turbidity, and total solids in such an environment leading to a decrease in dissolved oxygen with microbial bloom from rich nutrients (nitro-groups, sulfur-groups, and phosphors) [8–12].

Heavy metals are also released from these effluents. Studies have shown that long-term exposure to low concentrations of some heavy metallic anions can result in the development of sub-chronic to chronic illnesses and diseases in a given population, usually between 1 in 1000 to 1 in 1,000,000 as institutionalized by the US. Environmental Protection Agency (EPA) [13, 14].

The forms in which metal pollutants exist in wastewater discharges determine their release into the aquatic ecosystem. Some metals become bio-available when soluble airborne solids are dissolved by weak acids such as carbonic acid. Their concentration became enhanced by the abundance of metals in road dust and tire residues [15].

The physicochemical properties and selected heavy metals of industrial effluents from consolidated Breweries in Benue State, Nigeria were studied. Therefore, the study aimed to assess the concentration levels, and ecological and health risks of industrial effluents discharged daily into the nearby River, a primary source of fishing activity and domestic purposes in the neighboring community. Information from the present study will be helpful to the relevant government agencies and policymakers in preparing preventive action to control the direct discharge of effluents from chemical industries, agro-based activities, and domestic waste to the rivers and the sea.

2. Materials and method

2.1 Location/study area

Consolidated breweries plc Makurdi, Benue state is located at kilometer 5 Gboko Road, pm 102,339 Makurdi, Benue State, Nigeria. The locality is predominantly an agrarian community where farming and animal husbandry takes place with a few industrials such as small and medium scale enterprise (mechanic workshop, local markets, construction, and mining) activities. **Figure 1** shows the map of the study location.

2.2 Sample collection and preparation

Four samples of effluent were collected with a cleaned plastic container at a different location in the brewery. The plastic containers used were carefully washed with 1% HNO_3 acid, rinsed with tap water, and then distilled water. The samples were labeled appropriately; Sample A-untreated effluent, B- treated effluent, C- contact point of the treated effluent with the river, and D- 10 kilometers away from the contact point. These samples were transported to the laboratory for analysis in an ice-packs container and protected from direct sunlight. They were stored in the refrigerator at 20°C.

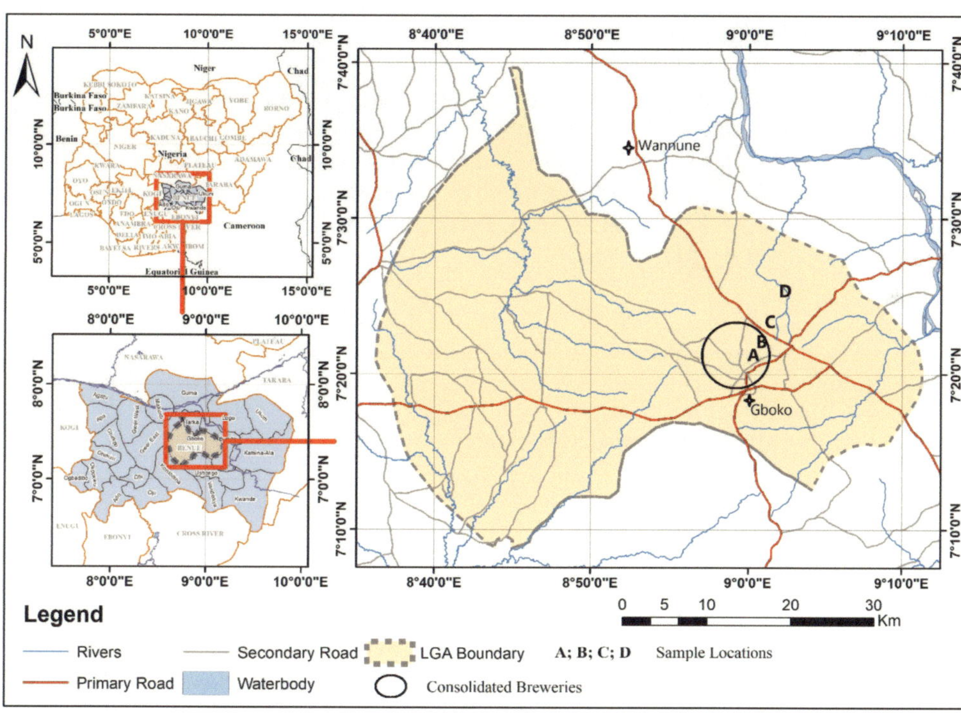

Figure 1.
Study map showing the study locations.

2.3 Physicochemical analysis

2.3.1 pH

The pH values of the samples were determined at the point of sampling using a portable pH meter after calibrating against buffer solution (pH 4.7 and 9.2).

2.3.2 Turbidity

The turbidity of the samples was determined using the turbidity meter (Labtech digital model), and EPA 180 was selected as the measurement mode.

2.3.3 Total dissolved solids (TDS)

TDS was determined by an electrometric method using a TDS meter (Jenway, model 4076).

2.3.4 Dissolved oxygen (DO)

DO is the amount of gaseous oxygen dissolved in the water. It was determined using a dissolved oxygen meter (Model H1 9146, HANNA) as described by AOAC [16].

2.3.5 Biological oxygen demand (BOD)

BOD measures the amount of dissolved oxygen used by microorganisms in the oxidation of organic matter in a water sample. The BOD was determined by collecting the water sample in a sealed bottle, incubating for a standard period in the dark (usually 5 days at 200°C), and determining the residual oxygen in the water at the end of incubation (Model H1 9146, HANNA). BOD was determined using the following formula as described by AOAC [16].

$$BOD = (DO_1 – DO_5)\ mg/l \tag{1}$$

Where:
DO_1 = Sample before incubation.
DO_5 = Sample after 5 days of incubation.

2.3.6 Chemical oxygen demand (COD)

COD is the amount of oxygen consumed under specified conditions of organic and oxidizable inorganic matter in water and wastewater. The COD is determined first by pipetting 10 ml of the water sample into a conical flask and adding 5 ml of 0.025 N potassium dichromate ($K_2Cr_20_7$). 15 ml of sulfuric acid (H_2S0_4) was added to it, and the solution was diluted with 40 ml of distilled water to get a 70 ml solution. Seven drops of phenolphthalein ferrous sulfate indicator were added, and the solution was allowed to cool. The solution was titrated with 0.025 N ferrous ammonium sulfate. A blank solution was also titrated. COD was determined by using the following formula:

$$\frac{(T_1 - T_2) \times N \times 8000 \times C}{Volume\ of\ water\ sample\ used}(\text{mg/l}) \tag{2}$$

Where:
T_1 = Titer value for blank.
T_2 = Titer value for effluent sample.
N = Normality of the ferrous ammonium Sulfate used is 0.025.
C = Chloride correction which is in Milligram per liter of chloride × 0.03.
8000 = Milliequivalent weight of oxygen x 1000 mL/L.

2.3.7 Chloride

Ten ml of the water sample was pipetted into a conical flask with 3 drops of potassium chromate (K_2Cr0_2) and the solution was titrated with 0.1 N silver nitrate ($AgNO_3$). Chloride was determined by using the following formula:

$$\frac{(\text{T}_v\ \text{x}\ 0.003546\ \text{x}\ 105)\text{mg}}{1} \tag{3}$$

Where:
Tv = Titer value of sample.
0.003546 = equivalent weight of chloride

2.3.8 Sulfate

Ten ml of water sample was pipetted into a conical flask plus 5 ml of 2 N HCl and 0.05 N $BaCl_2$. The solution was boiled for 5 minutes and allowed to cool. Two ml of ammonium (NH_4^+) and 5 ml of 0.01 N EDTA were added to the solution and boiled for 5 minutes. Five ml of pH buffer 10 and 3 drops of Eriochrome black) indicators were added. The solution was titrated with 0.01 N $MgCl_2$. Sulfate was determined by using the following formula:

$$\left(\frac{10 - (\text{T}_v \times 0.93) \times 96.01484}{10}\right) \text{mg/l} \tag{4}$$

Where:
Tv = Titrate value of the sample.
96.01484 = molecular weight of sulfate.
0.93 = Constant.
10 = Volume of water sample used.

2.3.9 Analysis of effluent samples for heavy metals concentration

An effluent sample of 200 mL was measured into a 500 mL beaker, and 5 mL of concentrated nitric acid was carefully added. This solution was concentrated to 20 mL by heating in a water bath for a few hours. The concentrated extract was cooled and transferred into a 50 mL standard flask, then made up to mark with distilled water. Heavy metal (Pb, Cd, Zn, As, Cr, Fe, Mn, Co, and Ni) contents of the samples were determined using atomic absorption spectrophotometer (Spectra AA Varian 400 plus) involving direct aspiration of the aqueous solution into air-acetylene flame. A reagent blank was prepared and analyzed. Heavy metal concentrations of a series of standards were determined and a calibration graph was developed. From the graph, the concentrations of heavy metals in the sample were calculated as described by Braid *et al.* [17].

2.4 Ecological risk assessment

2.4.1 Contamination factor (CF)

CF is the extent of pollution of the contaminant of interest, it is expressed:

$$CF = \frac{\text{chemical contaminant of interest}}{\text{background value using WHO standard}} \tag{5}$$

The following terminology was used to describe the contamination factor:

i. $CF < 1$: low contamination factor;

ii. $1 \leq CF < 3$: moderate contamination factor;

iii. $3 \leq CF < 6$: considerable contamination factor;

iv. $CF \geq 6$ _ very high contamination factor.

2.4.2 Degree of contamination (Cdeg)

This is the summation of the contamination factor of all chemical contaminants in the study site. It is calculated as follows:

$$C_{deg} = \sum(CF) = CF_1 + CF_2 + CF_3 + \cdots + CF_n \tag{6}$$

For the description of contamination degree (Cdeg), the following terminologies have been used:

i. Cdeg <8: low degree of contamination;

ii. 8 ≤ Cdeg <16: moderate degree of contamination;

iii. 16 ≤ Cdeg <32: considerable degree of contamination;

iv. Cdeg ≥32: very high degree of contamination.

2.4.3 Modified degree of contamination (mCdeg)

This is the average effect of all chemical contaminants of interest, the advantage of mCdeg is that it quantifies the chemical contaminants into a composite aggregate to derive salient information about the study site using the formula:

$$mC_{deg} = \frac{1}{n}\sum(\mathrm{CF}) \tag{7}$$

Where: n is the total chemical contaminant and CF is the contamination factor. mCdeg is classified as:

i. mCdeg <4: low moderate contamination;

ii. 4 ≤ mCdeg <16: medium moderate contamination;

iii. 12 ≤ mCdeg <20: high moderate contamination;

iv. mCdeg ≥20: extreme moderate contamination.

2.4.4 Pollution load index (PLI)

This is the geometric mean of CF value to the nth number of chemical contaminants of interest, it is given as described by Tomlinson *et al.* [18]:

$$\mathrm{PLI} = (\mathrm{CF}_1 \times \mathrm{CF}_2 \times \mathrm{CF}_3 \times \cdots \times \mathrm{CF}_n)^{1/n} \tag{8}$$

Where:
n is the total chemical contaminant.
CF is the contamination factor.
The PLI gives the level of pollution classified as:

i. PLI < 1: no pollution;

ii. 1 < PLI < 2: modest pollution;

iii. 2 < PLI < 3: high pollution;

iv. 3 < PLI: extremely high pollution.

2.4.5 Potential ecological risk index (PERI)

Assesses the toxicity factor of a particular chemical contaminant of interest, where the definite contamination status is evaluated concerning the ecosystem. It is expressed as shown in Eq. (9) and (10). A methodology to assess ecological risks for aquatic pollution was developed by Hakanson [19]. The ecological risk index (R_I) is calculated as a sum of eight elements of heavy metals (As, Cd, Cr, Ni, Mn, Pb, Co, and Zn).

$$\mathrm{PERI} = \sum \mathrm{E_r} = \mathrm{TF} \times \mathrm{CF} \tag{9}$$

$$\mathrm{R_i} = \sum \mathrm{E_r} \tag{10}$$

Where:
Er is the ecological risk index of different chemical contaminants,

TF is the toxicity factor of each chemical contaminant of interest as described by Hussain et al. [20] and Umeh et al. [21].

CF is the contamination factor in Eq. (5),

RI is the risk index calculated as the sum of the potential ecological risk factors for heavy metals in the wastewater.

Er and RI values are categorized using:

i. Er <40 and RI <150: low ecological risk;

ii. 40 < Er ≤80 and 150 < RI <300: moderate ecological risk;

iii. 80 < Er ≤160, appreciable ecological risk;

iv. 160 < Er ≤320 and 300 < RI <600: high ecological risk;

v. Er > 320 and RI ≥ 600: extremely high ecological risk.

3. Results

The values of the physical and chemical parameters of effluent samples from the Brewery are presented in **Table 1** and **Figure 2** respectively. The pH values were 6.98, 6.20, 6.48, and 6.72 for sampling points A, C, D, and E respectively. In all sampling locations, the highest pH value of 6.98 was obtained at sampling point A and the lowest value of 6.20 was obtained at sampling point C.

The TDS values in the present work ranged from 4.0 to 48.0 mg/l. Dissolved oxygen values ranged from 4.70 to 23.60 mg/l for all sampling points. The highest (15.90) and lowest (0.40) biological oxygen demand values were recorded at sampling points C and A, respectively. The chemical oxygen demand in the present work ranged from 18.88 to 19.24.32 mg/l. The values of nitrate varied between 1.28–1.95 mg L-1. The highest (7.16) phosphate value was recorded at sampling point C and the lowest (6.93) at sampling point A. Sulfate values ranged from 31.69 to 35.39 mg/l for all sampling points. Chloride values ranged from 28.00 to 53.00 mg/l for all sampling points.

Figure 3 shows the results of heavy metals concentration analyses of effluent samples across the study area. Among the 9 elements studied, concentrations of As, Pb, and Ni were higher than WHO recommended limits. In contrast, lower concentrations of Co, Mn, and Zn were observed in the different sampling locations. Elements displayed wide variations in their distribution, suggesting control of

Parameters	Sample A	Sample B	Sample C	Sample D	WHO
pH	6.20	6.98	6.48	6.72	6.50–8.50
Turbidity (NTU)	102.70	115.00	39.90	30.70	50. 00
TDS (mg/l)	48.00	4.00	16.00	10.00	500.00
Conductivity (μS/cm)	272.00	273.00	178.00	187.00	1000.00

**A-un treated effluent, B- treated effluent, C- contact point of the treated effluent with the river and D- 10 kilometers away from the contact point.*

Table 1.
Physical parameters of the studied effluents.

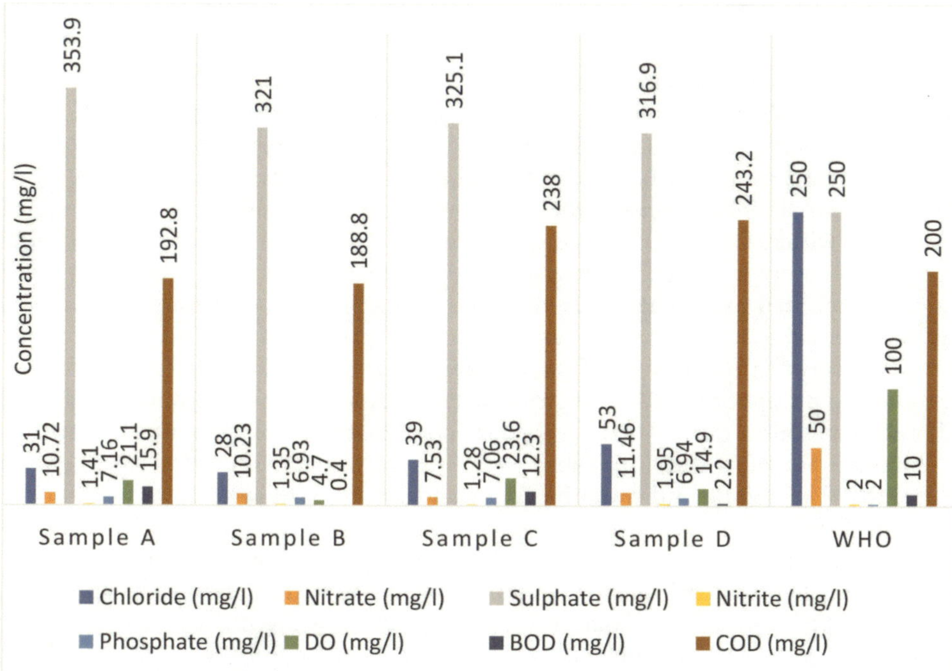

Figure 2.
Chemical parameters from the studied effluents. DO: Dissolved oxygen; BOD: Biological oxygen demand; COD: Chemical oxygen demand.

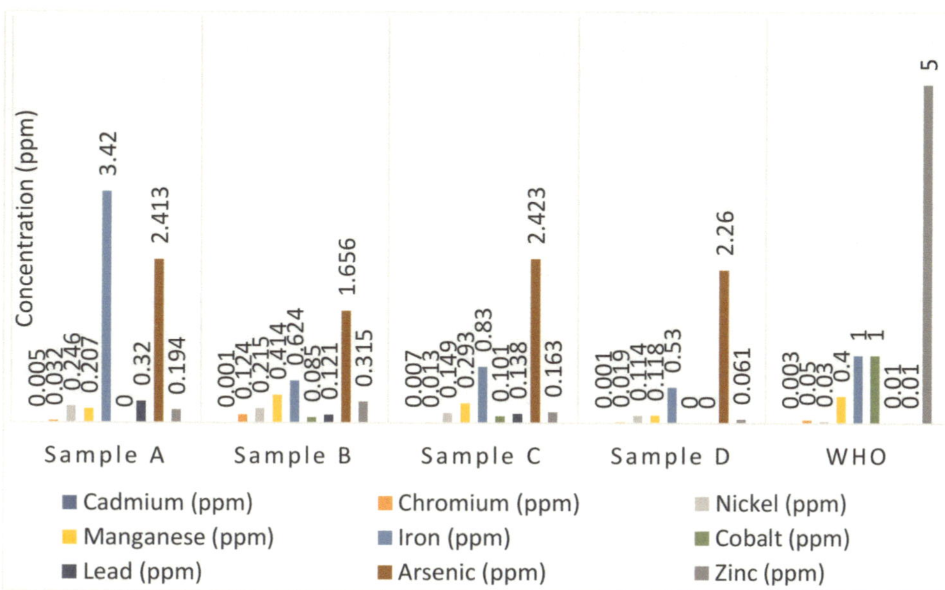

Figure 3.
Heavy metal concentrations.

anthropogenic activities on water chemistry. Overall, concentration of the study elements followed the order: As >Fe > Mn > Zn > Ni > Pb > Cr > Co > Cd. The sampling site C recorded the highest concentration of Fe and As.

4. Discussion

Water quality in an aquatic environment is very important for the survival of its flora and fauna. Water pH can affect aquatic organisms as their metabolic activities are pH-dependent [22, 23]. The pH across the sampling points ranges from 6.20 in sample A to 6.98 in sample B, indicating slight acidity. A significant ($P \leq 0.05$) difference was observed between the pH values for each sampling point, although, B and D were within the WHO [24] guideline regulatory limit of 6.5–8.5 set for drinking water, while samples A and C were a little below the standard limit. The slight acidity could be attributed to the chemicals used in the treatment processes and the water may serve as a sink for various wastes and chemical preservatives used in the brewery such as oxides of sulfur, nitro, carbon, and phosphor in turn form sulfuric, nitric, carbonic and phosphoric acid on reaction with water leading to microbial bloom from rich nutrients source thereby causing reduction in dissolved oxygen, increase turbidity, conductivity, odor and diminish aquatic esthetic respectively. Water pH helps to control metal solubility, and water hardness and serves as an indicator of water pollution [7, 9].

Nitrate in the present study was all below-recommended limit when compared to the WHO [24] standard for safe drinking water. Nitrate is alleged to be an indicator of pollution in the public water supply [25]. It is the stable form of nitrogen that plays a significant role in the process of eutrophication. The conductivity range of the various sampling points varied considerably across the study area. Point B showed the highest value and, therefore, decreased along with the sampling points, most likely due to the effect of dilution and removal of soluble salts by biological utilization.

The biological oxygen demand (BOD) and chemical oxygen demand (COD) are useful parameters in water quality analysis. The highest and lowest BOD values were recorded at sampling points A and B, respectively. Biological oxygen demand is the amount of oxygen required by aerobic microorganisms to stabilize the organic material of wastewater at a standardized temperature (20°C) and time of incubation (usually 5 days). It is used to indicate the organic strength of water. When BOD is less than 4 mg/l, water is deemed to be reasonably clean and unpolluted, while a BOD level greater than 10 mg/l indicates pollution [26].

Chemical oxygen demand is a measure of organic contamination in water. It is the amount of dissolved oxygen required to cause chemical oxidation of the organic material in water and is a key indicator of the environmental health of surface water [18]. There was a gradual increase in chemical oxygen demand from point B to point D. Chemical oxygen demand values were below the WHO recommended value of 200 mg/l [24]. High chemical oxygen demand COD values indicate pollution due to oxidizable organic matter [27].

Phosphate concentration was high in all sampling points and greater than the WHO recommended value of 2.0 mg/l. Phosphate is known as a limiting nutrient in the aquatic ecosystem [28]. There is little variation in dissolved oxygen values of effluent samples across the study areas. The dissolved oxygen concentration is a function of temperature, pressure, salinity, and biological activities in the water body. The tropical aquatic ecosystem should have a dissolved oxygen concentration of at least 5 mg/l in other to support diversified biota, including fish [29–32]. The level of 4.70 mg/l for point B was within the WHO, [24] standard value of 5 mg/l necessary for aquatic productivity, while other points were above the standard limit of 5 mg/l.

The highest value of sulfate was observed in point C (35.39 mg/l). This value is far below the permissible limit stipulated by World Health Organization WHO, [24]. The present work was in line with the work of Alao [30], who also reported low sulfate levels in the water receiving brewery effluent in Majawe Ibadan.

The values of chloride and iron obtained from point B, C, and D falls below the WHO permissible limit, while point C was within the 1 mg/l desirable level from WHO. The result of chloride agrees with Imoobe and Koye [33], who reported the value of chloride in Eruvbi Stream to be below the permissible limit stipulated by the World Health Organization [24]. The discharge of industrial effluents into receiving water bodies invariably results in the presence of a high concentration of pollutants in the water and sediments.

The pollutants are present in concentrations that may be toxic to different organisms [34–36]. The concentration of Cadmium across the study area ranged from 0.001 at sampling points B and D to 0.007 at sampling point C. The values recorded were lower than (0.043 mg/l) and (0.072 mg/l) in the water reported by Oguzie and Okhagbuzo [37]. The value of all samples assessed was above the permissible limit of 0.003 ppm set by WHO [24] for drinking water except for sampling points B and D. High concentrations of Cadmium (Cd) have been reported to inhibit the bio-uptake of Phosphorus and Potassium by plants [38].

Specific industries involved in electroplating, pigments production, chemicals, and alloy processing are sources of cadmium to the urban environment. Chromium (Cr) levels in the effluents were relatively low across the different sampling points. The concentrations of chromium in effluents were below the 0.050 mg/l value recommended by the World Health Organization (WHO) [24] in industrial effluents except for sampling point B.

A high concentration of nickel (Ni) was recorded in the effluent samples ranging from 0.114 ppm in point D to 0.246 ppm in point A. The concentrations of nickel in effluents are higher than the <1 mg/l value recommended by the WHO [24] in industrial effluents. Ni has wide applications in the manufacture of batteries, fertilizer, welding products, electroplating, and household appliances and has essential functions in every area of industrial activity [2].

Lead (Pb) and Arsenic (As), a major environmental pollutant is a multi-organ poison that, in addition to well-known toxic effects, depresses immune status and causes damage to the central nervous system, kidney, and reproductive system [39]. The lead (Pb) values were quite low in all the sampling points except in point E where it was not detected. All the points showed a lead value above the maximum acceptable concentration.

4.1 Contamination factor/pollution index

The contamination factor (CF) values were revealed in **Table 2**. Arsenic (As) can be categorized as a very high contamination factor across all the sampling locations. The highest values of CF of As at location C (48.46) and the lowest at location B (33.12), indicating severe anthropogenic contribution to the contamination load of rivers at this site. The CF of Cd (Cadmium) can be categorized as low to moderate. Two locations (B and D) can be categorized as having low CF of Cd, and two locations (A and C) can be categorized as having moderate CF of Cd. Lead (Pb) can be categorized as a very high CF across all the sampling locations. The highest values of CF of Pb at location A (32) and the lowest at location B (12.1). The CF of Ni (Nickel) can be categorized as considerable to very high contamination. Two locations

Sample locations/elements	A	B	C	D
Arsenic (As)	48.26	33.12	48.46	45.20
Cadmium (Cd)	1.67	0.33	2.33	0.33
Cobalt (Co)	0.00	0.085	0.10	0.00
Chromium (Cr)	0.64	2.48	0.26	0.38
Iron (Fe)	3.42	0.624	0.83	0.53
Manganese (Mn)	0.51	1.035	0.73	0.30
Nickel (Ni)	8.20	7.17	4.96	3.80
Lead (Pb)	32.00	12.1	13.8	0.00
Zinc (Zn)	0.063	0.038	0.033	0.012
PLI	2.10	1.25	1.16	0.70
C.deg	94.7	25.14	71.50	50.55
M-C.deg	40.17	21.05	29.56	5.616

**A-un treated effluent, B- treated effluent, C- contact point of the treated effluent with the river and D- 10 kilometers away from the contact point. PLI: pollution load index; C.deg.: degree of contamination; M-C.deg.: modified degree of contamination.*

Table 2.
Contamination factor, pollution index and contamination index.

(C and D) can be categorized as having considerable CF of Ni, and two locations (A and B) can be categorized as having very high CF of Ni. The CF of cobalt and Zinc can be categorized as low contamination factors of Co and Zn, respectively. Other elements such as Cr (0.26–2.48), Mn (0.29–1.035), and Fe (0.62–3.42) can be categorized as low to moderate CF. The result indicates that contamination of effluents from Nigeria Brewery contributed to As and Pb [21].

The pollution Load Index (PLI) is a resourceful tool to measure and compare contamination. Analyzed effluents samples discharged into rivers at locations A (2.1), B (1.25), and C (1.16) displayed higher PLI values (PLI > 1) and progressive deterioration in quality. Location D was observed to have a low pollution index value of 0.7. The order for PLI was A > B > C > D. Higher PLI values in rivers demonstrated substantial anthropogenic impacts on the river quality signifying the need for immediate intervention to prevent pollution. In contrast, lower PLI values pointed to no considerable anthropogenic activities, signifying no need for intervention but requiring constant monitoring [31].

Degrees of contamination (Cdeg) values of effluents from Nigeria's brewery are revealed in **Table 2**. The degree of contamination across the sampling locations can be categorized into four categories according to the Patil *et al.* [28] classification. Sampling locations A, C, and D can be categorized as having a very high degree of contamination (Cd value = 94.7, 71.50, and 50.54, respectively), this indicates very severe anthropogenic pollution at these sampling sites. Location B indicates a considerable degree of contamination with a Cd value of 25.137. The present study revealed Pb and Ni as the most severe component causing moderate to very high river contamination. A similar pattern was noted for contamination degree (Cdeg), where sampling locations having dominant anthropogenic activities displayed a high Contamination degree. Regular monitoring of the river for the presence of trace elements, especially Arsenic, lead, and nickel, is required [34].

Sample Locations/Elements	A	B	C	D	Mean
Arsenic (As)	482.6	331.2	484.6	452	437.6
Cadmium (Cd)	50.1	9.9	69.9	9.9	34.9
Cobalt (Co)	0	0.425	0.5	0	0.231
Chromium (Cr)	3.2	12.4	1.3	0.76	4.42
Manganese (Mn)	2.55	5.175	3.65	1.475	3.212
Nickel (Ni)	41	35.83	24.8	19	30.15
Lead (Pb)	160	60.5	69	0	72.37
Zinc (Zn)	0.038	0.063	0.0326	0.012	0.036
Risk index	739.48	455.5	653.78	483.14	582.9
Risk grade	Extremely High	High	Extremely High	High	High

A- un treated effluent, B- treated effluent, C- contact point of the treated effluent with the river and D-10 kilometers away from the contact point.

Table 3.
Potential ecological risk index.

4.2 Potential ecological risk index method

The evaluation results on the potential ecological risk factor (Eir) and the potential ecological risk index (RI) are summarized in **Table 3**. The order of potential ecological risk coefficient (Eir) of heavy metals in discharge effluents was As > Pb > Cd > Ni > Cr > Mn > Co > Zn. The mean potential ecological risk coefficient of Cd, Cr, Mn, Ni, Co, and Zn were all lower than 40, which is low ecological risk. At the same time, the mean potential ecological risk coefficient of Pb and As were greater than 80 and 320, respectively, which indicates moderate to extremely high ecological risk. All the sampling locations were at High to very high-risk levels where the RI values were much greater than 600. However, because most samples are contaminated with As, Pb, and Ni, their impact on the ERI became very obvious and predominant. Therefore, the present study indicates that As, Pb, and Ni were the major heavy metal posing an ecological risk in the study area [21, 31, 35].

5. Conclusion

The laboratory analysis results of the effluent samples indicated that metals and other contaminants from the effluents have compromised the River quality. The results of the physicochemical analysis showed that sulfate, phosphate, COD, and heavy metals such as Cr, Ni, Pb, Mn, As, and Cd were slightly higher than WHO and FEPA standards for drinking water, while the pH, Chloride, Nitrate, Nitrite, TDS, Conductivity, BOD, COD and some heavy metals such as Fe, Zn, Co were within the standard of WHO and FEPA, set for drinking water. The study, however, showed that some contaminants sampled were within statutory limits. It was also observed that sample A (Untreated effluent) and sample B (Treated effluent) had lower mean differences than sample C and sample D. Contamination factors follow a similar trend in metal contamination. At the same time, PLI index models confirmed that the effluents from the different sampled locations were polluted, except for location D, which is unpolluted. The mean anthropogenic input for the sampled effluents for the individual metals followed the order As> Fe > Mn > Zn > Ni > Pb > Zn > Cr > Co > Cd. The ecological risk assessments for the heavy metals were at high ecological risk. Furthermore, the potential ecological index depicts As at extremely high risk, Pb at appreciable risk, Ni at moderate risk, and Cr, Mn, Co, Zn, and Cd at a low ecological risk level. Hence any significant increase would persuade environmental challenges. However, the present study recommends proper treatment of effluents before discharging to reduce their mean difference from the WHO standard and protect the health of the local population.

Acknowledgements

The authors acknowledge and extend their sincere gratitude to all who assisted in realizing this present work.

Conflict of interest

All the authors declare no conflict of interest regarding this manuscript.

Notes/thanks/other declarations

We are thankful for the invitation to contribute to River Basin Management and other anonymous reviewers for their assistance. This research work was self-funded, as the authors declare no conflict of interest.

Author details

Eucheria N. Nweke[1,2], Victor U. Okechukwu[1], Daniel O. Omokpariola[1*], Theresa C. Umeh[1] and Nwanneamaka R. Oze[3]

1 Department of Pure and Industrial Chemistry, Nnamdi Azikiwe University, Awka, Nigeria

2 Department of Chemistry, College of Education Umunze, Nigeria

3 Department of Chemistry, Federal University of Technology, Owerri, Nigeria

*Address all correspondence to: omeodisemi@gmail.com

References

[1] Fakayode SO. Impact assessment of industrial effluent on water quality of the receiving Alaro River in Ibadan, Nigeria. The Journal of African Medical Association. 2005;**10**:1-13

[2] Onwukeme VI, Okechukwu VU. Leaching matrix of selected heavy metals from soil to groundwater sources in active dumpsites: A case study of southern Nigeria. IOSR Journal of Environmental Science, Toxicology, Food Technology. 2021;**14**(4):1-18

[3] Lin Q, Liu E, Zhang E, Li K, Shen J. Spatial distribution, contamination and ecological risk assessment of heavy metals in surface sediments of Erhai Lake, a large eutrophic plateau lake in Southwest China. Catena. 2016;**145**: 193-203

[4] Phiri O, Mumba P, Moyo BHZ, Kadewa W. Assessment of the impact of industrial effluents on water quality of receiving rivers in urban areas of Malawi. International Journal of Environmental Science and Technology. 2005;**2**:237-244

[5] Malu SP, Andrew C, Abah J, Oko OJ. Determination of heavy metals in Brewer's spent grains obtained from Benue brewery limited (BBL), Makurdi, north Central Nigeria. Journal of Natural Science Research. 2014;**4**(1):119-122

[6] Ogbu KC, Ebenebe CI, Abaijue MC. Physico-chemical characteristics of Ama Brewery effluent and its receiving Ajali River in Udi, Enugu state, Nigeria. Animal Research International. 2016; **13**(2):2392-2399

[7] Ikhajiagbe B, Kekere O, Omoregbee O, Omokha FI. Microbial and physiochemical quality of effluent water from a Brewery in Benin City, Midwestern Nigeria. Journal of Scientific Research and Reports. 2014;**3**(3):514-531

[8] Teshome JY, Gelanew A, Abayneh B. Impact of Dashen brewery effluent on irrigation water quality of Shinta River in Gondar, Ethiopia. Journal of Academia and Industrial Research. 2019;**8**(1):11-15

[9] Abrha BH, Chen YJ. Analysis of Physico-chemical characteristics of effluents from beverage industry in Ethiopia. Journal of Geoscience and Environment Protection. 2017;**5**:172-182. DOI: 10.4236/gep.2017.56016

[10] Alemayehu HG. Physico-chemical characterization of commercial local alcohol beverages available in south nations, nationalities and People's regional state, Ethiopia. International Journal of ChemTech Research. 2018; **11**(8):227-231

[11] Naibaho J, Korzeniowska M. The variability of physico-chemical properties of brewery spent grain from 8 different brewerie.s. Heliyon. 2021;**7**:e06583. DOI: 10.1016/j.heliyon.2021.e06583

[12] Pérez-Aguilar J, Ramos-Barrales R. Study of physico-chemical parameters in smallscale craft beer production. Biosaia (revista de los másteres de Biotecnología Sanitaria y Biotecnología Ambiental, Industrial y Alimentaria de la UPO). 2021;**10**:1

[13] USEPA. Regional Screening Levels (RISLS)-user's Guide-Risk Assessment. Washington DC, USA: United States Environmental Protection Agency; 2017. Available from: https://www.epa.gov/risk/regionalscreening-levels-rsls

[14] USEPA. Integrated Risk Information System (IRIS). Washington, DC, USA:

United States Environmental Protection Agency; 2010. Available from: www.epa.gov/ncea/iris/index.html

[15] Lagerweff JV, Specht AW. Contamination of roadside soil and vegetation with Cd, Ni, Pb, and Zn. Environmental Science & Technology. 1970;**5**:483-586

[16] AOAC. Official Methods of Analysis. 18th ed. Marylland, U.S.A: Association of Official Analytical Chemists; 2011

[17] Braid RN, Eaton AD, Rice EW, Bridgewater L. Standard methods for the examination of water and wastewater. In: United States: Water Environment Federation. 23rd ed. Washington, DC: American Water Works Association; 2017

[18] Tomlinson DL, Wilson JG, Harris CR, Jeffrey DW. Problems in the assessment of heavy-metal levels in estuaries and the formation of a pollution index. Helgoland Marine Research. 1980;**33**:566-575

[19] Hakanson L. An ecological risk index for aquatic pollution control. A sedimentological approach. Water Research. 1980;**14**:975-1001

[20] Hussain R, Khattak S, Shah M, Ali L. Multistatistical approaches for environmental geochemical assessment of pollutants in soils of Gadoon Amazai industrial estate, Pakistan. Journal of Soils and Sediments. 2015;**15**(5): 1119-1129

[21] Umeh CT, Nduka JK, Omokpariola DO, Morah JE, Mmaduakor EC, Okoye NH, et al. Ecological pollution and health risk monitoring assessment of polycyclic aromatic hydrocarbons and heavy metals in Enugu River, South-Eastern Nigeria. SSRN. 2022;**406**:1-30. DOI: 10.2139/ssrn.4067308

[22] Asare-Donkor NK, Boadu TA, Adimado AA. Evaluation of groundwater and surface water quality and human risk assessment for trace metals in human settlements around the Bosomtwe crater Lake in Ghana. Springer Plus. 2016;**5**:1812-1821

[23] Wang WN, Wang AL, Chen L, Liu Y, Sun RY. Effects of pH on survival, phosphorus concentration, adenylate energy charge, and Na(+)-K(+) ATPase activities of Penaeus chinensis Osbeck juveniles. Aquatic Toxicology. 2002;**60** (1–2):75-83

[24] WHO. Guidelines for Drinking-Water Quality: Incorporating the First Addendum. 4th ed. Geneva. Available from: https://apps.who.int/iris/rest/bitstreams/1080656/retrieve;: World Health Organization (WHO); 2018

[25] Ijeh BI, Onu NN. Assessment of pollution levels of groundwater in parts of Imo River basin, south eastern Nigeria. International Journal of Water Research and Environmental Engineering. 2013;**5**:194-202

[26] Mocuba JJ. Dissolved Oxygen and Biochemical Oxygen Demand in the Waters Close to the Quelimane Sewage Discharge [Thesis]. Bergen, Norway: University of Bergen; 2010

[27] Atobatele OE, Morenikeji OA, Ugwumba OA. Spatial variation in physical and chemical parameters of benthic macroinvertebrate fauna of river Ogunpa, Ibadan. Zoology. 2005;**3**:58-67

[28] Patil PN, Sawant DV, Deshmukh RN. Physico-chemical parameters for testing of water – A review. International Journal of

Environmental Sciences. 2012;**3**(3): 1194-1207

[29] Jha P, Barat S, Nayak CR. Fish production, water quality, and bacteriological parameters of koi carp ponds under live-food and manure-based management regimes. Zoological Research. 2008;**29**(2):165-173

[30] Alao O, Arojojoye O, Ogunlaja O, Famuyiwa A. Impact assessment of brewery effluent on water quality in Majawe, Ibadan, southwestern Nigeria. Resources. 2010;**2**(5):21-28

[31] Ojaniyi OF, Okoye PAC, Omokpariola DO. Heavy metals analysis and health risk assessment of three fish species, surface water and sediment samples in Ogbaru Axis of river Niger, Anambra state, Nigeria. Asian Journal of Applied Chemistry Research. 2021;**9**(1): 64-81. DOI: 10.9734/AJACR/2021/ v9i130205

[32] Omokpariola DO, Omokpariola PL. Health and exposure risk assessment of heavy metals in rainwater samples from selected locations in Rivers state, Nigeria. In: Ramasami P editor. Vol. 2. Chemical Sciences in the Focus. 2021. pp. 16-23. ISBN 978-3-11-072659-6. DOI: 10.1515/9783110726145-002

[33] Imoobe TOT, Koye PIO. Assessment of the impact of effluent from a soft drink processing factory on the physicochemical parameters of Eruvbi stream Benin City, Nigeria. Bayero Journal of Pure and Applied Sciences. 2010;**4**(1):126-134

[34] Matsuo T, Kumazawa K, Ishii R, Ishihara K, Hirara H. The Science of the Rice Plant Physiology. Tokyo, Japan: Food and Agriculture Research Centre; 1995

[35] Okafor VN, Omokpariola DO, Igbokwe EC, Theodore CM, Chukwu NG. Determination and human health risk assessment of polycyclic aromatic hydrocarbons (PAHs) in surface and ground waters from Ifite Ogwari, Anambra state, Nigeria. International Journal of Environmental Analytical Chemistry. 2022;**22**:2038587. DOI: 10.1080/03067319.2022.2038587

[36] Okechukwu VU, Omokpariola DO, Onwukeme VI, Nweke EN, Omokpariola PL. Pollution investigation and risk assessment of polycyclic aromatic hydrocarbons in soil and water from selected dumpsite locations in Rivers and Bayelsa state, Nigeria. Environmental Analysis Health and Toxicology. 2021;**36**(4):e2021023. DOI: 10.5620/eaht.2021023

[37] Oguzie FA, Okhagbuzo GA. Concentrations of heavy metals in effluent discharges downstream of Ikpoba river in Benin City, Nigeria. African Journal of Biotechnology. 2010; **9**(3):319-325

[38] Anetor JI, Oleleye AFAA, SR. Molecular epidemiology: A better approach for the early detection of pathophysiologic response to environmental toxicants and diseases. African Journal of Biomedical Research. 2003;**6**:113-118

[39] Omokpariola DO, Nduka JK, Omokpariola PL, Omokpariola ECO. Ionic composition of rainwater from different sampling surfaces across selected locations in Rivers state, Nigeria. World Science News. 2020;**150**: 132-147

Chapter 11

Evaluation of Water Quality Using Physicochemical Parameters and Aquatic Insects Diversity

Muhammad Xaaceph Khan and Abida Butt

Abstract

Biomonitoring studies focus on the component of biodiversity, its natural habitats, and species populations which display the ongoing variations in ecosystem and landscape. Physicochemical parameters are important water quality parameters of river water i.e., pH, temperature, turbidity, conductivity, total dissolved solids, total suspended solids, total alkalinity, sulfate, nitrate, heavy metals, and phosphate. This chapter focuses on assessing water quality through Physicochemical Parameters and Aquatic Insects Diversity. The case study investigated the effect of pollutants produced by the human dwelling, agricultural and industrial activities on aquatic invertebrate communities of water of part of Soan River, Pakistan. Four sites were selected based on variation in micro-habitat accessibility to examine the pollution in water. Samples were collected from these sites during spring, 2015. Water samples for physio-chemical analysis and macroinvertebrates were collected from all sites. Results showed that conductivity, dissolved oxygen, sodium, and cadmium at all sites were higher than the drinking water quality of WHO standards while potassium, chromium, and manganese were higher in concentration at most downstream sites. However, all other studied parameters were within recommended range of WHO standards. A total of 412 individuals of aquatic insects were collected from the studied sites, belonged to 6 orders and they were the most abundant in April. Total abundance was used to estimate the quality of water at the sites. Most biotic indexes showed that water was of good quality at upstream stations rather than downstream stations, while water quality index (WQI) showed fair water quality at downstream sites. This study showed that aquatic insects could be useful as bioindicators for biomonitoring of water quality along with physiochemical parameters.

Keywords: Biomonitoring, Pakistan, Physicochemical, Soan River, WQI

1. Introduction

1.1 Physicochemical problems and emergences of biomonitoring

Common methods that rely on chemicals to monitor river pollution are increasingly suitable for monitoring systems as they can detect physical and environmental pressures occurring over time and on multiple scales [1–5]. However, concepts, and principles of biomonitoring, which are more efficient and effective than traditional methods, have been developed and widely used worldwide to monitor river pollution. Compared to common and uncommon species, bioindicators are more tolerant to environmental change. They are sensitive enough to detect environmental change thanks to their tolerance, but they are also resilient enough to deal with some variability and represent the overall biotic response [6]. However, this new initiative has liberated tropical areas by allowing the model and adaptation of current geologically developed non-tropical species using natural freshwater organisms. These biomonitoring indicators are often developed for specific regions to respond to regional variables using local biotic collections that reflect regional variability based on sensitivity or biological tolerance. Such variations may affect the strength, performance, and reliability of biomonitoring indicators developed in non-tropical areas (mean temperature between 21 and 30°C and rainfall 100 inches a year) when used in tropical rivers (mean temperatures above 18°C and monsoonal patterns rainfall 79 to 394 inches) [7]. Similarly, modification of non-thermal bio-monitoring indicators used in tropical areas is often captured by incomplete taxonomical resolution and unknown levels of tropical taxa [4]. Abiotic variables or physiochemical samples are problematic in identifying a change or impact in some environmental conditions. For example, contamination can be present in toxic quantities or bio-accumulated, which causes adverse biological deterioration. However, contaminated concentrations may be too small to be detected using this procedure [8]. Consequently, changes in behavioral or pathological responses, population dynamics, environmental pollution, and impacts have been measured using biological rather than physic-chemical indicators by many scientists because of the direct interaction of an organism with the ecosystem [9–11].

1.2 Importance of biologic index

The use of biological indicators to assess the health of the river ecosystem has become increasingly important because the function of life, biodiversity, population density, human settlement, and the activity of aquatic organisms are affected by all changes in the water ecosystem. River life decisions can be made based on biodiversity and quantity. Many aquatic species such as fish [12, 13], algae [14], plankton [15], and benthic macroinvertebrates [16, 17] are common biologic indicators of water pollution and are used in biotic reliability for the aquatic ecosystem [18–21]. The types of indicators are those taxes that are known to be more sensitive to certain environmental factors so that changes that occur or in large quantities can directly reflect local change [22]. The usage of biomonitoring techniques in river ecosystems have many advantages compared to physiochemical techniques [23]. Freshwater organisms play an important role in the continuously monitoring water quality and pollutants that enter at different time intervals. In most cases, the disorder occurs during at least one stage (egg, larva, caterpillar, adult) of the invertebrate animal life cycle. If this category is affected by the disruption, changes will be seen in the community structure if sampled over time [24]. Macroinvertebrates are also sensitive to stress; it can be natural or human-based. This change will lead to an impaired community. The aquatic insects show the effect of point and non-point contaminants, physical habitat alteration, and pollutant accumulation over the life cycle [25].

2. Estimation of physiochemical and biological index, a case study of Soan River

2.1 Study site

Sampling sites have been selected that vary in their physiochemical characteristics of the Soan River, Pakistan, at the time of spring 2015. These sites were named A (N 33°43.120, E 073°20.44), B (N 33°37.133, E 073°17.88), C (N 33°33.174, E 073°08.547), and D (N 33°32.906, E073°05.844) which was starts from site A and end at D site as mentioned in [26]. Their positions are shown on the map (**Figure 1**). The starting point was Simply dam, and the sampling sites were selected thereafter. Photos for Site A (**Figure 2**), Site B (**Figure 3**), Site C (**Figure 4**), and Site D (**Figure 5**) are presented. In addition, the site descriptions are listed in **Table 1**. Site A was considered upstream, site B and C considered mid-stream, while all the sites were categorized

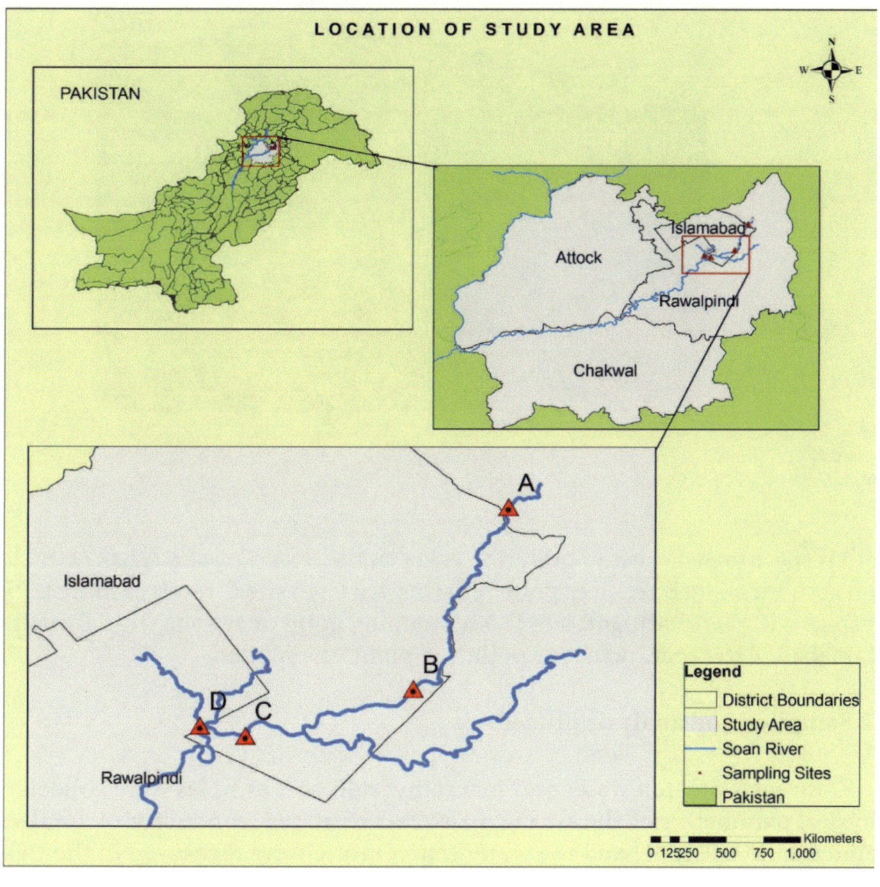

A: Simly Dam; B: Aari Syedan; C: Hummak; D: Soan Adae

Figure 1.
Map of study area and location of sampling sites at Soan River, Pakistan.

Figure 2.
Site A.

with respect to major water pollution entry in the river. i.e., site A has no major pollution activities; site B receives poultry farms wastage; site C receives industry pollution wastage without treatment; site D is a dumping point of sewage from Rawalpindi and Islamabad. After that, no major pollution point was spotted.

2.2 Sampling and analysis procedure

From each site, ten water and macroinvertebrates samples were collected. Physical parameters of the area was also recorded. pH, conductivity, total suspended solids (T.D.S), and the water temperature were measured in the field using a portable instrument (HANNA HI 9811-5), and dissolved oxygen (DO) was calculated by using DO-510. The other parameters recorded in **Table 2** were determined in the laboratory using a flame (air-acetylene) atomic absorption spectrophotometer (Hitachi, Model Z-5000) and tritremetric methods following standard procedure [27–31].

Figure 3.
Site B.

2.3 Sampling of aquatic insects

Macroinvertebrates were collected by D-net of 30 cm × 30 cm × 60 cm with 0.5 mm mesh. The net was held vertically right angle to the current. The collection was done by shaking the net at a depth of the river for 1 minute. The collection was done in the morning for 1 week. The collection was done from upstream (A) to downstream (D). For maximizing the complete assemblage of samples, ten replicates from each site were selected every day according to the geographic conditions of the surrounding, i.e., anthropogenic interference and natural causes. After the collecting of water and macroinvertebrates, samples were carried out into 500 ml plastic bottles with enough water, so that samples were not damaged during transportation.

In the laboratory, macroinvertebrates were preserved within a few hours of collection in 90% alcohol. The collection was sorted within a month by RIVPACS (River

Figure 4.
Site C.

Figure 5.
Site D.

Study site	Location	Site description
A	Simly Dam	Rich fauna with a greater number of riparian places
B	Aari Syedan	A moderate number of riparian and fast-flowing of water
C	Hummak	No riparian, industrial dumping, and waste product lying on the bank of the river
D	Soan Adae	No riparian, the waste of Rawalpindi and Islamabad was dumped into that point

Table 1.
The geographical position of sampling sites at Soan River.

Tests	Sampling sites				WHO Standards
	A	B	C	D	
pH	7.75(0.05)	7.55(0.15)	7.85(0.05)	**8.25(0.15)**	6 to 8
Conductivity (μS)	**451.5(1.5)**	**427.5(5.5)**	**477.0(2.0)**	**437.0(1.0)**	200 to 400
T.D.S (mg/l)	673(2)	677.5(2.50)	682.0(2.0)	653.5(3.5)	500 to 1000
Water Temperature (°C)	26.45 (0.54)	26.04(0.33)	27.13(0.66)	25.21(1.53)	
D.O (mg/l)	**1.88(0.035)**	**1.93(0.01)**	**1.56(0.025)**	**2.255(0.025)**	>4
Chloride (mg/l)	26.50(1.50)	28.5(1.50)	26.50(0.50)	24.5(1.5)	< = 250
Sulfate (mg/l)	32.0(2.0)	35.0(1.00)	27.50(1.50)	32.0(2.00)	< =250
Nitrate (mg/l)	2.06(0.05)	1.81(0.08)	2.20(0.08)	2.25(0.04)	<=10
Calcium (mg/l)	26.85(0.05)	25.6(0.30)	27.60(0.40)	26.60(0.20)	< = 100
Magnesium (mg/l)	18.95(0.15)	18.80(0.20)	19.0(0.10)	18.65(0.15)	< = 50
Total Hardness (mg/l as $CaCO_3$)	312.50(2.50)	322.5(2.50)	317.50(2.5)	302.0(1.0)	100 to 500
Sodium (mg/l)	**20.85(0.05)**	**20.70(0.20)**	**20.50(0.10)**	**23.0(2.0)**	< = 20
Potassium (mg/l)	8.25(0.15)	7.75(0.15)	**10.35(0.25)**	**11.55(0.15)**	< = 10
Aluminum (μg/ml)	0.056(0.003)	0.056(0.002)	0.087(0.0015)	0.012(0.001)	< 0.1–0.2
Cadmium (μg/ml)	**0.145(0.025)**	**0.167(0.002)**	**0.180(0.01)**	**0.173(0.002)**	< 0.003
Chromium (μg/ml)	0.047(0.002)	0.046(0.002)	**0.0530(0.002)**	**0.064(0.001)**	< 0.05
Copper (μg/ml)	0.405(0.015)	0.375(0.015)	0.455(0.017)	0.548(0.018)	< 2
Manganese (μg/ml)	0.394(0.003)	0.374(0.002)	**0.565(0.015)**	**0.589(0.002)**	< 0.5

A: Simly Dam; B: Aari Syedan; C: Hummak; D: Soan Adae: WHO: World Health Organization; S.E in brackets. Bold indicates that the values didn't follow the WHO Standards.

Table 2.
Physio-chemical analysis of water samples collected from Soan River.

Invertebrate Prediction and Classification System) standard procedure (Environment Agency, 1997) using taxonomic keys [32]. Small aquatic insects were sorted under a dissecting microscope, whereas large insects were sorted with naked eyes. All samples were then kept in properly labeled vials containing 90% alcohol.

Biotic Index, Biological Monitoring Working Party-Average Score Per Taxon (BMWP-ASPT), The HBI (Hilsenhoff Biotic Index) or Family Biotic Index (FBI), Ephemeroptera, Plecoptera, and Trichoptera (EPT) Index and Water Quality Index were assessed using the data gathered. Small description for each of the indexes are also given as follows.

3. Biotic index

Aquatic insects of the freshwater river and stream ecosystems have frequently been examined to assess the species-habitat relationship concerning the water quality of the habitat [33]. Aquatic insects can indicate the water quality of streams, rivers, and lakes. Once aquatic insects are collected, and after analysis, the data can be compared between different sites by using four standard indices, i.e., Hilsenhoff's Biotic Index [34], EPT (Ephemeroptera, Plecopterans, and Trichopterans) Index [35], the Benthic Index of Biotic Integrity [25] and Beck's Biotic Index [36]. The EPT Index stands for Ephemeroptera, Plecopterans, and Trichopterans, three orders of Class Insecta which can easily be sorted, identified, and commonly used as water quality indicators. The EPT index is based upon a high-quality stream ecosystem with a great species richness. Biotic Index shows the quality of the environment by the presence of different organisms present in it, this index is also known as the "Family Biotic Index". This index is commonly used for river water quality. Biotic Index shows four basic water quality, i.e., (Excellent, Good, Fair, or Poor) measured as 1 to 10. 1 is good, and 10 is poor.

3.1 Biological monitoring working party-average score per taxon (BMWP-ASPT)

Biological Monitoring Working Party-Average Score Per Taxon (BMWP-ASPT) is a biotic index method. That index also estimates the diversity of organisms concerning pollution level. According to different indexes, specimens are placed at various levels from 1-to 10. Plecoptera (rock fly larvae), Ephemeroptera (mayfly larvae), and Crustacea (pole shrimp) are on level 10, Gastropoda (freshwater limpet), Odonata (kini – kini) are at level 8, Trichoptera (caddisfly larvae) at level 7, Odonata (dragonfly larvae), Crustacea (freshwater shrimp) and Bivalvia (shell) at 6 levels, Hemiptera (backswimmer), Diptera (fly larvae), Trichoptera and Coleoptera (water scorpion, diving beetle) at 5, Arachnida (water mite) and Platyhelminthes (flatworm) at four levels, Syrphidae (rattail maggot), Hirudinea (leech), Gamaridae (water pig bug), Gastropoda (snail) and Bivalvia (shell) at 3, Chironomidae (mosquito larvae) at level 2 and Oligochaeta (worm) at level 1. The sequence starts from 10 as excellent and 1 as poorer. Many aquatic species are pollutants intolerant, i.e., levels 10–7 and absent in polluted water bodies. The greater the pollution, the lower the number of insects because few species are tolerant to pollutants, i.e., level 1. The BMWP score equals the sum of the tolerance score of all families in the sample. ASPT was calculated by taking the average number of the tolerance scores of all families of macroinvertebrates which varies from 0 to 10.

3.2 The HBI (Hilsenhoff biotic index) or family biotic index (FBI)

The HBI (Hilsenhoff Biotic Index), also known as the family biotic index (FBI) calculates the level of tolerance in the community of a specific area and the categorization of each taxonomic group by relative abundance. Organisms are grouped with a tolerance number 0–10, and 10 is the most tolerant, while 0 is the most sensitive to organic pollutants [1, 37–39]. The tolerance values were modified for macroinvertebrates for application in the Modified Family Biotic Index.

The family biotic index (FBI) is calculated as [37].

$$FBI = \Sigma(xi^{*}ti)/(n),$$

Where a *xi* is the number of individuals within a taxon, *ti* is the tolerance value of a taxon, and *n* is the total number of organisms in the sample. If the value is between 0.00–3.75, then excellent, 3.76–4.25 very good, 4.26–5.00 good, 5.01–5.75 fair, 5.76–6.50 fairly poor, 6.51–7.25 poor, and 7.26–10.00 very poor.

3.3 Ephemeroptera, Plecoptera, and Trichoptera (EPT) index

EPT index was estimated by summing a total number of individuals group Ephemeroptera (Mayflies), Plecoptera (Stoneflies), and Trichoptera (Caddisflies). If the EPT number is greater than 7 per sample, then it is considered excellent, between 2 and 7 is good, and below 2 is considered poor [40].

4. Water quality index

Water quality was identified based on the water quality index (WQI) [41].

$$WQI = 100 - \sqrt{}\left(F_1^{\,2} + F_2^{\,2} + F_3^{\,2}\right)/1.73$$

Where,
F_1 = [Water variables that do not meet objectives/Total number of water variables]*100
F_2 = [Number of tests that do not meet objectives/Total number of tests]*100
F_3 = nse/0.01 nse + 0.01
Where,
nse (normalized sum of excursions) = Σ^n departure/number of tests.

The variables that do not meet objectives are those parameters that exceed the WHO permissible limits. At the same time, the number of tests that do not meet objectives is the number of replicates that do not fall between permissible limits. WQI, the score was scaled between 0 to 100; higher values represent better quality of water, e.g., excellent >80, good 60–80, fair 60–40, and poor <40.

5. Results

The water quality parameters of the upstream section of Soan river tested in this study are summarized in **Table 2**. The pH was normal in all sites except in site D. Conductivity, D.O, sodium, and cadmium had high values concerning the standards. While, T.D.S, chloride, sulfate, nitrate, calcium, magnesium, total hardness, potassium, aluminum, and copper did not exceed the limit, and fell within the normal range. Potassium, chromium, and manganese show the normal range at sites A & B while surpassing ranges at sites C & D. The water temperature was 25–27°C from upstream to downstream.

Order	Tolerance value	A	B	C	D
Ephemeroptera	5	6.61	0	0	0
Plecoptera	2	1.61	0	0	0
Trichoptera	4	3.22	0	0	0
Hemiptera	5	22.84	45.62	56.8	0
Coleoptera	5	65.72	54.38	35.2	0
Diptera	8	0	0	8	100

Table 3.
Percent composition (%) of aquatic insects in four different stations of Soan River, Pakistan.

Study site	Biotic index								Physicochemical index	
	FBI		BMWP		ASPT		EPT index		WQI	
	Value	Class	Value	Class	Value	Class	Value	Class	Value	Class
A	4.6	Good	54	Fair biological quality	6	Good	7	Good	71.28	Good
B	4.49	Good	30	Poor biological quality	5	Fair	0	Poor	68.36	Good
C	5.23	Fairly	17	Poor biological quality	4.25	Poor	0	Poor	59.82	Fair
D	10	Very poor	0	Very poor biological quality	0	Very poor	0	Poor	55.57	Fair

Notes: FBI = Family Biotic Index; BMWP = Biological Monitoring Work Party; ASPT = Average Score Per Taxon; EPT = Ephemeroptera, Plecoptera and Trichoptera; Water Quality Index = WQI.

Table 4.
The water quality of the Soan River is based on biological indices and water quality index.

A total of 412 individuals of macroinvertebrates were captured, which represents 7 orders. Site A has shown more variety of insects than all the other sites. Total abundance was higher at site C. The variety of insects drops from upstream to downstream stations and only limits to order Diptera (**Table 3**). In the biotic index, the FBI shows good water quality in upstream stations (sites A & B) while quality decreases fairly in site C and becomes very poor at the last site. The BMWP shows the same trend just like FBI but starts from fair biological quality to very poor biological quality from upstream to downstream. The ASPT represents good water quality at site A and decreases the status to very poor at site D. The EPT only shows good water quality at site A and poor in all remaining sites. In the Physicochemical index, the WQI shows good water quality at sites A & B while fair in sites C & D (**Table 4**).

6. Discussion

In this study, water quality was examined using chemical, physical, and bio-monitoring methods. In chemical analysis, different parameters were tested, such as magnesium, chromium, aluminum, copper, cadmium, dissolved oxygen (DO), nitrogen (N), and phosphorous (P) [42]. Physical parameters include pH, total suspended solids (T.D.S), conductivity, chloride, sulfate, nitrate, calcium, total hardness, sodium, and potassium. In bio-monitoring FBI, EPT, ASPT, and BMWP were calculated to determine water quality.

The concentration of metals was almost normal but some of them were high such as cadmium, sodium, potassium, and chromium. The pH values were low at the start and exceeded in site D. This shows that Organic influx wastage in the monsoon season may lower the water pH at that site D. The pH affects the biochemical process as well. It also indicates water quality and the extent of pollution in the watershed [42]. Among the different heavy metals, Cd shows a higher level of concentration which did not support any life or for drinking purposes (3 $\mu g\ L^{-1}$) [43]. Cd is the typical anthropogenic metal affected by human activities [44] and showed enrichment. It is shown that Cd is associated to a greater extent with colloidal materials in surface run-off which can easily be transported into river flow [45]. Metals, such as Cu, Zn, and Pb, have a high affinity to human substances present in organic matter. The presence and quality of organic matter differentially influence the binding of metals within the sediments, reducing the adsorption of Cd and Co and increasing the adsorption of Zn [46]. Discharge of industrial, sewage, and poultry waste largely untreated forms may cause the elevation of metals in the water [47–49]. The water quality shows poor quality at those sites which drain sewage of the twin cities, while at downstream sites, the natural process shows some recovery from stress conditions due to the huge amount of sewage waste from the urban [50]. Activities of humans can change the smallest change in the ecosystem, especially downstream of the Soan River. The poor quality of water at downstream rather than upstream stations can result from several human activities, sewage, nutrient, sedimentation, and agriculture pesticides residue run-off. Wahizatul et al. [51] also studied in the Sekayu stream and found that agricultural and recreational activities were directly related to the destruction of aquatic species diversity in the Sekayu recreational forest. The higher organism abundance at site A is related to greater availability of coverage of riparian vegetation, which offers them a great supply of hiding places, allochthonous material, and food availability. Roque et al. [52] pointed out that the area with greater vegetation coverage has greater taxonomic richness. Although, at sites C and D, low diversity is found which could be related to the loss of riverbank vegetation and replaced by waste material, shrubby, exotic vegetation, and a lower quantity of heterogeneous substrate. This phenomenon was noted by [53]. Adamu Mustapha and Geidam [54] reported that high nutrients loading at urban sites are due to discharge of sewage wastewater of Rawalpindi and Islamabad.

Chironomidae (Diptera) were most abundant at downstream of the Soan River. They show no variation and are found in all stations. Yule and Sen [32] reported that in Malysia, Chironomidae is probably the most abundant and diverse group of all macroinvertebrate's streams. The sandy or muddy areas and slow-flowing or standing streams with a high number of sediment particles are the best areas where

Chironomidae can excel [32, 55]. Due to heavy rainfall, the flood affected the macroinvertebrates from all the sites. Thus, effect seasonal taxa richness. The member of Chironomidae was most affected by the flood. The mayflies (Ephemeroptera) and true bug (Hemiptera) did not show any response to heavy rainfall because they are morphologically better adapted, attachment abilities to stones, mobility in water and behavioral pattern during mating. Holomuzki and Biggs [56] studied the behavioral pattern in response to the flood. The fluctuation of water level in the winter season remained very low. This is the major stressing factor for littoral organisms. Nairn et al. and Gopal [57, 58] also recorded a similar finding on littoral destruction. The macroinvertebrate density in the Soan River was found to be the lowest when the monsoon season starts and increases when the monsoon stops. Wallace et al. and Jakob et al. [59, 60] pointed out that monsoon floods decrease macroinvertebrates' density, especially Chironomid species known as two-winged flies. EPT was not found abundantly in any collection points, especially downstream. EPT members are known to be the most sensitive insects to environmental stress. Therefore, the presence of EPT upstream indicates a relatively clean environment [61, 62]. Therefore, the EPT can be used for potential bioindicator purposes. The BMWP index shows poor results for all the sampling sites, but some sites show fair biological quality at site A as well as ASPT [62]. Therefore, the presence and absence of macroinvertebrates along with water physicochemical analysis at upstream and downstream shows the influence of anthropogenic and natural influences. This suggests that aquatic insects can be used to access the water management in Pakistan as the role of potential bio-indicators.

7. Conclusion

The biological and WQI index shows water quality was good at upstream rather than at downstream stations. The biotic index shows a clear variation throughout the sites. While the physicochemical index shows the same trend at site D and it shows fair water quality despite the biotic index is Poor. The biotic index is detailed and efficient while the individual can gather information on the spot, but physicochemical parameters are costly, laboratory intensive, and time-consuming. These two indexes can be used alternatively concerning the situation.

Acknowledgements

This study was sponsored by the World Wildlife Fund (WWF) Pakistan through the young Ecologist Scholarship. Special funds for a multipurpose water analysis kit were provided by the University of the Punjab, Lahore. The authors are extremely thankful to both organizations.

Author details

Muhammad Xaaceph Khan* and Abida Butt
Institute of Zoology, University of the Punjab, Lahore, Pakistan

*Address all correspondence to: xaaceph@gmail.com

References

[1] Barbour MM, Fischer RA, Sayre KD, Farquhar GD. Oxygen isotope ratio of leaf and grain material correlates with stomatal conductance and grain yield in irrigated wheat. Functional Plant Biology. 2000;**27**(7):625-637

[2] Böhmer J, Rawer-Jost C, Zenker A. Multimetric Assessment of Data Provided by Water Managers from Germany: Assessment of Several Different Types of Stressors with Macrozoobenthos Communities. Integrated Assessment of Running Waters in Europe: Springer; 2004;**514**. pp. 215-228

[3] Chaves MLC. Spatio-Temporal Dynamics of Undisturbed Macroinvertebrate Communities in the Mondego River Basin: Contribution to the Ecological Assessment of Streams. Universidade de Lisboa, Portugal: ProQuest Dissertations Publishing; 2008

[4] Jacobsen D, Cressa C, Mathooko JM, Dudgeon D. Macroinvertebrates: Composition, life histories and production. Tropical Stream Ecology: Elsevier; 2008. p. 65-105.

[5] Elias J. The Assessment of Benthic Macroinvertebrate Structures in some Tanzanian Rivers as an Indicator for Water Quality Biomonitoring Programme. Stockholm, Sweden: Unpublished report submitted to IFS; 2009

[6] Holt E, Miller S. Bioindicators: Using organisms to measure environmental impacts. Nature Education Knowledge. 2010;**3**(10):8

[7] Syvitski JP, Kettner AJ, Overeem I, Hutton EW, Hannon MT, Brakenridge GR, et al. Sinking deltas due to human activities. Nature Geoscience. 2009; **2**(10):681-686

[8] Suter GW II. Applicability of indicator monitoring to ecological risk assessment. Ecological Indicators. 2001;**1**(2):101-112

[9] Cairns J Jr, van der Schalie WH. Biological monitoring part I—Early warning systems. Water Research. 1980;**14**(9):1179-1196

[10] Landres PB, Verner J, Thomas JW. Ecological uses of vertebrate indicator species: A critique. Conservation Biology. 1988;**2**(4):316-328

[11] Carignan V, Villard M-A. Selecting indicator species to monitor ecological integrity: A review. Environmental Monitoring and Assessment. 2002;**78**(1):45-61

[12] Harris J, Silveira R. Large-scale assessments of river health using an index of biotic integrity with low-diversity fish communities. Freshwater Biology. 1999;**41**(2):235-252

[13] Belpaire C, Smolders R, Auweele IV, Ercken D, Breine J, Van Thuyne G, et al. An index of biotic integrity characterizing fish populations and the ecological quality of Flandrian water bodies. Hydrobiologia. 2000;**434**(1):17-33

[14] Stevenson RJ, Smol JP. Use of algae in environmental assessments. In: Freshwater Algae in North America: Classification and Ecology. 2003;**1**. pp. 775-804

[15] Reynolds C. Planktic community assembly in flowing water and the ecosystem health of rivers. Ecological Modelling. 2003;**160**(3):191-203

[16] Weigel BM, Henne LJ, Martinez-Rivera LM. Macroinvertebrate-based

index of biotic integrity for protection of streams in west-Central Mexico. Journal of the North American Benthological Society. 2002;**21**(4):686-700

[17] Silveira M, Baptista D, Buss D, Nessimian J, Egler M. Application of biological measures for stream integrity assessment in south-East Brazil. Environmental Monitoring and Assessment. 2005;**101**(1):117-128

[18] Butcher JT, Stewart PM, Simon TP. A benthic community index for streams in the northern lakes and forests ecoregion. Ecological Indicators. 2003;**3**(3):181-193

[19] Davis NM, Weaver V, Parks K, Lydy MJ. An assessment of water quality, physical habitat, and biological integrity of an urban stream in Wichita, Kansas, prior to restoration improvements (phase I). Archives of Environmental Contamination and Toxicology. 2003;**44**(3):0351-0359

[20] Wang B, Yang L, Hu B, Shan L. A preliminary study on the assessment of stream ecosystem health in south of Anhui Province using benthic-index of biotic integrity. Acta Ecologica Sinica. 2005;**25**(6):1481-1490

[21] Zhang Y, Xu C, Ma X, Zhang Z, Wang J. Biotic integrity index and criteria of benthic organisms in Liao River basin. Acta Scientiae Circumstantiae. 2007;**27**(6):919-927

[22] New TR. Insect Conservation. An Australian Perspective: Dr. W. Junk; 1984

[23] Rosenberg DM. Introduction to freshwater biomonitoring and benthic macroinvertebrates. Freshwater Biomonitoring and Benthic Macroinvertebrates. 1993;**488**:1-9

[24] Relyea CD, Minshall GW, Danehy RJ. Stream insects as bioindicators of fine sediment. Proceedings of the Water Environment Federation. 2000;**2000**(6):663-686

[25] Fore LS, Karr JR, Wisseman RW. Assessing invertebrate responses to human activities: Evaluating alternative approaches. Journal of the North American Benthological Society. 1996;**15**(2):212-231

[26] Butt A, Khan MX, Aihetasham A, Khan MA, Nazli H, Ramzan A. Assessment of water quality of Soan River using physicochemical parameters and aquatic insects diversity. The Biological Bulletin. 2021;**48**(6):813-820

[27] Korkmaz D. Precipitation titration:"determination of chloride by the Mohr method". Methods. 2001;**2**(4):1-6

[28] Dedkov YM, Korsakova N, Sychkova V. New metallochromic indicator for barium: Determination of sulfate in water and soil extracts. Journal of Analytical Chemistry. 2006;**61**(12):1154-1162

[29] Nguyen HT, Vo KTK, Bui LTT, Hua HH, Oko GE. Gries-Ilosvay spectrophotometry for determination of nitrite in water and vegetables in Vietnam. Asian Journal of Chemical Sciences. 2018;**5**:1-9

[30] Tucker B, Kurtz L. Calcium and magnesium determinations by EDTA titrations. Soil Science Society of America Journal. 1961;**25**(1):27-29

[31] Davies T. Chemistry and pollution of natural waters in western Kenya. Journal of African Earth Sciences. 1996;**23**(4):547-563

[32] Yule CM, Sen YH. Freshwater Invertebrates of the Malaysian Region. Malaysia: Academy of Sciences; 2004

[33] Azrina M, Yap C, Ismail AR, Ismail A, Tan S. Anthropogenic impacts on the distribution and biodiversity of benthic macroinvertebrates and water quality of the Langat River, peninsular Malaysia. Ecotoxicology and Environmental Safety. 2006;**64**(3):337-347

[34] Hilsenhoff WL. Use of Arthropods to Evaluate Water Quality of Streams [Wisconsin]. USA: Technical Bulletin-Wisconsin Dept of Natural Resources, Division of Conservation; 1977

[35] Lenat DR. Water quality assessment of streams using a qualitative collection method for benthic macroinvertebrates. Journal of the North American Benthological Society. 1988;**7**(3):222-233

[36] Terrell CR, Perfetti PB. Water Quality Indicators Guide. Washington, DC: Surface waters Terrene Institute; 1996

[37] Hilsenhoff WL. Seasonal correction factors for the biotic index. The great lakes entomologist. 2017;**21**(1):3

[38] Bode RW, Novak MA, Abele LE, Heitzman DL, Smith AJ. Quality Assurance Work Plan for Biological Stream Monitoring in New York State: Stream Biomonitoring Unit. Bureau of Monitoring and Assessment, New York: Division of Water Albany; 1996

[39] Plafkin JL. Rapid Bioassessment Protocols for Use in Streams and Rivers: Benthic Macroinvertebrates and Fish. Washington, DC: U.S Environmental Protection Agency, Off. Water Regulations and Standards; 1989

[40] Bonner L, Hayes R, Lister J, Myer P, Wolf J. Rapid bioassessment of Crabtree Creek (Wake County, North Carolina) using macroinvertebrate and microbial indicators. Journal of Freshwater Ecology. 2009;**24**(2):227-238

[41] Lumb A, Halliwell D, Sharma T. Application of CCME water quality index to monitor water quality: A case study of the Mackenzie River basin, Canada. Environmental Monitoring and Assessment. 2006;**113**(1-3):411-429

[42] Kannel PR, Lee S, Lee Y-S, Kanel S, Pelletier G. Application of automated QUAL2Kw for water quality modeling and management in the Bagmati River, Nepal. Ecological Modelling. 2007;**202**(3-4):503-517

[43] Muhammad Ekramul Mahmud HN, Huq A. Removal of heavy metal ions from wastewater/aqueous solution using polypyrrole-based adsorbents: A review. RSC Advances. 2016;**6**(18):14778-14791

[44] Zhang H, Shan B. Historical records of heavy metal accumulation in sediments and the relationship with agricultural intensification in the Yangtze–Huaihe region, China. Science of the Total Environment 2008;**399**(1-3):113-120

[45] Wakida F, Lara-Ruiz D, Temores-Pena J, Rodriguez-Ventura J, Diaz C, Garcia-Flores E. Heavy metals in sediments of the Tecate River, Mexico. Environmental Geology. 2008;**54**(3):637-642

[46] Tomlinson D, Wilson J, Harris C, Jeffrey D. Problems in the assessment of heavy-metal levels in estuaries and the formation of a pollution index. Helgoländer Meeresuntersuchungen. 1980;**33**(1):566-575

[47] Tao Y, Yuan Z, Wei M, Xiaona H. Characterization of heavy metals in water and sediments in Taihu Lake, China. Environmental Monitoring and Assessment. 2012;**184**(7):4367-4382

[48] Kansal A, Siddiqui NA, Gautam A. Assessment of heavy metals and

their interrelationships with some physicochemical parameters in eco-efficient rivers of Himalayan region. Environmental Monitoring and Assessment. 2013;**185**(3):2553-2563

[49] Ali Z, Malik RN, Qadir A. Heavy metals distribution and risk assessment in soils affected by tannery effluents. Chemistry and Ecology. 2013;**29**(8):676-692

[50] Reza R, Singh G. Heavy metal contamination and its indexing approach for river water. International Journal of Environmental Science and Technology. 2010;7(4):785-792

[51] Wahizatul A, Amirrudin A, Raja Noor Balqhis R, editors. Diversity of aquatic insects in relation to water quality in stream of Sekayu recreational Forest, Terengganu. In: Proccedings of the National Seminar in Science, Technology and Social Sciences. Kuantan, Pahang: Science and Technology; 2006

[52] Roque F, Trivinho-Strixino S, Strixino G, Agostinho R, Fogo J. Benthic macroinvertebrates in streams of the Jaragua State Park (southeast of Brazil) considering multiple spatial scales. Journal of Insect Conservation. 2003;7(2):63-72

[53] Bueno AA, Bond-Buckup G, Ferreira BD. Community structure of benthic invertebrates in two watercourses in Rio Grande do Sul state, southern Brazil. Revista Brasileira de Zoologia. 2003;**20**:115-125

[54] Adamu Mustapha AA, Geidam AL. The influence of land use and land-cover changes on surface water quality variation in the Jakara basin North-Western Nigeria. Nigeria: IJAIR; 2013

[55] Marques M, Barbosa F, Callisto M. Distribution and abundance of Chironomidae (Diptera, Insecta) in an impacted watershed in south-East Brazil. Revista Brasileira de Biologia. 1999;**59**:553-561

[56] Holomuzki JR, Biggs BJ. Taxon-specific responses to high-flow disturbance in streams: Implications for population persistence. Journal of the North American Benthological Society. 2000;**19**(4):670-679

[57] Nairn R, Zuzek P, Bachmann R, Jones J, Peters R, Soballe D. Artificial beaches as shoreline treatment. Lake-reserve Mmt. 1995;**11**:174-175

[58] Gopal B. The Role of Ecotones (Transition Zones) in the Conservation and Management of Tropical Inland Waters. Internationale Vereinigung für Theoretische und Angewandte Limnologie: Mitteilungen. 1994;24(1):17-25.

[59] Wallace ID, Wallace B, Philipson G. A Key to the Case-Bearing Caddis Larvae of Britain and Ireland 1990

[60] Jakob C, Robinson CT, Uehlinger U. Longitudinal effects of experimental floods on stream benthos downstream from a large dam. Aquatic Sciences. 2003;**65**(3):223-231

[61] Armitage P, Moss D, Wright J, Furse M. The performance of a new biological water quality score system based on macroinvertebrates over a wide range of unpolluted running-water sites. Water Research. 1983;**17**(3):333-347

[62] Czerniawska-Kusza I. Comparing modified biological monitoring working party score system and several biological indices based on macroinvertebrates for water-quality assessment. Limnologica. 2005;**35**(3):169-176

Section 5

Community's Role in River Basin Management

Chapter 12

Community Participation in River Basin Management

Benny Syahputra, Berkah Fajar and Sudarno

Abstract

River basin management can increase biodiversity conservation, land productivity, ecosystem restoration, rehabilitation, and land reclamation. River basin management cannot be handled by one institution alone but requires cooperation and coordination with various parties. Community involvement in river basin management has a strategic role and is carried out based on the principle of sustainability that combines a balance between productivity and conservation to achieve river basin management goals. Community participation has the power to make decisions autonomously in order to be able to solve the needs and interests of life and improve the standard of living by utilizing the resources that must be owned. In addition, it is also necessary to harmonize structural relationships between institutions in government both at the center and at the local level so that internal factors can be controlled, and programs and activities do not overlap in the management of natural resources and air. One of the impacts of climate change is the occurrence of river basin damage. Upstream river basins as buffer areas, water catchments, and sources of water filters will be damaged. Sensitive people will be more vulnerable, while people who can adapt will survive.

Keywords: community, river basin, stakeholders, management, environment

1. Introduction

Community empowerment in the management of river basins is getting more and more attention, as seen by the number of authors who state the importance of community participation and independence in river basin management [1]. Several researchers have developed a transdisciplinary, scientific, and explicit-scale watershed system model jointly designed by the multidisciplinary community [2]. The authors put a massive water reform program into effect, which resulted in considerable institutional, social, and economic changes. Policies must be enforced around a scientific basis, with a management focus on incorporating new knowledge [3]. Other authors proposed that for watershed management, direct community participation for conservation, and overcoming the risk of watershed damage are required [4] coordination and consultation between stakeholders for each policy [5].

Community empowerment is an effort to enable and empower the community to complete the needs and interests of their lives and improve their standard of living by utilizing the resources they have [6]. Thus, community empowerment emphasizes initiative and autonomy in decision-making by the community, therefore emphasizing the importance of the learning process in empowerment to equip the community towards sustainable change [7].

There are three important stages in the empowerment process, namely: the awareness stage, the stage of capacity building, which include increasing human, organizational, and value system capacities, and the empowerment stage, namely the granting of power, authority, or opportunity [8]. In the empowerment process, counseling and mentoring activities must be carried out. This is also stated in the articles on community empowerment in the laws and regulations relating to river basin management.

Mentoring and counseling are often seen as separate or distinct activities. In the past, counseling was only considered an effort to convey information and technology to the public. In its development, counseling is not only defined as a process of disseminating information and technology but also a non-formal education process, a capacity-building process, and a behavior change process so that people can help themselves, and improve their welfare [9], meaning that counseling is also a process of community empowerment. While mentoring means the existence of assistance from outside parties to increase public awareness and capacity to understand problems and look for alternative solutions to problems, so as to achieve sustainable development, empowerment, and community welfare. Thus, counseling and mentoring are activities that cannot be separated from the community empowerment process. Both are learning processes to increase community capacity which will continue throughout the community empowerment process. In addition, to achieve an empowered society, there are several efforts that also need attention, namely:

a. Creating an atmosphere that allows the potential of the community to develop. Community empowerment requires a strong commitment from the government and other related parties. These parties are required to create a supportive atmosphere or climate so that the potential of the community develops. Community participation must be encouraged as widely as possible through mentoring programs toward their independence.

b. Strengthening the potential of the community (empowering). This strengthening includes concrete steps and involves the provision of various inputs, as well as opening access to various opportunities that will make the community more empowered such as information, markets, and capital.

Apart from existing efforts, community participation is an important element in the process of community empowerment [8], because, without community participation, community empowerment efforts will not be achieved. In the community empowerment process, there are several typologies of participation as follows:

a. Manipulation, namely participation, is not based on individual participation but only on representation in groups and the community does not get any information.

b. Passive (receiving information) i.e., outsiders take decisions, and the public is only given information without a response from the community about the decision.

c. Consultation is community participation that is only consulted and answered; outsiders define problems oversee analysis, and make decisions, while the community does not take part in decision making.

d. Advices namely outsiders submit plans and ask for community feedback. If necessary, it will make changes to the plan by taking into account and considering the views of the community.

e. Functional (co-planning) is outside the temporary exhibition and discusses it with the community and provides the opportunity for the community to participate in final decision making.

f. Delegating power (interactive) i.e., external parties regarding identification and presentation of problems to the community, analyzing with the community for planning development, but decision making is carried out by the community, which means that decision making is local by the community.

g. Independent (self-mobilization): The community identifies problems and decides on goals and how to achieve them. Outside parties only act as assistants who provide information on the necessary resources and techniques, but the community has control over the use of resources [10].

Community empowerment is essentially about enabling and self-reliant communities so that more emphasis is placed on decision-making autonomy from a community group based on the resources they have [11]. Thus, community empowerment should place more emphasis on the process of positive change that occurs as well as the improvement and sustainability of community empowerment. In practice, many community empowerment activities are not in accordance with the concept of community empowerment. Community empowerment is often trapped in a "project" logic that emphasizes results and administrative responsibilities such as the size of the budget, the number of activities carried out, and the assistance provided [10]. Assistance makes community empowerment efforts tend to be participation mobilized by material incentives. Instead of creating community independence, it actually causes community dependence on the government and other outside parties.

This chapter aims to explain the current condition of the river basin and its conservation efforts under changing climate. River basin conservation requires community participation and the implementation of its empowerment in river basin management under changing climate, both individually and institutionally.

2. River basin conservation under changing climate

The river basin conservation can be achieved if every activity is carried out based on the principle of sustainability that combines a balance between productivity and conservation to achieve the river basin management goal of improving water management; improving soils; controlling land degradation processes; increasing farmers' income; and encouraging the community towards conservation activities that control runoff and flooding [12].

Broadly speaking, the river basin system can be divided into three parts, namely upstream, middle, and downstream. The upstream river basin ecosystem is very important in the river basin system because it functions as an overall river basin protection system. The upstream area is characterized as a rural ecosystem with four main components, namely: villages, cropland, rivers, and forests. Thus, upstream river basin management is not only to maintain the river basin function but also to improve livelihoods and improve the economy of local communities sustainably [13]. It emphasizes that the balance between meeting the needs of local communities and preserving natural resources is a prerequisite for achieving the goals of sustainable river basin management. The strategy that is seen as an approach to community participation in river basin management in community-based natural resource management is known as Community-Based Natural Resource Management. This approach began to develop in the late 1990s, along with the passing of the era of decentralization and democracy.

The rapid increase in population and economic development activities in the river basin causes changes in land use and very high use of fossil fuels. These two activities are the largest sources of greenhouse gas (GHG) emissions in the river basin. The continuous increase in GHG emissions causes global warming, which affects the climate in the river basin.

Global GHG emissions are projected to peak between 2020 and at the latest before 2025 in global modeled pathways that limit warming to 1.5°C (>50%) with no or limited overshoot and in those that limit warming to 2°C (>67%) and assume immediate action [14]. Global warming is a problem that humans must face now and is no longer a future problem. This event impacts the long-term accumulation of atmospheric pollution due to human activities, causing the release of GHG into the atmosphere at a very high rate which then impacts climate and the environment. Climatic events such as floods, long droughts, and strong winds have been happening more frequently lately with high intensity [15]. This incident is increasingly causing a greater impact with the high level of environmental damage.

Global warming caused by GHG has affected the world's climate. IPCC (2001) in Climate change 2001: Impacts, Adaptation, and Vulnerability: Contribution of Working Group II to the third assessment report of the 74 Intergovernmental Panel on Climate Change states that climate change refers to the variation in the average climatic conditions of a place or its statistically significant variability over a long period of time, at least 30 years. It is further said that over the last 100 years (1906-2005), the earth's average surface temperature has risen by about 0.74°C, with greater warming on land than in the oceans.

Climate change has positive and negative impacts on all sectors of human life. However, most of the impacts are negative. To assess the impacts of climate change, it is necessary to estimate how the climate is changing at local and regional levels and how these changes affect ecosystems and human life. Most scientists use Global Circulation Models (GCMs). GCMs have been used to assess the impact of climate change on all sectors of life in Indonesia.

Climate change, especially temperature and rainfall, causes changes in discharge fluctuations in the river basin. For example, in Indonesia, it shows an increase in temperature of around 0.1°C–0.5°C in 2010 and in 2070 around 0.4°C–3.0°C, while globally there is an increase in temperature between 0.6°C–1.7°C by 2030 and 1.0°C–1.7°C by 2070. Naylor et al. have projected that until 2050 April, May, and June, there will be an increase in rainfall in Central Java, while in July, August, and September, the conditions are projected to be extraordinarily dry [16].

Changes in debit fluctuations in the river basin that are getting higher greatly affect the lives of people in the river basin. Sensitive communities will respond to this condition and cause an increase in community vulnerability in the river basin. However, people who have the ability to adapt will survive with changes or hydrological conditions in this river basin. The sensitivity and adaptability of the community can be assessed from five aspects of life, namely: physical/technology, social/institutional, economic, human resources, and nature.

The level of community vulnerability is influenced by the amount of exposure, the sensitivity of the community, and the adaptability of the community. The higher the exposure and sensitivity of the community, the higher the level of community vulnerability will be. Meanwhile, the higher the adaptability, the lower the level of community vulnerability. In other words, the level of vulnerability is a positive function of community exposure and sensitivity and a negative function of the community's adaptive capacity.

The impact of climate change will occur slowly and continuously. Therefore, adaptation to climate change is very important. One of them is using nature, especially River Basin ecosystems, as one of the adaptation strategies to climate change. River Basin is an area bounded by a ridge that drains water from upstream to downstream. The river basin is divided into upstream, middle, and downstream areas, so the river basin has the function of regulating the hydrological, economic, ecological, and social aspects of an ecosystem.

Based on the results of a search on the International Disaster Database, 345 natural disasters fall into the global disaster category. Around 60% of these disasters are natural disasters due to extreme climate events such as floods, droughts, forest fires, strong winds/storms, landslides, high tidal waves, and outbreaks of disease [15]. This finding is in line with the results of a study by the Intergovernmental Panel on Climate Change [17] that global warming will increase the frequency and intensity of extreme climate events.

For example, extreme climate events in Indonesia are often associated with the El Niño Southern Oscillation (ENSO) phenomenon. There is a strong correlation between ENSO incidence and rainfall variability in Indonesia, namely long dry spells in El Niño years and far above normal rainfall in La Niña years [15]. Global warming will impact increasing the incidence of droughts and floods. Besides that, the beginning of the season and the length of the season will also shift.

Naylor et al. found an effect of global warming on seasonal changes in Java. Their research showed that areas south of the equator would experience a decrease in rainfall while those north of the equator will experience an increase in rainfall. The results of the projections of Naylor et al. showed that in the next 40 years, global warming would cause the beginning of the rainy season in Central Java to experience a setback with rainfall that tends to fall, while the end of the rainy season will be faster with rainfall that tends to increase. This has implications for the increasing risk of drought in the dry season and the risk of flooding or landslides in the rainy season. WWF Annual Review 2007: A watershed year states that the change in the distribution of rainfall causes various potential natural disasters triggered by rainfall to become higher, such as floods, landslides, overflowing rivers, and the spread of disease vectors, while in conditions of reduced rainfall, potential disasters such as drought, crop failure, lack of clean water, and various social problems that may arise.

River Basins provide natural resources that offer many benefits to the surrounding population, including agricultural resources, clean water sources, freshwater fisheries resources, and other water uses. Various community activities along the river basin affect the quality and quantity of the river basin. In addition, the issue of climate change has exacerbated the condition of the River Basin. The increase in temperature and rainfall affects the hydrological conditions of the River Basin.

River Basin degradation causes ecosystems to not optimally provide functions and services that are very important for human life. This condition causes a decrease in the level of community welfare and increases the level of community vulnerability. Community vulnerability is a condition of society that cannot adapt to changes in the ecosystem caused by a certain threat [18]. Vulnerability is a function of three components: exposure, sensitivity, and adaptive capacity (IPCC 2001). The main impact of climate change is the level of community welfare in the upstream, middle and downstream areas.

The shift of seasons in the River Basin area impacts agricultural activities by the community. Agriculture that relies on water supply from the River Basin is done in an ineffective and unprofitable way. The unpredictable climate is also one of the reasons the rice fields are not productive. Water production at some points decreased and even disappeared. The number of dangerous and vulnerable areas has increased. Landslides have increased in the last 2 years [15]. The increasingly critical upstream River Basin degradation causes the River Basin not optimally to provide its functions and services for the community. Upstream River Basins as buffer zones, water catchments, sources of water filters, and carbon sinks will be damaged. Consequently, it will lose water supply and in the rainy season will result in flooding. In a matter of years, this region will become critical. Sensitive people will be more vulnerable, while people who can adapt will survive. Based on these conditions, it is necessary to study the level of community vulnerability to climate change in the upstream River Basin. The existence of climate change and River Basin damage that occurs in the upstream River Basin requires an adaptation strategy. River Basin-based adaptation strategies are important considering that River Basins are providers of ecosystem services for the sustainability of upstream River Basin functions.

3. Law enforcement and institutional arrangements for integrated river basin management

Natural resources and the environment must be maintained to avoid environmental damage or environmental disasters so that development and environmental sustainability can run synergistically. One of the uses of natural resources which is also used for environmental conservation is river basin management.

The utilization of natural resources in river basins can have positive and negative impacts for various purposes. The positive impact is indicated by an increase in economic growth and community welfare. The community can utilize the river basin for irrigation canals for agriculture, excavation of soil and sand that can be used for building materials, tourism objects, and many other benefits. The negative impact is a decrease in the quality and quantity of the river basin environment caused by sediment sourced from erosion and industrial waste caused by densely populated slums. Besides that, another negative impact is the continuous excavation of soil and sand to form basins in the river. River basins that can damage the shape of the land and facilitate landslides, especially on the left and right of the river and riverbed, become rougher. It can increase the erosion and carrying capacity of the river. These negative impacts can change the condition of rivers and river basin ecosystems.

The complexity of the problems of the artificial environment along the river basin requires multidimensional and comprehensive problem-solving. One of the determining factors for the success of efforts to solve these problems is the participation of all levels of society. At this time, community involvement in environmental management along the river basin, from the planning, and implementation to monitoring stages, is still relatively low due to:

a. Understanding the low level of awareness and the problems of the artificial environment.

b. The weak role of social institutions and the business world in supporting the artificial environment management program.

c. Limited community income causes participation capacity to be not optimal.

Furthermore, there seems to be no awareness that actions taken in one river basin will be linked to what happens in another. The level of community education is one aspect that determines the extent to which people have environmental concerns on a wider scale than the environment in which they live.

There are river basin sustainability problems related to the community's social conditions around the river basin and the management of the river basin itself institutionally. A comprehensive, integrated river basin approach requires open management that ensures the continuity of the coordination process between related institutions. In addition, it is also necessary to consider the importance of community participation in river basin management, starting from planning, policy formulation, implementation, and utilization. River basin planning cannot be carried out alone through a sectoral approach. Still, there must be inter-sectoral linkages both in the planning of the state revenue and expenditure budget, work programs, and coordination of implementation. In addition, there are social factors that influence river basin management, including population density, conservation behavior, customary law, traditional values, institutions, and a culture of cooperation or mutual cooperation.

The harmonization of structural relations between institutions in government in the era of regional autonomy, both at the central and local levels, was disrupted due to uncontrollable internal factors, such as sectoral egos and regional egos. This is further complicated by the imbalance in the potential of natural resources and the financial condition of each region. The reality on the ground shows that governmental institutions have overlapping programs and activities in managing natural and water resources.

The reasons behind River Basin management regulations include national economic development, which is carried out based on the principles of sustainable development and a sound environment. The spirit of regional autonomy in governance has brought about changes in the relationship and authority between the central government and regional governments, especially in the field of environmental protection and management.

For example, in Indonesia, having Law Number 32 of 2009 concerning Environmental Protection and Management, it is expected that the use of natural resources must be in harmony and balance with environmental functions. Consequently, development policies, plans, and/or programs must be imbued with the obligation to preserve the environment and realize sustainable development goals included in the management of river basins. The Law of the Republic of Indonesia, Number 32 of 2009 concerning Environmental Protection and Management Article 70 paragraph (1) states that the community has the same and widest possible rights and opportunities to play an active role in environmental protection and management. In addition, based on the Government Regulation of the Republic of Indonesia Number 37 of 2012 concerning River basin Management Article 57, which states:

a. The community can participate in river basin management.

b. The participation of the community, as referred to in paragraph (1) can be carried out either individually or through a coordination forum for river basin management.

c. The river basin management coordination forum, as referred to in paragraph (2), assists in supporting the integrated implementation of river basin management.

Through these laws and regulations, the government gives comprehensive authority to regional governments in protecting and managing the environment in their respective regions.

The regulation provides space for the community to reduce the possibility of the community refusing to accept decisions. Providing access to information on environmental management is also an integral part of community participation in environmental management.

The purpose of community participation since the planning stage is to generate useful inputs and perceptions from citizens and interested communities to improve the quality of environmental decision-making. Because by involving potentially affected communities and interest groups, decision-makers can capture the views, needs, and expectations of these communities and groups and translate them into concepts. The views and reactions of the community, on the other hand, will help decision-makers determine priorities, interests, and positive directions from various factors.

The process of community participation must be open to the public; community participation will affect the credibility (accountability) of the agency concerned. By documenting the actions of this state agency's decisions, to be able to provide a satisfactory means if the public and even the courts feel the need to examine the considerations that have been taken when making the decision, which in the end will be able to force the responsibility of the state agency for the activities it carries out.

Planning institutions that enable integrated or coordinated planning are important and necessary, including institutions for collecting and presenting data and information. The institution in question is the organization and regulation of the mechanism of the relationship between components within the organization and between related organizations. Establishing institutions does not always have to form new organizations but strengthening the role of existing organizations and clarifying relationships between existing organizations.

The main principle in river basin management, such as in Selat Village, Jambi Luar Kota Regency, is to synergize sectoral programs with river basin resource management objectives based on the issue of water resources in the river basin. As a framework for a working approach, every sectoral plan and activities of related parties need to be monitored by the competent authorities in responding to water resources issues and communicating with related parties to resolve problems that cause these issues to occur within an integrated river basin management framework.

Institutional arrangements determine how individuals interact with other individuals and between organizations and other organizations using river basins. This arrangement is important to ensure various stakeholders can accept that river basin management from their respective goals and interests. Three factors can be identified from this institutional arrangement, namely: coherence of interests and activities among stakeholders; strength of local institutions; and benefits to local communities within the river basin.

River basins can be managed well if their resource potential is high and social arrangements and external factors can create a good balance between incentives and controls. Communities will act in the context of rehabilitation and conservation of river basins if they can share the benefits of their actions.

To be able to realize the above-described river basin management, it must first be realized that the performance of river basin management is largely determined by the performance of many institutions/organizations, each of which has different interests, roles, and functions according to the sector, resource, and region.

The initial stage in the search for institutional innovation for integrated river basin management is initiating a dialogue process that aims to identify and, without concern, concepts, programs, and actions based on views in river basin management. In principle, the initiation process can be carried out by individuals or community groups, community organizations, local, regional and national NGOs, district, provincial, or national government agencies, or private institutions, either voluntarily or after a public mandate. Initiation by public institutions with main tasks and functions directly related to river basin resources is expected to maintain a continuous process so that river basin management objectives can be achieved through overall institutional strengthening. The dialog process is carried out in a participatory manner and should be facilitated by an independent facilitator who understands the context of integrated river basin management.

The participation process for integrated river basin management at the initiation stage encourages the formation of a collective decision-making mechanism that is oriented towards solving the main problems in the field, so it is necessary to apply a bottom-up approach to ensure its efficiency and effectiveness. The principles of various capacities, resolutions and building conventions need to be mutually agreed upon in every process implemented. Consensus building is the most effective method of decision-making in the integrated river basin management process. Although it takes a relatively long time and requires commitment, decisions are made based on a contributive approach (sharing capacity and resources) and are determined by mutual agreement. All participants are responsible for implementing any agreed decisions.

The final crystallization of the series of initiation processes is formulating an integrated river basin management plan that is still macro in nature but lays the foundation for establishing a solid social contract. Integrated river basin management planning starts with a basic information review, condition assessment, problem definition, priority setting, objective analysis, alternative analysis, and a joint plan.

If the above agreements/outcomes can be reached, the door for capacity building and division of roles of each involved institution/organization has been opened and can be followed up with efforts to operationalize all agreements through structuring better institutional relations and being accountable to the public, strengthening legal aspects and implementation at the program level. Every collective agreement must ultimately be adopted constitutionally/legally by the institution that has the authority.

Water and river basin management institutions are essentially a social construction that constantly changes and develops according to social and ecological evolution. The form of interaction and the direction of socio-ecological change is highly dependent on the dynamics of the interplay between social and ecosystem elements. In the process of change, there is a multi-way interaction between the river basin management system as the designer and decision-maker; members of the river basin management organization as the main stakeholders for the sustainable use of the river basin, and other ecological and social elements involved.

The concept of water and river basin management institutions includes formal and informal regulations, norms, cognitive bases, and structured symbolic systems to regulate use and distribution and determine the status of water resources within a community group. The concepts mentioned above can be broadly divided into aspects of policy, law, and administration, including formal and informal elements. Water law issues refer to the legal status of water, water rights, conflict resolution and mechanisms, possible conflicts between laws, legal diversity, and the presence or absence of administrative regulations to implement these laws. Policy aspects include the priority of use, cost, ability to decentralize or centralize, and participation and coordination with other policies. The administrative aspect is the organizational structure of water management, including financing, staffing, capacity, and fundraising.

4. Performance indicators and role of stakeholders in river basin management

Stakeholder collaboration and synergy are needed in river basin management to maintain the formalization and implementation of activities or programs that manipulate natural and human resources found in river basins to obtain production and service benefits without causing damage to water and soil resources. It is important to manage and allocate natural resources for river basins, including prevention of flooding and erosion, as well as protection of the aesthetic value associated with natural resources. River basin management should include strategies for identifying linkages between land use, land and water, and linkages between upstream and downstream areas of a river basin. River basin management needs to consider the social, economic, cultural, and institutional aspects operating within and outside the river basin concerned.

Ecosystems must be viewed holistically, namely by identifying the key components that make up the ecosystem and examining the interactions between these components. A holistic approach is carried out so that the utilization and conservation of natural resources can be carried out efficiently and effectively, which is a requirement for the realization of the use of natural resources for sustainable development. River basin management is one of the government's authorities that can be decentralized based on authority and function. The form can refer to the division model, namely deconcentration, delegation to parastatal or semi-autonomous organizations, devolution, privatization, or transfer of affairs from the government to non-governmental institutions [19]. Successful intergroup performance is a function of a number of factors. The broader concept (umbrella concept) that overrides these factors is that the coordination of each of the following factors can affect coordination efforts.

To achieve sustainable river basin development, economic development activities and environmental protection must be harmonized. In this case, it is necessary to unify the two perspectives realistically by adjusting river basin management activities and conservation of upstream areas into economic and social realities. Therefore, it is necessary to know the respective roles of the parties who will collaborate in river basin management, and it is necessary to know how to identify performance indicators.

Stakeholders have the authority and responsibility in terms of managing natural resources around the river basin. Stakeholders are expected to contribute directly or indirectly to river basin management. There are five measurement indicators obtained based on the results of interviews and assessment of the performance scores of activities carried out by various stakeholders, namely:

a. Policy Determination: The role of Stakeholders has a policy in determining the target of activities in the river basin ranging from central policies to regional policies. This policy determines the sustainability of the criteria in river basin management.

b. Activity Goals and Objectives: Each stakeholder must have a set of goals and objectives for activities in the river basin that cannot be contradictory so that the goals and objectives of river basin management can be achieved. Case in point: the Agriculture Service may not clear agricultural and plantation land in the upstream area; the upstream area should be managed by the Forestry Service.

c. Activity Planning: In planning river basin management activities, all stakeholders must be involved in planning activities to establish good coordination and cooperation between stakeholders. These good coordination and cooperation are because each stakeholder has a clear job description.

d. Implementation of Activities: Implementation of activities must be in accordance with the plan of activities that can benefit the community, such as in the development of agriculture, plantations, and animal husbandry. The successful implementation of this activity occurs when each stakeholder collaborates and synergizes in river basin management.

e. Activity Monitoring and Evaluation: This activity must be carried out regularly periodically so that weaknesses in river basin management can be addressed immediately.

5. Conclusions

Based on the description above, it can be concluded:

a. Community empowerment in river basin management must have the power to make decisions autonomously so that community participation has a major role. The current condition is that community empowerment is still consultation participation and participation mobilized by incentives.

b. There is a need for harmonization of structural relationships between institutions in government both at the central and local levels so that internal factors can be controlled, and programs and activities do not overlap in the management of natural resources and water.

c. River basin management starting from upstream, middle, and downstream, needs to involve stakeholders in order to build an integrated river basin management pattern in collaboration and synergy of stakeholders. So that through this river basin, management will create biodiversity conservation, increase land productivity, ecosystem restoration, rehabilitation, and land reclamation.

Acknowledgements

We express our deepest gratitude to the Dean of the Faculty of Engineering at the Sultan Agung Islamic University, who has greatly supported our research costs, and also we do not forget to express our gratitude to all our Doctoral friends in environmental science at Diponegoro University who have discussed and provided input on our writing.

Author details

Benny Syahputra[1*], Berkah Fajar[2] and Sudarno[2]

1 Sultan Agung Islamic University, Semarang, Indonesia

2 Diponegoro University, Semarang, Indonesia

*Address all correspondence to: benny@unissula.ac.id

References

[1] Ohno T, Tanaka T, Sakagami M. Does social capital encourage participatory watershed management? An analysis using survey data from the Yodo River Watershed. Society and Natural Resources. 2010;**23**:303-321. DOI: 10.1080/08941920802078224

[2] Li X, Cheng G, Lin H, Cai X, Fang M, Ge Y, et al. Watershed system model: The essentials to model complexhuman-nature system at the river basin scale. Journal of Geophysical Research-Atmospheres. 2018;**123**:3019-3034. DOI: 10.1002/2017JD028154

[3] Thompson RM, Bond N, Poff NL, Byron N. Towards a systems approach for river basin management-Lessons from Australia's largest river. River Research and Applications. 2019;**35**:466-475. DOI: 10.1002/rra.3242

[4] Withanachchi SS, Kunchulia I, Ghambashidze G, Al Sidawi R, Urushadze T, Ploeger A. Farmers' perception of water quality and risks in the Mashavera River Basin, Georgia: Analyzing the vulnerability of the social-ecological system through community perceptions. Sustainability. 2018;**10**:3062

[5] Angriani P, Ruja IN, Bachri S. River management: The importance of the roles of the public sector and community in river preservation in Banjarmasin (A case study of the Kuin River, Banjarmasin, South Kalimantan, Indonesia). Sustainable Cities and Society. 2018;**43**:11

[6] Chandrakar B, Dewangan NP, Verma S, Mishra A. Empowering community for river basin management. In: Sarma A, Singh V, Kartha S, Bhattacharjya R, editors. Urban Hydrology, Watershed Management and Socio-Economic Aspects. Cham: Springer; 2016. pp. 323-329

[7] Norton BG. Sustainable Values, Sustainable Change: A Guide to Environmental Decision Making. Illinois, Chicago: University of Chicago Press; 2015. p. 344. DOI: 10.1086/692192

[8] Conger JA, Kanungo RN. The empowerment process: Integrating theory and practice. The Academy of Management Review. 1988;**13**(3): 471-482. DOI: 10.2307/258093

[9] Mcwhlrter EH. Empowerment in counseling. Counseling and Development. 1991;**69**:222-227. DOI: 10.1002/j.1556-6676.1991.tb01491.x

[10] Ife JW, Tesoriero F. Community Development: Community-based Alternatives in an Age of Globalisation. 3rd ed. Australia: Pearson; 2006. p. 374

[11] Hales C. Management and Empowerment Programmes. Work, Employment and Society. 2000;**14**(3):501-519. DOI: 10.1177/09500170022118545

[12] Wulandari C. Penguatan Forum DAS sebagai Sarana Pengelolaan DAS secara Terpadu dan Multipihak. In: Proceeding of the Lokakarya Sistem Informasi Pengelolaan DAS: Inisitif Pengembangan Infrastruktur Data. 2007. pp. 171-183

[13] Debnath R. A review of the sustainability of recent watershed management programmes in Bangladesh. Lakes and Reservoirs. 2016;**21**:152-161. DOI: 10.1111/lre.12129

[14] Climate Change. Mitigation of Climate Change [Internet]. 2022.

Available from: http://www. https://report.ipcc.ch/ar6wg3/pdf/IPCC_AR6_WGIII_SummaryForPolicymakers.pdf. [Accessed: May 24, 2022]

[15] de Boer J, Wardekker JA, van der Sluijs JP. Frame-based guide to situated decision-making on climate change. Global Environmental Change. 2010;**20**:502-510. DOI: 10.1016/j.gloenvcha.2010.03.003

[16] Naylor RL, Battisti DS, Vimont DJ, Falcon WP, Burke MB. Assessing risks of climate variability and climate change for Indonesian rice agriculture. In: Proceedings of the National Academy of Sciences (PNAS). 2007. pp. 7752-7757

[17] IPCC. Climate Change 2007: Synthesis report. Contribution of Working Groups I, II and III to the Fourth Assessment Report of the Intergovernmental Panel on Climate Change [Core Writing Team)]. In: Pachauri, RK, Reisinger A, editors. Geneva, Switzerland: IPCC; 2007. p. 104

[18] Füssel HM. Vulnerability: A generally applicable conceptual framework for climate change research. Global Environmental Change. 2007;**17**:155-167. DOI: 10.1016/j.gloenvcha.2006.05.002

[19] Raharja SJ. Paradigma Goevernance dalam Penerapan Manajemen Kebijakan Sektor Publik pada Pengelolaan Sungai. Ilmu administrasi dan organisasi. 2009:82-86

Chapter 13

Incentives for Managing Water Demands: Lessons from the Umgeni River Basin, KwaZulu-Natal, South Africa

Quinex W. Chiluwe, Bimo A. Nkhata and Dev Tewari

Abstract

This paper examines the incentives for managing water demands from a catchment or basin perspective by focusing on defined property rights dimensions. Using property rights theory, the paper has investigated the existence of relationships between attributes of property rights and intentions of water users to conserve water. A case study was used to test whether property rights can be used as incentives in the management of water demands. The results from the analyses that were conducted using IBM SPSS indicated that property rights would be very significant in curtailing water demands in a catchment by acting as incentives in water resource utilisation, specifically by motivating water user users to conserve water. This is an important finding because it would thus help water resource managers to use a properly defined property rights system (better duration and secure tenure) to enable water users curtail the ever-increasing water demands in the river basins.

Keywords: property rights, water demand management, water licences, water conservation, water scarcity

1. Introduction

At a global level, water has been declared to be a social and an economic good by *fiat* by the definition of water in The United Nations Conference on Environment and Development Agenda 21 "*as an integral part of the ecosystem, a natural resource and a social and economic good…*" [1] (Chapter 18) and by the fourth Dublin Statement Principle "*Water has an economic value and should be recognized as an economic good, taking into account affordability and equity criteria*" [2] since 1992. As seen in [3] it also satisfies *de facto* the Robbins conditions of the definition of economy as "*the science which studies human behavior as a relationship between ends and scarce means which have alternative uses*" [4] since scarcity is accepted by OECD in [5] pp. 18,129 and finiteness by the first Dublin Statement Principle "*Water is a finite, vulnerable and essential resource which should be managed in an integrated manner*" [2].

Booker and his colleagues further assert this by stating that water is used in the production of virtually all economic goods and services; and above all, plays a vital role in the provision of basic ecosystem services for human beings and organisms [6]. The importance of water in the economy can also be seen via the Water-Energy-Food (WEF) Nexus as seen in general in [7] and for South Africa in particular in [8–10]. In addition to this, other scientists have recognised the influence that water has on development [11]. For example, it has been claimed by the World Water Assessment Programme [12] that proper management of water resources brings the prospects of poverty reduction and economic growth to developing economies. Brown and Lall [13] further add that the production of food and most infrastructural development initiatives across the world have been affected by the amount of rainfall received and its variability. Rainfall variability plays an important role seen in general in [14], for South Africa in [15–17] and for KwaZulu-Natal Province in particular [18] while extreme rainfall is seen in general in [14] and for South Africa in [19–21]. This is particularly true in Sub-Saharan Africa where infrastructure development in the water sector still lags behind, and storage of the available water is a challenge such that farmers are not able to continue food production without depending entirely on rainfall [22]. As a consequence, it has been argued by Ward [23] that the need to satisfy the growing human demands for water while protecting the aquatic ecosystems on whose products and services economies and life itself depend has emerged to be a significant challenge for 21st Century water policy especially as the demand for freshwater sources continue to increase worldwide.

It is further noted that literature is awash with evidence on the relationship between resource use and incentives to manage the resource. Musole [24] has argued in his paper that resource users tend to increase resource use efficiency when they have stake in the resource. In addition, some scholars [25–27] posit that by properly defining the rights of use of the water resource, there are high chances that a water user will invest in the improvement of the resource and hence ensure its efficient use. However, despite the growing body of knowledge on property rights and natural resource use [28–30], little research has been conducted to examine how the property rights definition would help in managing the increasing water demands at the scale of a river basin. While numerous studies have been conducted to examine the role that property rights play in creating incentives for investment in land use rights and conservation of fish and forestry resources [31–37], there is scarce literature to indicate the existence of similar research in water demand management. In addition to that, while research efforts have advanced in water demand management, most of this research places its focus on residential or domestic water demands and economic instruments like pricing. Efforts to study the response of water users to property rights institutions in water resource utilisation have been insignificant.

The purpose of this paper is therefore to explore how a property rights system can enhance incentives towards managing water demands by luring users to conserve water on their properties. The argument is that property rights can serve as both incentives and disincentives towards the actions of water users and those actions may either reduce or increase water demands. We learn from Bruns and Meinzein-Dick [38] that property rights can secure access to water for existing users and offer equitable ways to meet additional water needs/demand, including urban expansion, economic growth and environmental protection. However, in order to advance the understated aim, an understanding of water availability and scarcity, demand management and property rights theory is required.

The paper is organised as follows; following hereafter is the background to debates on water availability and scarcity bringing out the rationale for improved water management efforts. After this section, the paper gives a brief overview of the current understanding in demand management efforts in the water sector. A discussion of the property rights theory and its applicability in water resources management has been presented next. This section is followed by a methods section which precedes the results and discussion section. The paper finally closes with some significant conclusions that have been generated from the results of the study.

2. Water availability and scarcity

There are various definitions for water availability divided into blue and green parts e.g., "*blue water availability is defined as total natural runoff net of 20% assigned to environmental flow requirements*" [39], "*Green water availability is defined as total rainfall infiltration in agricultural land minus runoff from this area multiplied by a reduction factor for minimum evaporation losses in agriculture of 0.85*" [ibid]. The same authors argue that water accessibility is even more important as water sources are often far from their point of use due to issues of spatial population or/and productive land distribution and industry. For instance, it is reported that the Amazon river has a 95% flow inaccessibility [40] while only less than 50% of the Congo River's flow can be assessed by the population due to infrastructure challenges despite being one of the largest rivers in Africa [41].

As seen in [42] water scarcity is divided into physical scarcity and economic scarcity. Physical scarcity occurs when, due to the global interconnectedness of the hydro-climatic system [43], water fails to satisfy consumption demand as well environmental flows [44]. Economic scarcity occurs when failure to satisfy the aforementioned demands is attributed to the socio-economic system's failure in water utilisation [ibid] either due to inadequacies in storage, timely distribution and access (infrastructure development) [45] or as seen in [46] in case human/ institutional actions or lack of capital place limits to water access.

An apt definition of water scarcity, among many seen in the relevant literature, is the one employed by the EU "*water demand exceeds the water resources exploitable under sustainable conditions*" [47]. According to Shiklomanov [48], 75% of the earth's surface is covered with water, but only 3 per cent of the earth's water is available as freshwater for human use with the other fraction locked up in oceans and hence salty. The quantity of usable water available is further constrained by a number of factors exacerbated by continued economic growth, population growth, climate change and rapid urbanisation that have increased pressure on the resource [49].

Figure 1 shows water withdrawals across the globe as projected from 1995 to 2025 [50]. It is interesting to note from the figure that water withdrawals continue to rise especially in Africa, China and South Asia and South East Asia. Alcamo, and his colleagues [ibid] report that water withdrawals are prone to grow in these regions due to rapid population and economic growth. For instance, Hoekstra, Mekonnen [51] allege that the increased need for food to feed the growing population will lead to more water withdrawals in the agriculture sector which is the driving force behind economic development in these regions. At the same time, the rate of urbanisation especially in the developing world has led to increased domestic water demand, a situation described by Serageldin [52] as worrisome due to multifaceted implications this has on the social well-being of urban populations. Consequently,

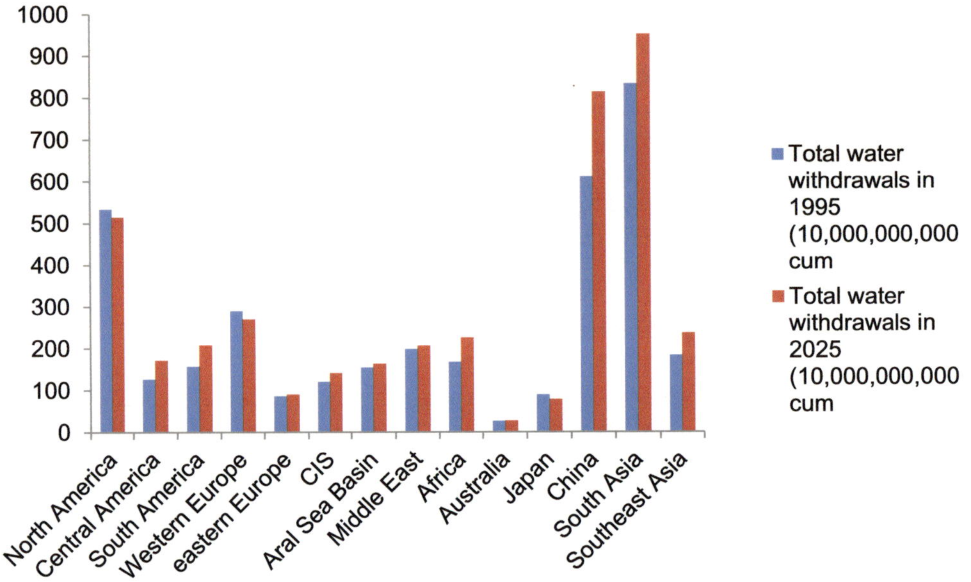

Figure 1.
Graph showing Water Withdrawals projected from 1995 to 2025 [50].

the increase in population coupled with increased water withdrawals has seen approximately 2.1 billion people living in severely water stress basins [50] with Dzikus [53] warning that a total of 1.1 billion people in African countries will be greatly affected by the dwindling water availability if the status quo is maintained.

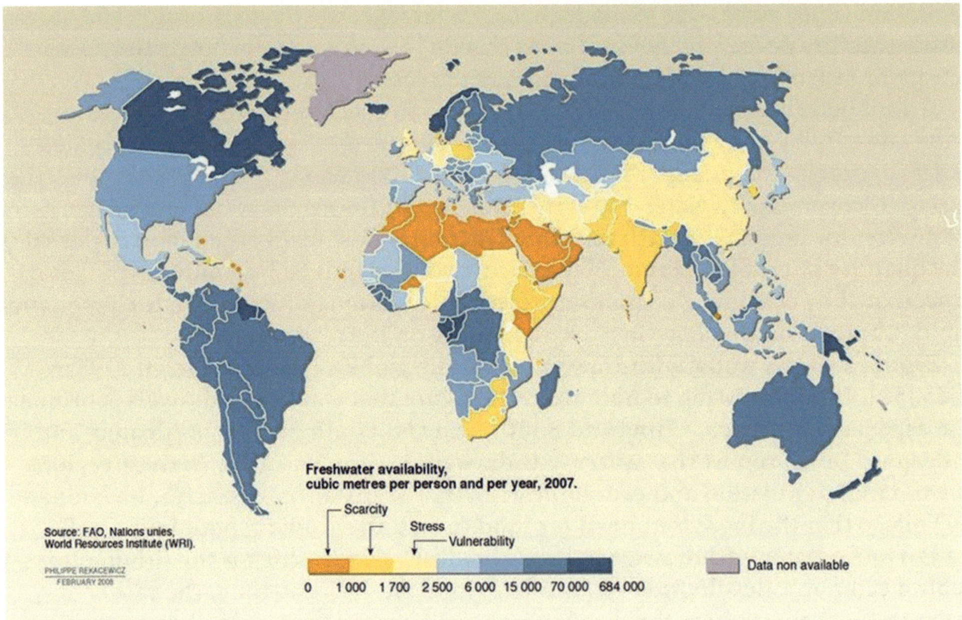

Figure 2.
Map showing freshwater availability across the globe.

The question that remains is whether the available water will be able to meet the growing demands if the situation remains the same [51], and this has led to some authors claiming of an impending global water crisis. However, Lall and Heikkila [54] has acknowledged in their report that the existence or emergence of a global water crisis still remains a topic of controversy among scientists, with some scholars claiming that it is being overstated while raising are questions regarding available data [55]. This not-withstanding, Brown and Lall [13] argues that it is the ability of states to manage their available water resources that affect economic development and social well-being of the society.

Figure 2 is the map showing the availability of freshwater across the globe. While **Figure 1** indicates that global water withdrawals continue to rise in developing countries, it can be observed from **Figure 2** that most African countries including South Africa are heading towards stress levels with some countries especially in North Africa in critical situations of water scarcity.

All this points to the fact that increased water management efforts are imperative and more proactive measures should be sought [56]. However, it has to be admitted that managing the scarce water resources for increasing demands in an equitable and sustainable manner is one of the greatest challenges facing the world in the 21st Century [57, 58].

3. Managing water demands

While the conventional approach to deal with increasing water demands has been to increase supply through infrastructure development for dams and new water supply schemes (29), this has become pecuniary expensive alternative over time as water resources have been affected by multifaceted challenges which include climate change, demographic changes and pollution. It must be highlighted that while managing water demands has been exhorted as probably a more beneficial alternative to supply side management, some authors have found otherwise [56]. For the water users, it has been contended that effective water demand management would enable equity among them and aide in financial savings that would emanate from water use bills [59]. In addition, water supply and management institutions would be saved from making huge infrastructure investments like dam constructions, new water schemes and inter-basin transfers. On the other hand, studies conducted in Iran found that while the adoption of trickle irrigation methods led to improved water use efficiency, there was a significant reduction of the downstream return flows leading to less water available for ecological purposes and those users reliant on these flows [56]. It has been argued by Molle that while there are indeed significant savings by various commercial farmers by employing improving methods of water application and changes in the crop husbandry practices in order to save water, the reality has been that water users tend to utilise every drop of their allocated water by even expanding their current farm coverage [ibid]. This notwithstanding, water demand management strategies should be designed in such a way that they are not a means to themselves but rather with downstream measures to ensure there are water savings that can be reallocated to other users in cases of closed basins as well as environmental uses.

There are thus various types of instruments that have been used to curtail water demands across the globe. These include legal instruments (institutions), economic/financial/market-based instruments, technical instruments and social-political arrangements [59–61]. For the purposes of this study, this discussion will focus on property rights as a form of institutions that can be used to manage water demands.

3.1 Property rights theory

There has been a lot of disagreements on how to define the concept of property rights among scholars in literature [24]. The differences are prominent among legal, economic and social scholars. For example, Furutbotn and Pejovich [62] as cited by Musole [24] defined property rights from a legal perspective as the claims, entitlements and related obligations among people regarding the use and disposition of a scarce resource. However, Barzel [63] later in 1989 cited by Musole [24] contested from an economic view that a person's property rights consists of the rights or the power to consume, obtain income from and alienate the property to another person. On the other hand, Wiebe and Meinzen-Dick [64] in their study on property rights as policy tools in resource use defined property rights as formal and informal institutions and arrangements that govern access to resources, as well as the resulting claims that individuals hold on those resources and on the benefits they generate. Irrespective of the fact that these scholars come from different backgrounds, however, these definitions share some common features. More generally, they all point to the fact that property rights determine what can be done with a resource, by whom, at what time and in what manner, and a permit or a licence system is used for administering or allocating the resource to the user. For the purposes of the current study, the definition proposed by Wiebe and Meinzen-Dick [64] was adopted due to its applicability in the context of water resources management [65]. It must be pointed out that as some authors argue, property rights affect economic outcome in various ways e.g., the resource use pattern, the goods and services produced quantity and mixture as well as the resulting income and wealth distribution but they do not determine it as it can affect in its turn the structure of property rights themselves [66].

3.2 The attributes of property rights

According to Scott and Coustalin [65], rights to natural resources such as water have prescribed attributes such as duration, exclusivity, quality of title, flexibility and transferability that make the property rights structure effective or otherwise. In addition to these, other scholars point out that a successful property rights system needs to have enforcement mechanisms [24, 67, 68]. The way these attributes have been defined represents the quality of the property right system and to an extent may lead to the success or failure of the system in sustainable management of the natural resources. Some authors have argued that a property rights system would be ill-defined if these attributes are not considered in the design of the system leading to the increase in transaction costs and uncertainty among resource users [62, 63, 69–71]. The discussion in this study has, however, been limited to duration, flexibility, enforcement and transferability of the property rights; while the discussion of enforcement has been combined with exclusivity because they quite often share similar traits.

3.2.1 Duration

Duration of a property right is defined by Crase and Dollery [72] as a representation of the period or length of the right possessed by an individual. The length of property rights is an important element in determining water use. As pointed out by several scholars [27, 67, 73], property rights of long duration encourage water users to invest in water saving technologies as well as infrastructure improvements on the resource. This would enhance efficient use of the water resource thereby leading to sustainable utilisation as well as curbing the insurgent demands of water in water stressed river basins.

3.2.2 Exclusivity and enforcement

Exclusivity is the description of the extent to which other resource users can be prevented from accessing the resource and enjoying the benefits of the resource [72]. Most scholars agree that this attribute is an important dimension that determines the success or failure of most property right regimes [27]. According to Musole [24], the right of the resource users to the resource would be deemed exclusive if there are adequate enforcement mechanisms in place. This suggests that the enforceability of the property right is an important prescription that needs to be considered if property rights regimes are to be successful. Understanding the enforceability of the property rights structure would help in understanding the behaviour of water users in the way they use the resource.

Several authors assert to the need for exclusive property rights by stating that they tend to internalise resource depletion [24, 74, 75]. Furthermore, well enforced property rights have been upheld by Kemper and Olson [76] in that they lead to emergence of water markets in water scarce areas. This means that water would be allocated to its highest values within the water scarce river basins. In addition, Rosegrant and Binswanger [77] bring about another significant contribution of exclusive property rights systems particularly in water demand management which was also later echoed by Bruns and Meinzen-Dick [38]. They claim that having an enforceable property right could motivate long-term investments in water saving technologies (an important element in demand management) among water users, cause users to consider the opportunity costs of water and to use it efficiently, and gain additional income from the sale of water and internalise externalities.

3.2.3 Flexibility

Crase and Dollery [72] defines flexibility of property rights as the extent to which the right permits an alteration to the pattern of use without forfeiting the right. Flexibility of the right has some intriguing consequences on the sustainability of water resource utilisation. It has been argued that placing exclusive flexibility on property rights especially in water resources implies that a right-holder can alter the pattern of water usage without regard to the impacts on other users. In view of this, therefore, many scholars recommend that flexibility need to be attenuated to limit the extent to which right-holders can modify usage [64, 68, 78, 79].

3.2.4 Transferability

The transferability of property rights has been defined by Veettil, Speelman [80] as the ability of the resource user to transfer the individual property right to another user either temporarily or permanently in line with the specified rules of the governing body. Authors such as Crase and Dollery [74] consider property rights especially in water resources that are not transferable as ill-defined property rights. This is in agreement with several other scholars who argue that transferability of property right enables resource users to get incentives to invest in the resource and hence improve resource use efficiency [27, 38, 75]. Further than that, transferable property rights in water have been deemed important in the emergence of water markets whereby underutilised and low productive resources can be allocated to higher productive uses [80, 81].

4. Methods

The data used in this study was collected in the Umgeni River Catchment in the year 2013 as part of a larger research program. It focused on water users in the basin who are individual farmers (commercial and small scale), industrial companies and forestry companies in the study area such that the unit of analysis was the individual and corporate water user. According to the National Water Act of 1998, all users of water in a specified area including individuals (farmers, smallholders, landowners or lessees), communities, companies or businesses, water users associations and water service providers are regarded as 'water users' [82] and are required by law to obtain a licence to use the water. The Umgeni River catchment has the population of about 1.6 Million with total urban population at 74% and 26% rural population [83]. This population encompasses a total of 368,250 households. According to the Department of Water Affairs (DWA), the area has higher per capita income as compared to other areas within the province reaching as high as R15,100 while the average lies around R11,000.

A probability sampling technique was utilised in order to determine the respondents that made up the required sample from the population of registered water users in the river catchment. This was deemed appropriate because this would validate the generalisation that can be made from the sample about the population [84] and every member of population had equal possibility of being part of the sample. The sampling technique used here enabled the study to imply the results that emanated from the sampled water users to the rest of the users within the Umgeni River Catchment area.

Questionnaires were distributed to a random sample of 351 out of 818 users that was drawn from an extensive database of registered water users in the study area referred to as the Quaternary catchment U provided by the Department of Water Affairs (DWA). Out of this sample, 146 water users returned the questionnaires. This sample size was calculated using the online sample size calculator [85] at 95% Confidence Level and 6% margin of error or Confidence Interval. The Online Sample Calculator is a tool used for determining sample size in survey research developed by the Creative Research System group of America. The tool uses the target population, Confidence Interval and Confidence Level to determine appropriate sample size for survey research.

The collected data was analysed using Statistical Package for Social Scientists (SPSS) version 20. Four hypotheses were developed to facilitate the investigation of relationships between property rights attributes and intentions by water users to conserve water.

5. Results and discussion

It has been argued in paper that the attributes of property rights affect the incentives for utilising and sustaining water resource base over a period of time. As demands for water continue to increase across the globe particularly due to social and economic dynamics like population growth (and urbanisation) and of course compromised resource quality, it is increasingly becoming important that water users reduce the amount of water they use by practising water conservation on their properties. One of the ongoing challenges is, however, finding the best approach to encourage water users conserve water and curtail the increasing demands. As was reported previously, a number of studies have been conducted aimed at expanding our knowledge on incentives that can enhance water conservation behaviour among water users. The evidence in this paper agrees with Becker and Gibson [86] who argues that formulating and defining property rights to natural resources is one of the fundamental requirements that is necessary for ensuring that resource users have the incentives to conserve the resources and avoid degradation.

5.1 Duration of property rights and water conservation

The importance of the duration of property rights is highlighted by the fact that water users know the period by which they would continue to benefit from the resource using their right. In accordance with the property rights institutions theory, this would have an impact on the behavioural actions among water users in the catchment [74, 87]. The results from the analysis using SPSS Kruskal Wallis Test are presented in **Table 1** showing the relationship between intention to water conserve water and the duration or tenure of their property right.

From the results of the study, it has been evident that there is a significant direct relationship between duration of the right and water conservation intentions among water users. This result denotes that the duration of the property right can be an incentive to enhance conservation of water among water users in the Umgeni River Basin. Nonetheless, this finding does not in itself clarify the extent to which duration would influence the intentions to conserve water as the results indicated a difference in the influence between five to 10 and 21 to 30 durations and 21 to 30 and 31 to 40 year duration (See **Table 2**).

By looking at the mean scores, the results revealed that water users were more motivated to conserve water with the lower durations and this motivation continued to drop as the duration increased. This finding would suggest that shorter water durations are key if water users are to be more conservative in water use, and this would help in reducing water demands in water stressed basins. According to Adhikari [88], resource users with shorter property rights durations attach more importance to optimising their benefits from the resource within the given period. The implication of this proposition would be two-fold. On the one hand, resource users would expropriate more resources from the source as they have no concerns for

Test Statistics[a,b]	
Test Statistics	**Intention to conserve water**
Chi-Square	19.518
Df	5
Asymp. Sig.	.002

[a]*Kruskal Wallis Test.*
[b]*Grouping Variable: What is the length of your water licence?*

Table 1.
The relationship between intention to water conserve water and the duration or tenure of their property right.

Dependent variable	Length of property right	N	Mean rank
Intention to conserve water	5 years	14	92.64
	6 to 10 years	15	90.47
	11 to 20 years	10	76.95
	21 to 30 years	23	49.61
	31 to 40 years	23	86.22
	Greater than 40 years	47	62.55
	do not know	14	88.82
	Total	**146**	

Table 2.
Intention to conserve water vs length of property right.

longer term availability of the resource thereby leading to degradation of the resource and increased conflicts among water users due to increased demands. The resource users may, nevertheless, be conservative with the available resource supply if they have a shorter duration so as to maximise their benefits from the supply. However, other studies indicate that the ability to conserve and sustainably use the resource with respect to the duration of the property rights depends more on the certainty of whether they can easily renew their licence after the expiry of the current allocation [27, 89]. Further than that, although Nikouei, Zibaei [90] argues that resource users are believed to be conservative if they are certain that they will continue benefiting from the resource, Hasan [36] reports in her study that some resource users will reduce their commitment to preserve the integrity of the resource in the long run.

5.2 Flexibility of property rights and water conservation

There is huge evidence in literature that resource degradation and sustainable management is dependent upon the efficient institutional arrangements [71, 91]. It was argued in the preceding sections that placing flexibility on property rights can

Variables	Test statistics	Intention to conserve water	Overall flexibility of property right
Intention to conserve water	Correlation Coefficient	1.000	**.190****
	Sig. (2-tailed)	.	.022
	N	146	146
Overall flexibility of property right	Correlation Coefficient	**.190****	1.000
	Sig. (2-tailed)	.022	.
	N	146	146

***Correlation is significant at the 0.05 level (2-tailed). Bold shows statistically significant results.*

Table 3.
Flexibility of property right vs water conservation.

have both positive and negative impacts on the resource. The results from the correlation analysis between flexibility and intention to conserve water are presented in **Table 3**.

The results of this research indicated that the flexibility of the property rights correlated significantly with water conservation intentions among water users implying that low flexibility may lead to low conservation intentions while higher flexibility may lead to higher conservation intentions. However, the results revealed that the relationship was very weak (as observed from a correlation coefficient of 0.19). Since the flexibility of one's property right would affect the behaviour of another resource user within the resource regime [72], Corral-Verdugo, Frias-Amenta [92] suggest that the intentions of resource users to conserve water would be affected by the actions of those other users in the catchment. Corral-Verdugo and his colleagues argue that water users that are affected by other rights holders who change the pattern of water use are more likely going to be demotivated to conserve water on their property. In addition, these authors claim that flexible property rights would also reduce the likelihood of collective action towards conservation and sustainable utilisation of the resource. From this observation, it can be suggested that there is need to ensure that property rights should exhibit some form of flexibility in order to motivate water users conserve water while at the same time protecting the interests of neighbouring water users.

5.3 Enforceability of property rights and water conservation

Well defined property rights entails having a good compliance monitoring and enforcement mechanisms either done by the community of resource users themselves or indeed by the state [93]. Ostrom and several other colleagues argue that the behaviour of actors lean to a large extent onto the enforceability of the institutions within the context [78, 87, 94, 95]. An analysis on the relationship between enforceability and intentions of water users to conserve water using SPSS Spearman's Rho Correlations has been presented in **Table 4**.

In this study, it has been established that there is a positive correlation between enforcement of property rights and conservation of water by resource users. This result implies that any increases in enforcement of the property rights by the water management agency would result in increased water conservation efforts by water users. Although the current study finds a moderately strong correlation, there is agreement

Variables	Statistics	Intention to conserve water	Overall enforcement of property right
Intention to conserve water	Correlation Coefficient	1.000	.517**
	Sig. (2-tailed)	.	.000
	N	146	146
Overall Enforcement of property right	Correlation Coefficient	**.517****	1.000
	Sig. (2-tailed)	.000	.
	N	146	146

***Correlation is significant at the 0.01 level (2-tailed). Bold shows statistically significant results.*

Table 4.
Enforceability of property rights and water conservation.

with the findings from a study conducted by Yang, Zhang [96] in China. Yang and his colleagues reported that irrigation farmers were more motivated to conserve water by having enforceable water rights than changing the pricing incentives. From this finding, it can be concluded that the enforceability of the property right can be used as incentives towards the conservation of water by water users on either their farm or other properties. By having a well enforced property right system in place, water users would be able reduce water consumptions by adhering to their set water abstraction limits thereby reducing water demands. In so doing, water management agencies would be in a position to contain the surging water demands in stressed river basins.

5.4 Transferability of property rights and water conservation

Transferability of the property rights has received greater attention in property right literature especially by scholars looking water markets. Proponents of transferable property rights argue that transferable property rights are very important in water resource management because they aide in water being allocated to its highest value and efficiently utilised. In very water stressed basins, water can be transferred from one sector to another and from one user to another as long as the property rights institutions allow transferability of rights. In the light of escalating water demands, transferring the property rights from low value uses to high value uses may help manage the demands at basin level. An analysis on the relationship between transferability and intentions of water users to conserve water using SPSS Spearman's Rho Correlations has been presented in **Table 5**.

The results from this study suggest a significant relationship between transferability of the right and intentions of water users to conserve water although the correlation effect is not strong enough. This finding may imply that water users find the transferability of their right as incentives towards conservation of water but not to a larger extent. As discussed earlier in the beginning of the paper, even though transferability of property rights has some significant benefits in water allocation, some authors critiqued the concept of transferability.

Scholars such as Anderson have argued "*since rights cannot be perfectly enforced, ownership will always be probabilistic; but when the probability of capturing benefits from a use is low, it is less likely that the owner will devote the resource to that use*" [97].

Variables	Statistics	Intentions to conserve water	Transferability of property right
Intentions to Conserve Water	Correlation Coefficient	1.000	.580*
	Sig. (2-tailed)	.	.030
	N	146	146
Transferability of Property right	Correlation Coefficient	.580*	1.000
	Sig. (2-tailed)	.030	.
	N	146	146

**Correlation is significant at the 0.05 level (2-tailed).*

Table 5.
Transferability of property rights and water conservation.

In Pakistan a survey of watercourses (1990) showed the existence of 70% active trading between farmers [98] while market type trading of water rights is seen to result in social benefits via improvement of water resource allocation efficiency. Allocative efficiency maximisation which leads to net economic returns maximisation is rarely attained in practice as there are supply and demand imbalances [99]. Water intersectoral reallocation, as long as economic efficiency is attained via transfer to a use of higher value, is a positive process as seen in [100]. However, in this case, a framework of intersectoral trading is created where a competitive situation arises [101] unless the transfer is facilitated by central economic policy *fiat* as is the case in Jordan's shift from agricultural to industrial use [102]. Transferable property rights may lead to overutilization and/or overexploitation of the resource as seen in [103]. For instance, when water has been allocated to the highest efficiency sectoral value uses, like from agriculture production to industrial uses, there are high chances that the new right holders will aim to optimise production per unit water allocation within the particular industrial process employed which differs from agricultural use. The resulting consequence can be increased degradation of the water resource thereby creating water stress in the river basin.

6. Conclusions and implications for water demand management research

In the preamble of this paper, it was indicated that most demand management literature and research focus on the use of economic and market-based incentives to curtail water demands. Indeed, there is a lot of scholarly work that supports that water demand can better be managed by economic incentives like pricing [59, 76, 104–106]. However, there is also growing concerns on the applicability of these instruments in a river basin context [56, 80, 107] as most of this research is conducted with a primary focus on either residential water demands or irrigation water demands.

While there is acknowledgement of property rights institutions as an instrument that can be used to manage water demands [56, 59, 60, 108], it was observed that little research work was done to understand the potential and applicability of property rights in this respect. This research, therefore, aimed at examining the role of property rights in managing this water demand as an alternative incentive to the economic instruments. The main argument was that property rights institutions guide the behaviour of resource users towards resource utilisation [24, 64, 89, 95]. By utilising the property rights theory [29], a proposition was made that the attributes of property rights would act as incentives and influence the patterns of behaviour of water resource users resulting in either reduction or increase in water use. The interaction between these property rights attributes and water users would result in a well-managed water demand scenario or another tragedy of the water resources commons.

Using evidence obtained in this study, property rights would be very significant in curtailing water demands in a catchment by acting as incentives in water resource utilisation, specifically by motivating water user users to conserve water. This finding presents a new dimension in water demand management research. Although these results are mostly inconclusive as regards the amount water savings property rights would have, efforts would be done to delve into understanding more on the incentive structure of property rights as regards water resource utilisation across the sectors in a catchment. By focusing on catchment wide water users, research would be able to incorporate dynamics that affect water user behaviour often ignored in studies focusing on residential water demand management.

Acknowledgements

The author acknowledges the support of colleagues from the Research Directorate at Monash, Professor Charles Breen, Duncan Hay, Machaya Chomba and all participants who took their time to complete the questionnaires. Most importantly, recognises Mr. Tylcoat Cameron from the Department of Water and Sanitation in Durban for providing necessary information that was needed for the research. Thanks to Monash South Africa for funding this research.

Author details

Quinex W. Chiluwe[1*], Bimo A. Nkhata[2] and Dev Tewari[3]

1 University of Pretoria, Hatfield, South Africa

2 Water Research Centre, IIEMSA, Roodepoort, South Africa

3 University of Zululand, KwaDlangezwa, South Africa

*Address all correspondence to: qchiluwe@gmail.com

References

[1] United Nations. Protecting the Quality and Supply of Freshwater Resources: Application of Integrated Approaches to the Development, Management, and Use of Water Resources. In Agenda 21. New York, NY: United Nations Publication. 1992; Chapter 18

[2] ICWE. The Dublin statement on water and sustainable development. In: International Conference on Water and the Environment. Dublin, Ireland: ICWE; 1992

[3] Zisopoulou K, Zisopoulos D, Panagoulia D. Water economics: An In-depth analysis of the connection of blue Water with some primary level aspects of economic theory I. Watermark. 2022;**14**(1):103

[4] Robbins LRB. The Nature and Significance of Economic Science. London: Macmillan; 1932

[5] Herrington P. Pricing of water services. Organisation for Economic Co-operation and Development. Ottawa: Renouf; 1987

[6] Booker JF, Howitt RE, Michelsen AM, Young RA. Economics and the modeling of water resources and policies. Natural Resource Modeling. 2012;**25**(1):168-218

[7] Zisopoulou K, Karalis S, Koulouri M-E, Pouliasis G, Korres E, Karousis A, et al. Recasting of the WEF Nexus as an actor with a new economic platform and management model. Energy Policy. 2018;**119**:123-139

[8] Conway D, Van Garderen EA, Deryng D, Dorling S, Krueger T, Landman W, et al. Climate and southern Africa's water–energy–food nexus. Nature Climate Change. 2015;**5**(9):837-846

[9] Mabhaudhi T, Simpson G, Badenhorst J, Mohammed M, Motongera T, Senzanje A, et al. Assessing the State of the Water-energy-food (WEF) Nexus in South Africa. Vol. 76. Pretoria, South Africa: Water Research Commission (WRC); 2018

[10] Naidoo D, Nhamo L, Mpandeli S, Sobratee N, Senzanje A, Liphadzi S, et al. Operationalising the water-energy-food nexus through the theory of change. Renewable and Sustainable Energy Reviews. 2021;**149**:111416

[11] Lenton R, Muller M. Integrated Water Resources Management in Practice: Better Water Management for Development. Routledge, London; 2012

[12] World Water Assessment Programme (United Nations). Water: A shared responsibility. UNESCO, Paris; 2006

[13] Brown C, Lall U. Water and economic development: The role of variability and a framework for resilience. In Natural resources forum. Oxford, UK: Blackwell Publishing Ltd. 2006;**30**(4):306-317

[14] Panagoulia D, Vlahogianni EI. Nonlinear dynamics and recurrence analysis of extreme precipitation for observed and general circulation model generated climates. Hydrological Processes. 2014;**28**(4):2281-2292

[15] Mason SJ, Jury M. Climatic variability and change over southern Africa: A reflection on underlying processes. Progress in Physical Geography. 1997;**21**(1):23-50

[16] Fauchereau N, Trzaska S, Rouault M, Richard Y. Rainfall variability and changes in southern Africa during the 20th century in the global

warming context. Natural Hazards. 2003;**29**(2):139-154

[17] Reason CJ, Landman W, Tadross M, Tennant W, Kgatuke M. Seasonal to Decadal Predictability and Prediction of Southern African Climate. UCT, Cape Town. 2004; pp. 1-44

[18] Ndlovu M, Clulow AD, Savage MJ, Nhamo L, Magidi J, Mabhaudhi T. An assessment of the impacts of climate variability and change in Kwa Zulu-Natal Province, South Africa. Atmosphere. 2021;**12**(4):427

[19] De Waal JH, Chapman A, Kemp J. Extreme 1-day rainfall distributions: Analysing change in the Western cape. South African Journal of Science. 2017;**113**(7-8):1-8

[20] Landman WA, Botes S, Goddard L, Shongwe M. Assessing the predictability of extreme rainfall seasons over southern Africa. Geophysical Research Letters. 2005;**32**(23)

[21] Sikhwari T, Nethengwe N, Sigauke C, Chikoore H. Modelling of extremely high rainfall in Limpopo Province of South Africa. Climate. 2022;**10**(3):33

[22] Serdeczny O, Adams S, Baarsch F, Coumou D, Robinson A, Hare W, et al. Climate change impacts in sub-Saharan Africa: From physical changes to their social repercussions. Regional Environmental Change. 2017;**17**(6):1585-1600

[23] Ward FA. Decision support for water policy: A review of economic concepts and tools. Water Policy. 2007;**9**(1):1-31

[24] Musole M. Property rights, transaction costs and institutional change: Conceptual framework and literature review. Progress in Planning. 2009;**71**(2):43-85

[25] Ahmad M. Water pricing and markets in the near east: Policy issues and options. Water Policy. 2000;**2**(3):229-242

[26] Ray D. Introduction to development theory. Journal of Economic Theory. 2007;**137**(1):1-10

[27] Speelman S, Farolfi S, Frija A, Van Huylenbroeck G. Valuing improvements in the water rights system in South Africa: A contingent ranking approach 1. JAWRA Journal of the American Water Resources Association. 2010;**46**(6):1133-1144

[28] Meinzen-Dick R, Nkonya L. Understanding legal pluralism in water rights: Lessons from Africa and Asia. In African Water Laws Workshop: Plural Legislative Frameworks for Rural Water Management in Africa. IWMI. 2005; H038746

[29] Ostrom E. Doing Institutional Analysis Digging Deeper than Markets and Hierarchies. Boston, MA. Springer; 2005. pp. 819-848

[30] Schlager E, Ostrom E. Property-rights regimes and natural resources: A conceptual analysis. Land Economics. 1992:249-262

[31] Agrawal A, Ostrom E. Collective action, property rights, and decentralization in resource use in India and Nepal. Politics and Society. 2001;**29**(4):485-514

[32] Begossi A. Property rights for fisheries at different scales: Applications for conservation in Brazil. Fisheries Research. 1998;**34**(3):269-278

[33] Costello CJ, Kaffine D. Natural resource use with limited-tenure property rights. Journal of Environmental Economics and Management. 2008;**55**(1):20-36

[34] Guerin, K. Property rights and environmental policy: A New Zealand perspective. No. 03/02. New Zealand Treasury Working Paper; 2003

[35] Haddad BM. Property rights, ecosystem management, and John Locke's labor theory of ownership. Ecological Economics. 2003;**46**(1):19-31

[36] Hasan L. Property Regimes in Resource Conservation-A Framework for Analysis. Germany: University Library of Munich; 2000

[37] Heltberg R. Property rights and natural resource management in developing countries. Journal of Economic Surveys. 2002;**16**(2):189-214

[38] Bruns BR, Meinzen-Dick R. Frameworks for water rights: an overview of institutional options. Water rights reform: Lessons for institutional design. IFPRI; Washington. 2005:3-25

[39] Zisopoulou K, Panagoulia D. An in-depth analysis of physical blue and green water scarcity in agriculture in terms of causes and events and perceived amenability to economic interpretation. Watermark. 2021;**13**(12):1693

[40] Latrubesse EM. Patterns of anabranching channels: The ultimate end-member adjustment of mega rivers. Geomorphology. 2008;**101**(1-2):130-145

[41] Chishugi DU, Sonwa DJ, Kahindo J-M, Itunda D, Chishugi JB, Félix FL, et al. How climate change and land use/land cover change affect domestic water vulnerability in Yangambi watersheds (DR Congo). Landscape. 2021;**10**(2):165

[42] Seckler DW. World Water Demand and Supply, 1990 to 2025: Scenarios and Issues. IWMI; Colombo, Sri Lanka. 1998

[43] Alcamo JM, Vörösmarty CJ, Naiman RJ, Lettenmaier DP, Pahl-Wostl C. A grand challenge for freshwater research: Understanding the global water system. Environmental Research Letters. 2008;**3**(1):010202

[44] Ahopelto L, Veijalainen N, Guillaume JH, Keskinen M, Marttunen M, Varis O. Can there be water scarcity with abundance of water? Analyzing water stress during a severe drought in Finland. Sustainability. 2019;**11**(6):1548

[45] Nairizi S. Drought and Water Scarcity. International Commission on Irrigation and Drainage: New Delhi, India; 2017

[46] Molden D. Water for Food Water for Life: A Comprehensive Assessment of Water Management in Agriculture. Routledge, London; 2013

[47] European Commission. Communication from the Commission to the European Parliament and the Council. Brussels: European Commission; 2007

[48] Shiklomanov IA. Summary of the monograph "world water resources at the beginning of the 21st century" prepared in the framework of IHP. UNESCO. 1999;**12**(11):2010

[49] Rosegrant MW, Cai X, Cline SA. Global Water Outlook to 2025: Averting an Impending Crisis. IWMI; Colombo, Sri Lanka. 2002: No. 572-2016-39087

[50] Alcamo J, Döll P, Henrichs T, Kaspar F, Lehner B, Rösch T, et al. Global estimates of water withdrawals and availability under current and future "business-as-usual" conditions. Hydrological Sciences Journal. 2003;**48**(3):339-348

[51] Hoekstra AY, Mekonnen MM, Chapagain AK, Mathews RE, Richter BD. Global monthly water scarcity: Blue water footprints versus blue water availability. Plo S one. 2012;**7**(2):e32688

[52] Serageldin I. Water resources management: A new policy for a sustainable future. Water International. 1995;**20**(1):15-21

[53] Dzikus A. Managing water for African Cities: An introduction to urban water demand. In: Regional Conference on the Reform of the Water Supply and Sanitation Sector in Africa–Enhancing Public-Private Partnership in the Context of the Africa Vision for Water (2025). Uganda: Kampala; 2001

[54] Lall U, Heikkila T, Brown C, Siegfried T. Water in the 21st century: Defining the elements of global crises and potential solutions. Journal of International Affairs. 2008:1-17

[55] Dantas IRM, Delzeit R, Klepper G. Economic research on the global allocation of scarce Water resources needs better data. Water Economics and Policy. 2021;**7**(03):2150013

[56] Molle F, Turral H. Demand management in a basin perspective: is the potential for water saving overestimated?. In International Water Management Institute Conference . IWMI; Colombo, Sri Lanka. 2004: No. h034365

[57] Hoekstra AY. Sustainable, efficient, and equitable water use: The three pillars under wise freshwater allocation. Wiley Interdisciplinary Reviews Water. 2014;**1**(1):31-40

[58] Mekonnen MM, Pahlow M, Aldaya MM, Zarate E, Hoekstra AY. Sustainability, efficiency and equitability of water consumption and pollution in Latin America and the Caribbean. Sustainability. 2015;**7**(2):2086-2112

[59] Tate DM. Water Demand Management in Canada: Inland Waters Directorate. Water Planning and Management Branch; Quebec. 1990

[60] Mohamed AS. Water Demand Management: Approach, Experience and Application to Egypt. Delft: Technische University; 2001

[61] Savenije HH, Van Der Zaag P. Water as an economic good and demand management paradigms with pitfalls. Water International. 2002;**27**(1):98-104

[62] Furubotn EG, Pejovich S. Property rights and economic theory: A survey of recent literature. Journal of Economic Literature. 1972;**10**(4):1137-1162

[63] Barzel Y. Economic Analysis of Property Rights. Cambridge: Cambridge University Press; 1989

[64] Wiebe KD, Meinzen-Dick R. Property rights as policy tools for sustainable development. Land Use Policy. 1998;**15**(3):203-215

[65] Scott A, Coustalin G. The evolution of water rights. Natural Resources Journal. 1995;**35**:821

[66] Ditwiler CD. Water problems and property rights—An economic perspective. Natural Resources Journal. 1975;**15**(4):663-680

[67] Narain V. Institutions, Technology and Water Control: Water Users Associations and Irrigation Management Reform in Two Large-Scale Systems in. India: Wageningen University and Research; 2003

[68] Speelman S, Frija A, Buysse J, Van Huylenbroeck G. The importance of

irrigation water rights: Lessons from South Africa and Tunisia. Water Policy. 2011;**13**(5):663-676

[69] Bajt A. The property rights school: Is economic ownership the missing link? International Review of Law and Economics. 1993;**13**(1):85-97

[70] Bell D. The structure of rights in the context of private property. The Journal of socio-economics. 1995;**24**(4):607-622

[71] Fitzpatrick D. Evolution and chaos in property right systems: The third world tragedy of contested access. Yale LJ. 2005;**115**:996

[72] Crase L, Dollery B. Water rights: A comparison of the impacts of urban and irrigation reforms in Australia. The Australian Journal of Agricultural and Resource Economics. 2006;**50**(3):451-462

[73] Hodgson S. Modern Water Rights: Theory and Practice. Food & Agriculture Org; Rome. 2006

[74] Mwangi E, Meinzen-Dick RS. Understanding Property Rights in Land and Natural Resource Management. Institutional Economics Perspectives on African Agricultural Development; 2009. pp. 295-318

[75] Narain V. Water Rights System as a Demand Management Option: Potentials, Constraints and Prospects. Strategic Analyses of the National River Linking Project (NRLP) of India: Series 3. IWMI; Colombo, Sri Lanka. 2009:127-146

[76] Kemper KE, Olson D. Water pricing: the dynamics of institutional change in Mexico and Ceará, Brazil. In: The Political Economy of Water Pricing Reforms. 2000. pp. 339-357

[77] Rosegrant MW, Binswanger HP. Markets in tradable water rights: Potential for efficiency gains in developing country water resource allocation. World Development. 1994;**22**(11):1613-1625

[78] Orindi V, Huggins C. The Dynamic Relationship between Property Rights, Water Resource Management and Poverty in the Lake Victoria Basin. IWMI; Colombo, Sri Lanka. 2005: H038769

[79] White SE, Kromm DE. Appropriation and water rights issues in the high plains Ogallala region. The Social Science Journal. 1996;**33**(4):437-450

[80] Veettil PC, Speelman S, Frija A, Buysse J, Van Huylenbroeck G. Complementarity between water pricing, water rights and local water governance: A Bayesian analysis of choice behaviour of farmers in the Krishna river basin, India. Ecological Economics. 2011;**70**(10):1756-1766

[81] Tietenberg T. Tradable permits in principle and practice. Moving to Markets in Environmental Regulation. In: Moving to Markets in Environmental Regulation. Oxford: Oxford University Press Online; 2006

[82] Republic of South Africa. National Water Act No 36, Government Gazette. Cape Town. 1998; p 19182

[83] DWAF. Mvoti to Mzimkulu Water Management Area Situational Assessment. Kwa Zulu Natal Regional Office; Durban. 2001

[84] Bryman A. Social Research Methods. Oxford, UK: Oxford University Press; 2016

[85] American Marketing Association. Creative Research Systems, the Survey System-sample Size Calculator. 2007.

[86] Becker CD, Gibson C. The Lack of Institutional Supply: Why a Strong Local

Community in Western Ecuador Fails to Protect its Forest. FAO; Rome. 1998. pp. 111-133

[87] North DC. Institutions and the Performance of Economies over Time. Berlin: Springer; 2008. pp. 21-30

[88] Adhikari B. Literature review on the economics of common property resources. Report prepared for NRSP. 2001

[89] Snell M, Bell KP, Leahy J. Local institutions and Natural Resource Management. In 2010 Annual Meeting, July 25-27, 2010, Denver, Colorado. Agricultural and Applied Economics Association 2010 (No. 61880)

[90] Nikouei A, Zibaei M, Ward FA. Incentives to adopt irrigation water saving measures for wetlands preservation: An integrated basin scale analysis. Journal of Hydrology. 2012;**464**:216-232

[91] Gerber J-D, Knoepfel P, Nahrath S, Varone F. Institutional resource regimes: Towards sustainability through the combination of property-rights theory and policy analysis. Ecological Economics. 2009;**68**(3):798-809

[92] Corral-Verdugo V, Frias-Armenta M, Pérez-Urias F, Orduña-Cabrera V, Espinoza-Gallego N. Residential water consumption, motivation for conserving water and the continuing tragedy of the commons. Environmental Management. 2002;**30**(4):527-535

[93] Thobani M. Formal water markets: Why, when, and how to introduce tradable water rights. The World Bank Research Observer. 1997;**12**(2):161-179

[94] Brown KM. New challenges for old commons: The role of historical common land in contemporary rural spaces. Scottish Geographical Journal. 2006;**122**(2):109-129

[95] Ostrom E. Background on the institutional analysis and development framework. Policy Studies Journal. 2011;**39**(1):7-27

[96] Yang H, Zhang X, Zehnder AJ. Water scarcity, pricing mechanism and institutional reform in northern China irrigated agriculture. Agricultural Water Management. 2003;**61**(2):143-161

[97] Anderson TL. The new resource economics: Old ideas and new applications. American Journal of Agricultural Economics. 1982;**64**(5):928-934

[98] Thobani M. Tradable property rights to Water: How to improve water use and resolve water conflicts. The World Bank. 1995

[99] Vaux HJ. Economic factors shaping western water allocation. American Journal of Agricultural Economics. 1986;**68**(5):1135-1142

[100] Howe CW, Schurmeier DR, Shaw WD Jr. Innovative approaches to water allocation: The potential for water markets. Water Resources Research. 1986;**22**(4):439-445

[101] Farolfi S, Perret SR. Inter-Sectoral Competition for Water Allocation in Rural South Africa: Analysing a Case Study through a Standard Environmental Economics Approach. University of Pretoria; Pretoria, South Africa. 2002: No. 1737-2016-140373

[102] Carpio CE, Ramirez OA, Boonsaeng T. Potential for tradable water allocation and rights in Jordan. Land Economics. 2011;**87**(4):595-609

[103] Ajuzie EI, Altobello MA. Property rights and pollution: Their implications

for Long Island sound and the oyster industry. Applied Economic Perspectives and Policy. 1997;**19**(2):242-251

[104] Mwendera E, Hazelton D, Nkhuwa D, Robinson P, Tjijenda K, Chavula G. Overcoming constraints to the implementation of water demand management in southern Africa. Physics and Chemistry of the Earth. 2003;**28**(20-27):761-778

[105] Olmstead SM, Hanemann WM, Stavins RN. Water demand under alternative price structures. Journal of Environmental Economics and Management. 2007;**54**(2):181-198

[106] Winpenny J. Demand Management for Efficient and Equitable Use. Water: Economics, Management and Demand. E & FN Spon, Chapman Hall; London. 1997. pp. 296-303

[107] Mohamed A. Water demand management in Egypt: Policy objectives and strategy measures. Physics and Chemistry of the Earth, Part B: Hydrology, Oceans and Atmosphere. 2000;**25**(3):243-249

[108] Brooks B. Water demand management: Conceptual framework and pollicy implemation. In: Management of Water Demand in Africa and the Middle East: Current Practices and Future Need IDRC. 1997:1-10

Chapter 14

Water Control, Impacts and Sub-Regional Cooperation around a Transboundary Hydrological System – The Case of the Kayanga/Geba Catchment Area: (Guinea, Senegal and Guinea-Bissau)

Saly Sambou, Rene Ndimag Diouf and Joseph Sarr

Abstract

The Kayanga/Geba river basin is a transboundary basin shared between Guinea, Senegal and Guinea-Bissau. It concentrates important natural resources, notably water resources on which Senegal and Guinea-Bissau are particularly dependent. The drastic reduction of these water resources due to rainfall variability and climate change has had an impact on agricultural production in the basin; hence the hydro-agricultural developments, in Senegalese territory, boost socio-economic activities by increasing productivity in both the rainy and dry seasons. The negative effects of these developments go beyond administrative boundaries. The transboundary management of this basin is a real challenge because the dams built in Senegal do not have the legal status of common dams of the OMVG whose mission is to promote cooperation between its member states. This article first analyses water control and some of the negative impacts of hydro-agricultural developments, and then the cooperation initiatives that the OMVG is trying to implement for rational and harmonious exploitation of the common resources of this basin.

Keywords: water control, transboundary water system, sub-regional cooperation, Kayanga/Geba basin

1. Introduction

The catchment area becomes transboundary when it extends between two or more countries [1]. In addition to being numerous in West Africa, transboundary watersheds are often the primary water resources of the countries that border them [2]. The transboundary situation of these basins causes a fundamental problem when considering the management of shared natural resources. In many cases, an ecosystem divided by a political-administrative boundary is managed in a fragmented and sometimes contradictory way by the states that share it. This is due to the fact that sometimes states have different political priorities and environmental regulations [3].

At the end of the 1960s and the beginning of the 1970s, recurrent dry spells and climate change severely impacted agricultural production in the Kayanga/Geba basin, where the essential part of the population's income traditionally comes from rainfed crops in the uplands and rice cultivation in the lowlands, in addition to the livestock. To overcome this problem, the state of Senegal has adopted policies for the development and management of water resources in order to improve water management, boost socio-economic activities by increasing agricultural productivity, both in the rainy season and in the off-season, and promote local development [4, 5]. This is how the Agricultural and Industrial Development Company of Senegal (SODAGRI in French) created in 1974, was entrusted with the management of the three phases of the development of the Anambe basin, the central part of which is a vast flood basin of almost 16,000 ha [6]. The Anambe is the main tributary of the Kayanga/Geba in Senegal. The developments are for hydro-agricultural and pastoral purposes with integration of agriculture, livestock, and continental fishing. In total, two dams have been built for this purpose in addition to the one set up by the Local Small Scale Irrigation Support Project (PAPIL in French).

In Guinea-Bissau, no hydro-agricultural development has been carried out. The Kayanga/Geba and Koliba/Corubal rivers are the main source of surface water in this country. It shares the Koliba/Corubal with Guinea. The other rivers are deeply penetrated by the tide.

The developments in the Senegalese part of the basin may run counter to the benefits that can be derived from harmonious use of water resources and cooperative management at the scale of the hydrological system. Because managing at this scale means ultimately taking advantage of the comparative advantages of each part of the hydrological system and respecting its total productive capacity [2]. This article aims to analyse water control and transboundary cooperation around the Kayanga/Geba river basin. It is a modest contribution to the analysis of the challenges of transboundary basin management in West Africa.

2. Presentation of the Kayanga/Geba catchment area

The Kayanga/Geba watershed is a transboundary basin (**Figure 1**) shared by Guinea (1.3% of the total area), Senegal (34.3%) and Guinea-Bissau (64.4%) [7, 8]. The river has its source in the western part of the Badiar Plateau (Guinea), in swamps, at an altitude of about 90 m. It flows northwest for about 10 km, passes through Senegal as it flows west and then southwest, enters Guinea-Bissau after a course of about 150 km and takes the name Rio Geba [9]. The South Sudanese climate is characterised by a rainy season from May to October and a dry season for the rest of the year. The latitudinal configuration of the basin shows rainfall contrasts with an average annual rainfall of about 1000 mm in the north and 1500 mm in the south. Average monthly temperatures range from 23.7°C in December to 31.7°C in May. The average monthly temperature ranges from 23.7°C in December to 31.7°C in May at the Kolda station. In this basin, agriculture remains the main activity and source of income for the predominantly rural population. The other socio-economic activities such as the livestock, the continental fishing, the arboriculture, the market gardening and the trade are also

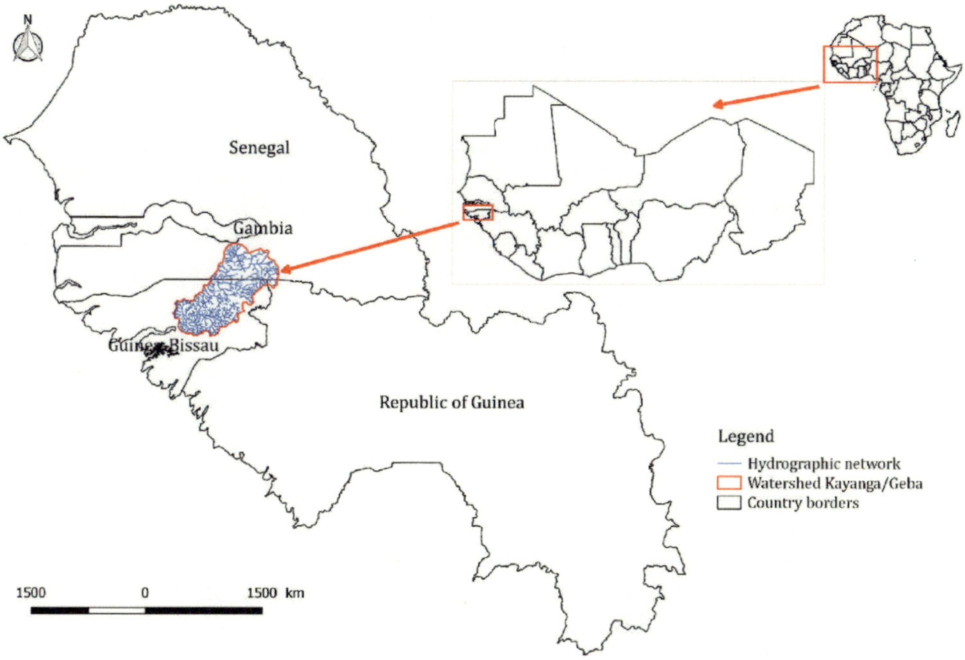

Figure 1.
Location of the Kayanga/Geba River watershed [7].

Basin	Countries sharing the basin	Area (km^2)	Population of the basin	% of the basin by country	% of the population by country
Kayanga/ Geba	Guinea	12,440	760,000	1.3	0.4
	Senegal			34.3	41.8
	Guinea-Bissau			64.4	57.8

Table 1.
Distribution of the basin population by country.

developed there. However, they are submitted to hydro-rainfall variations [7]. The total population of the basin is estimated at 760,000 residents, of which 0.4% live in Guinea, 41.8% in Senegal and 57.8% in Guinea-Bissau (**Table 1**) [10].

3. Water control in the basin

Water control generally involves the construction of dams, if necessary, which make it possible to compensate for the irregularity of seasonal or interannual water supplies by accumulating reserves. Dams are artificial structures built across the bed of a river to retain water or provide an artificial reservoir [11]. Thus, the benefits of an agricultural dam will be felt very gradually, as cultivation processes evolve, which we know are slow [12].

During the first phase of the development of the Anambe basin (Senegal), the Confluent dam-reservoir was built 300 m downstream from the Kayanga/Geba and the Anambe confluence. With a storage capacity of 59 million cubic meters, it was built in 1984 and diverts the flow of the river to fill the Anambe basin and irrigate 1365 ha [13]. In the non-rainy season, the water is blocked by the Kounkane weir. After more than a decade of exploitation, it was found that the additional water provided by this dam was not sufficient to achieve the objective of double cropping, due to rainfall deficits, coupled with significant water leakage downstream of the system, estimated at more than 50% of the runoff. These water leaks resulting from errors in the construction of the dam (defective retaining dyke), considerably reduce the possibilities of storing water [14]. It is in this context that the second phase was started with the construction of the Niandouba dam-reservoir in 1997, in addition to the development of new perimeters covering an area of 2805 ha. Located 10 km upstream of the Confluence reservoir, with a storage capacity of 85 million cubic metres, it provides a backup to this buffer reservoir. The main objectives are: to store water during the rainy season in order to provide the necessary complement to secure off-season crops in the Anambe basin; to allow the development of continental fishing in the reservoir in all seasons; and to contribute to the recharging of the underground water for human and animal water supply. This has resulted in improved hydrological conditions with over 100 million cubic metres of water available, allowing for relative water control in the Anambe-Kayanga/Geba system [13].

In 2012, PAPIL built the Velingara-Pakane dam-reservoir upstream of the Niandouba dam-reservoir to irrigate areas on the right bank of the river. The storage capacity is 1.5 million cubic metres.

Despite these dams, water control in this part of the basin is low, because development and production objectives have not yet been achieved due to, firstly, the rainfall deficit and, secondly, the production factors. However, a new dam project is envisaged at the confluence of the Niokolo-Koba and Koulountou rivers, tributaries of the Gambia River. It is expected to divert water from these tributaries, via a connecting channel, to reinforce existing reservoirs in order to expand the area cultivated and increase the production [15].

In Guinea-Bissau, on the contrary, there is no water control proper, because no hydro-agricultural development has been carried out, even though the country has a very old rice-growing tradition that is highly dependent on rainfall. According to STUDI International [10], rice production in mangrove and lowland areas is facilitated by the provision of salt dikes and water reservoirs. In the non-rainy season, small areas are irrigated from the river using motor pumps placed along the riverbanks. But irrigated agriculture is still in its embryonic stage. Enormous potential in developable lowlands has so far been little used for rice cultivation [16]. Today, despite the low level of development of the agricultural potential, the agricultural activities developed on the riverbanks are affected by the difficult climatic conditions combined with the hydraulic dams built in the Senegalese part of the basin. The difficulties in supplying drinking water to households are noted, and the strong degradation of ecosystems and the rise of the salt tongue are severely felt by the local populations [17].

4. Some impacts of developments in the catchment area

Since water is an economic good [18, 19], dams have contributed significantly to socio-economic development through their benefits as long as they are technically, economically and financially feasible [20]. It is a given [21] that dam development at the national and regional scale is the main factor of economic development, while basins with dams have an increased economic activity by 25% over those that have no dams [22]. In [23], it is seen that reservoir construction, groundwater exploitation and cropland irrigation are prime optimization instruments for water resource utilisation with particular success in Africa where dam construction balanced out water shortage.

However, the general picture of population increase for the three countries is as below (Base year 2022).

As dams are connected with social development, which includes population [25, 26]. **Table 2** shows that existing dam utilisation in these countries will come under rapidly increasing pressure. According to the World Bank Report [27], Senegal is classified as water-stressed country, as despite the fact that resources are plentiful there is high annual variability [28] with extreme precipitation [29] occurrences [30] and projections for 2035 show a withdrawal increase of 30–60%, the upper limit correlated with population increase. In the case of Guinea-Bissau, water resources are under pressure [31] from changing and variable climate, abstraction, the methodology for the disposal of wastewater and urbanisation rate while water stock depletion and water resource pollution are seen as being the major threats. Threats in Guinea are seen to materialise at the end of the century [32] where rainfall will be reduced by 26% in the Fouta Djallon Highlands, and the Konkouré River may see its flow reduced up to 50% while the Milo by up to 70%.

Although they are essential for development, dams must satisfy additional requirements, those of environmental, political, institutional and social acceptance [33]. Before dam construction resettlement costs and methodology play an important role as seen in [34] where a four-step method is described; and in a study of the Narmada Valley in India where cost-benefit analysis was applied including the assessment of the costs of displaced people [35]. Dams, '*the most cataclysmic event in the life of a riverine ecosystem*' [36], have environmental impact on water quality [37] which include the release of excessive sediment, eutrophication leading to anoxic conditions at the bottom [38] that reduces sulphate to acidic hydrogen [39], flooded [40] and pre-impoundment reduction of biomass which usually is burned, evaporation induced salt built-up [41] plus salt carried over by the rain due to proximity to an ocean [42]. As seen in [43], biophysical systems are impacted by dams primarily via the hydrograph change and river system fragmentation [44]. This leads [45] to changes in both sediment load and the morphology of the riverbed, as seen in [46] to the riparian areas' species composition and to aquatic biota in terms of both health and viability [47]. This is what the now extinct WCD [21] called double-edged developments. In this paper, only negative impacts affecting all or part of the basin are considered.

	Guinea-Bissau	Guinea	Senegal
2030	+19%	+23%	+22%
2040	+45%	+54%	+53%
2050	+72%	+87%	+88%
2100	+177%	+226%	+260%

Table 2.
Population increase in Guinea-Bissau, Guinea and Senegal [24].

From a hydrological point of view, the study of flows at the Wassadou and Sonaco downstream hydrometric stations showed a considerable decrease in flows which partly reflects the variability of rainfall [4]. River flow [48, 49] is one of the most determining factors of riverine ecosystems as seen in [50]. Regarding the effects of dams on flows, they alter flow quantity as well as water quality, and in terms of specific flow events, they impact all their main variables i.e., their change rate, seasonal timing as well as magnitude and duration [51, 52]. Moreover, the ecological instream flow is not respected despite the fact that it is very important [53] and according to Ennesser [54], there is, in fact, no instream flow. The analysis of the hydraulic distance separating the stations of the reservoir dams, using indices formulated by Payan [55], made it possible to deduce that the Confluent dam, the most downstream, is close to the Wassadou station. In other words, the surface of the intermediate zone acts more on the transfer of the flow than on the generation of flows. Thus, the flows are reduced by the dam. **Figure 2**, taken in the immediate surroundings of the Confluent dam, shows the small flow of water escaping from the structure. The lateritic track (red arrow) crossing the river just downstream of this dam is a shortcut to Pakour in the dry season. The stone in the middle makes it easier for pedestrians and people on bicycles to cross. In the rainy season, on the contrary, it is impracticable because of the significant flows that pass through the dam and possibly the spillway in case of spillage. At the Sonaco downstream station, there is a considerable distance between the reservoir and this station. The surface area of the intermediate zone being large, it is very likely that the phenomena linked to the generation of flows (lateral contributions, exchange with the aquifer, the contribution of rainfall on the intermediate zone, etc.) significantly support the flow [4]. A good monitoring of the hydrometric network and an in-depth analysis will make it possible to confirm or disconfirm the impact of the reservoirs on the flows downstream.

Moreover, on several occasions, the Guinea-Bissau authorities have complained to their Senegalese counterparts, and then to the OMVG, about the low quantities of water that arrive downstream because of the upstream reservoirs. Already in 1993 [56], before the put-in water of the Niandouba dam-reservoir in 1997, the French Society of Studies and Consultancy (SOFRECO in French), in the study of the Master Plan for the integrated management and development of the Kayanga/Geba and Koliba/Corubal river basins, pointed out that the river behaves like a lake downstream from the Sonaco Downstream station. From this station, a few dozen kilometres away, the riverbed is without water until the village of Fasse, where water is only found in the form of a lake. This situation is verified as far as the village of

Figure 2.
Weak flow downstream of the Confluence dam, June 2013 [4].

Sincha-Kagna. The little amount of water you see in these places comes from the aquifer that is outcropping.

From an ecosystem point of view, the vegetation formations of the Anambe watershed have disappeared, giving way to a mono-specific vegetation (rice cultivation). The pastoral space around the developed perimeters is reduced, in addition to the difficulty of access to certain water points by livestock. Unlike the Confluence reservoir, in the Niandouba reservoir, the standing timber was not fully recovered before the put-in water. Also, some plant species inside the reservoir have died of asphyxiation due to permanent submersion (**Figure 3**). These dead feet still present in the reservoir constitute an obstacle to fishing activity and are an integral part of the main objectives [4].

In some parts of the basin, there is a proliferation of invasive aquatic plants, such as Nymphea spp., Salvinia molesta, Typha australis and other unidentified species, linked to the lentic character of the waters. This proliferation will induce a modification of the environmental conditions: the formation of sandbanks caused by the root system and reduction of the oxygen available in the water. According to STUDI International [10], aquatic plants constitute microcosms where many vectors of waterborne diseases (bilharzia, filariasis, malaria, and onchocerciasis) live and feed.

Also, the decrease of the flow speed in the downstream part of the basin (Guinea-Bissau) has the effect of creating favourable conditions for sand deposits in the major river bed [17], as illustrated in **Figure 4**. The penetration of the tide over a length of more than 150 kilometres inland and the rising of the salty surface water are the cause of the salinisation of several rice fields in Guinea-Bissau [17].

The developments have caused an increase in conflicts related to access to water. The most frequent ones are related to livestock wandering, particularly in the developed areas. Indeed, in these perimeters, which more or less form a belt, there is no provision for grazing in accordance with the requirements of pastoralism [57] for livestock access to the confluence reservoir. In the dry season, when the water points dry up, herders are forced to make tracks through the irrigated areas to give their herds water. The ensuing conflicts may be reported to the gendarmerie or, more often, settled out of court. However, since the arrival of farmers from the groundnut basin of Senegal, relations with the breeders have become more strained, and complaints to the gendarmerie and impoundments have increased [57].

Land conflicts are most often between indigenous and non-indigenous people. The latter is considered by the natives as privileged competitors of SODAGRI, which was in charge of the exclusive distribution of developed plots. Between 2005 and

Figure 3.
Asphyxiated vegetation upstream of the Niandouba dam, June 2013 [4].

Figure 4.
Silting in the major bed of the Kayanga/Geba River downstream [17].

2009, due to the resurgence of conflicts, a forum on land management enabled the municipal councils to regain their prerogatives and to take charge of the allocation and decommissioning of developed plots of land under the approval of the Sub-Prefect and the technical assistance of SODAGRI [58].

In 2018, a conflict opposed the inhabitants of villages along the river in the commune of Pakour to a private promoter. At the origin of this conflict, the municipal council granted 1000 ha to the promoter for the establishment of a banana plantation. After about 10 hectares were cleared, the inhabitants of these villages mobilised to demonstrate their refusal. The establishment of this banana plantation would prevent a tenth of villages from accessing the river; their only source of life (fishing to meet their needs, market gardening, rice growing, and watering places for their cattle) and the loss of their production fields.[1] The event resulted in the arrest of 10 young people from the commune who were held in custody, referred to the prosecutor's office in Kolda and finally tried and sentenced to 2 months in prison.[2]

From a health point of view, the stagnant water has created conditions favourable to the development of numerous pathologies that affect the populations living in the Anambe basin. According to CSE [59], there are water-related vector diseases (malaria, onchocerciasis), water-borne diseases (diarrhoea, dysentery, typhoid fever) and water-borne diseases (urinary or intestinal bilharzia). Malaria is the primary reason for consultation in all health posts, both during the rainy and dry seasons. Urinary bilharzia is increasingly becoming a serious concern with cases detected in Kounkane, Wassadou and Medina Dianguette [59]. The cases of diarrhoeal diseases (dysentery and diarrhoea) detected in Kounkane over the period 2004–2009 include 801 cases in 2004, 993 in 2005, 1132 in 2008 and 1087 in 2009. This upward trend is observed at the Sare Coly Salle and Diaobe health posts. Dermatoses were diagnosed at the Kounkane health post, rising from 95 cases in 2004 to 125 in 2009, with a peak of 163 in 2007 [57].

The use of fertilisers and plant protection products in the irrigated areas of the Anambe zone risks polluting surface and groundwater in the long term. Indeed, the water taken from the Lake Waima in the Anambe basin for the irrigation of the

[1] https://www.koldanews.com/2018/03/29/pakour-velingara-les-populations-du-kayanga-contre-limplantation-dune-bananeraie-a820953.html

[2] http://www.enqueteplus.com/content/kolda-litige-foncier-dans-la-commune-de-pakour-10-jeunes-manifestants-arrêtés-seront-jugés

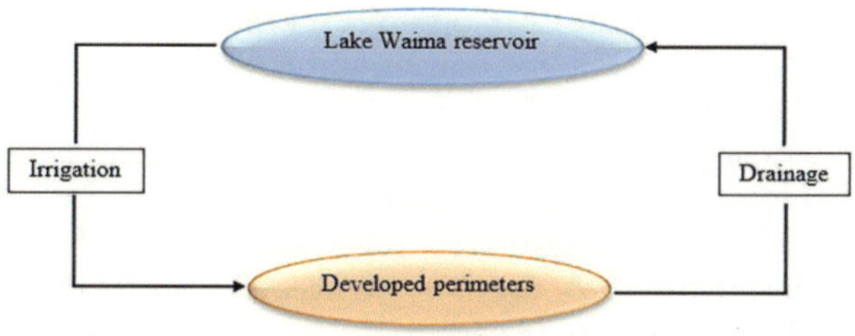

Figure 5.
Circuit schema of the irrigation water in Lake Waima (inspired by SODAGRI).

perimeters is returned directly to it via the drainage channels, without any treatment. This constitutes a closed circuit as illustrated in **Figure 5**. However, a study carried out by the CSE over the period 2004–2008 showed that, from a physicochemical point of view, surface water is poorly mineralised, given the chemical parameters measured, the levels of which are below the usual standards [57]. Therefore, the analyses do not reveal any pollution likely to alter the quality and suitability of the water. Concerning groundwater, nitrate levels that were limiting the probability of occurrence have increased significantly with a maximum value of 130.94 mg.1-1 in June 2008. The traditional wells most affected by the presence of nitrates are in the villages of Maoude (130.94 mg.1-1), Dialakegny (97.57 mg.1-1), Kandia (62.43 mg.1-1) and Sare Koutayel (58.94 mg.1-1) [57]. These excessive concentrations constitute risks for the populations that use the water from these wells without any pre-treatment. Also, the populations of the bordering villages complain about the resurgence of intoxications that affect them and their livestock [57, 60].

5. Sub-regional cooperation initiatives and limitations

The first cooperation initiatives were set up with the creation of the Organisation for the Development of the Gambia River (OMVG in French) by Senegal and the Gambia on 30 June 1978 in Kaolack. It is the successor to the coordinating committee for the development of the Gambia River Basin [61]. Its mission was to valorise the resources of the Gambia River. In 1981, it was enlarged with the adhesion of Guinea and in 1983, it was the turn of Guinea-Bissau. In February 1987 (Resolution No. 14 of the Conference of Heads of State), the territory covered by the OMVG was extended to the south to include the catchment areas of the Kayanga/Geba and the Koliba/Corubal, two rivers that have a common embouchure in Guinea-Bissau. Thus, the OMVG has become a sub-regional organisation comprising the four member states mentioned above, which speak three languages (French, English and Portuguese). From now on, its mission is to promote and undertake studies and development work in the three basins [62]. The specific objectives are the development of agriculture (fight against poverty in a context of sustainable development), the production of hydroelectric energy (potential energy to be developed estimated between 230 and 250 MW), the protection of the environment, the control of salinity in areas influenced by the tide, the improvement of existing waterways, the settlement of populations and the reduction of rural exodus, through programmes and projects common to the four countries. The organisation has organs such as the Conference of Heads of State and Government, the Council of Ministers, the Executive

Secretariat, the Permanent Water Commission and the Advisory Committee (States and funders) [63].

The member states are linked by four basic conventions: the convention relating to the status of rivers (no project likely to modify in a significant way the natural characteristics can be carried out without having been, as a preliminary, approved by the contracting States); the convention relating to the creation of the OMVG (defining the objectives, the attributions, the competences and the mode of functioning of the Organisation); the convention relating to the legal status of the common works (defining with precision the conditions of execution and exploitation of any work of common interest as well as the reciprocal obligations of the Member States); and the convention relating to the modalities of the financing of the common dams. They are largely consistent with the policy recommendations of the World Commission on Dams (CMB in French). It recommends that national water policies explicitly incorporate mechanisms for negotiation with other states affected by dam construction according to the principles of equitable and reasonable use, damage prevention and advance information [56].

In the case of the Kayanga/Geba basin, the first cooperation project (**Table 3**) concerns Integrated Water Resources Management (IWRM). The objective is to contribute to the improvement of the living conditions of the riparian populations and to the socio-economic development of the countries of the basin, in particular by: (i) improving knowledge of the resources and the rate of satisfaction of the demand for water for various uses; (ii) setting up a consultation platform for the harmonious management of water resources; (iii) building capacities for a better knowledge of the resources of the basin and (iv) increasing agricultural production. The project has created a favourable institutional and technical environment to organise integrated management and to foster the development of cooperation between the different users.

Project	Source of funding	General description of the project	Component	Scale
Integrated Water Resources Management Project in the Kayanga/Geba River Basin	African Water Facility OMGV Member States	This project focuses on improving knowledge of the resources of the Kayanga/Geba and setting up basic tools for integrated and concerted transboundary management of water resources in order to enable the riparian populations to live in harmony with their environment and to ensure sustainable development	Elaboration of the IWRM Plan in the river basin	Transboundary
			Preparation of studies for the hydro-agricultural exploitation of the Kayanga/Geba water resources in Guinea Bissau	
			Capacity building of OMVG and its member states in technical and institutional terms	
			Managing the implementation of the project	

Table 3.
Transboundary cooperation project in the Kayanga/Geba river basin.

It is important to note that long before this common project, there were hydro-agricultural developments in the Kayanga/Geba-Anambe system on Senegalese territory, which anterior the entry in the vigour of the Convention on the International Status of the Kayanga/Geba River, but post-date, the international treaties and agreements on shared basins. Indeed, the United Nations Convention on the Law of the Non-Navigational Uses of International Watercourses is an international instrument that focuses on shared water resources, with two key principles: equitable and reasonable use and the obligation not to cause significant damage [64]. The Confluence and Niandouba hydraulic dams built on the main course of the river and financed by Senegalese public funds with the support of funders, do not have the status of OMVG common dam, as they have been carried by Senegal until now. However, in their operational phase, they may require reciprocal obligations, notably on the conditions and modalities of management of the mobilised water resources. These dams are therefore not covered by this special convention. But if for any reason, the management of these dams should 1 day revert to the OMVG, this would be done by mutual agreement and after negotiations [57]. The water governance issues are not yet taken into account, and OMVG has for the time being no authority over the dams managed by SODAGRI. It does not control the number of releases planned in the dry season, nor the exact quantity of water that transits from the dams to Guinea-Bissau (downstream of the basin). According to the authorities of this country, the reduction in runoff is due to dams, a conflict that has become latent between Senegal and Guinea-Bissau [57]. The Velingara-Pakane dam, built in 2012 and financed by the African Development Fund, does not have the status of an OMVG common dam either, despite the entry in vigour of the convention [4]. This can be a major limitation for good transboundary cooperation.

In addition, the OMVG Executive Secretariat organised several field visits between 1989 and 2007, in which representatives of the two main states (Senegal, Guinea-Bissau) participated, to learn about dam management programmes and to observe the state of the river downstream of the water reservoirs. These visits made it possible to discuss the real causes of the drying up and to make recommendations for improving the water conditions of the river downstream of the dams. For its part, the OMVG believes that the problem should be solved at the end of the "Integrated Water Resources Management Project in the Kayanga/Geba river basin" [57]. So far, no concrete action has been taken. In Senegal, efforts are focused on immediate national concerns, without regard to regional benefits and ecological impacts, because the way dams are managed has not changed much.

6. Conclusion

The decrease of water resources linked to the dry spells of the 1960s and 1970s and climate change has impacted on socio-economic activities in the Kayanga/Geba basin, shared between Guinea, Senegal and Guinea-Bissau. This situation has incited the multiplication of hydro-agricultural dam projects in the Senegalese part of the basin, with the building of the Confluent, Niandouba and Velingara-Pakane dams, to increase agricultural production and fight against poverty. These dams have increased the pressure on water resources and raised the level of water withdrawals, hence the numerous complaints from the Guinea-Bissau authorities to their Senegalese counterparts and to the OMVG. These complaints relate to the low quantities of water that arrive downstream, especially in the dry season, due to the dams, a conflict that has become latent between the two countries. Nevertheless, many studies have shown that historically few tensions and disputes over water have led to open armed conflict [65–68]. The relations between countries bordering an international river may be tense, the disputes may arise, but in general, these countries almost always find a formula for cooperation rather than open confrontation [66, 68].

The OMVG as a basin organisation is trying to create a favourable environment to develop a common will to exploit and share the basin's potential together, with the implementation of common development projects such as the Integrated Water Resources Management project of the Kayanga/Geba river basin. But its mission in this basin is not easy because the dams do not have the status of common structures and it does not yet have a say in their management. Like the other basin organisations such as OMVS,[3] ABN[4] and CBLT,[5] it will have to demonstrate a great capacity for anticipation and adaptation in the face of the many changes that are on the horizon [69], in particular, the planned dam project at the confluence of the Niokolo-Koba and Koulountou rivers to strengthen the existing dams. The relevant suggestions made by [70] can help to prevent the risk of conflict and to manage it appropriately when it occurs.

[3] Organisation pour le Mise en Valeur du fleuve Sénégal.

[4] Autorité du Bassin du Niger.

[5] Commission du Bassin du Lac Tchad.

Author details

Saly Sambou[1*], Rene Ndimag Diouf[2] and Joseph Sarr[3]

1 Faculty of Letters and Human Sciences, Department of Geography, Cheikh Anta Diop University, Dakar-Fann, Senegal

2 Faculty of Science and Technology of Education and Training, Cheikh Anta Diop University, Dakar-Fann, Senegal

3 National Agency for Applied Scientific Research, Ministry of Higher Education, Research and Innovation, Dakar-Fann, Senegal

*Address all correspondence to: sambousaly@gmail.com

References

[1] Maganda C. Agua dividida, agua compartida Acuiferos transfronterizos en Sudamérica, una aproximación. Estudios Políticos. 2008;**32**:171-194

[2] Julien F. Maîtrise de l'eau et développement durable en Afrique de l'Ouest: de la nécessité d'une coopération régionale autour des systèmes hydrologiques transfrontaliers. VertigO. 2006;7:2. DOI: 10.4000/vertigo.2402

[3] Rodriguez T. Dynamiques de coopération transfrontalière sur la façade caraïbe du Costa Rica et du Panama: le cas du bassin du fleuve Sixaola. Études caribéennes. 2012:21. DOI: 10.4000/etudescaribeennes.5747. Available from: http://journals.openedition.org/etudescaribeennes/5747

[4] Sambou S. Contribution à la Connaissance des Ressources en eau du Bassin Versant du Fleuve Kayanga/Geba (Guinée, Sénégal, Guinée-Bissau) [Thèse de Doctorat]. Dakar, Sénégal: Université Cheikh Anta Diop; 2019. p. 314

[5] Sambou S, Diouf RN, Cisse B, Diouf I, Sarr J, Dacosta H. Evolution of Dam lakes in the Kayanga/Geba Basin: Contribution of remote sensing and GIS. Journal of Geoscience and Environment Protection. 2021;**9**:225-243. DOI: 10.4236/gep.2021.912014

[6] Dacosta H, Gomez R. Inventaire des Zones Humides des Bassins de la Casamance et de la Kayanga. Rapport RNZHS (Royal New Zealand Yacht Squadron). Dakar: UICN (Union Internationale pour la Conservation de la Nature); 1998. p. 33

[7] Sambou S, Dacosta H, Deme A, Diouf I. Contribution of TRMM 3B42 data to improve knowledge on rainfall in the Kayanga/Geba river basin (Republic of Guinea, Senegal and Guinea-Bissau). European Scientific Journal. 2018;**14**(9):260-275. DOI: 10.19044/esj.2018.v14n9p260

[8] Sambou S, Dacosta H, Paturel J-E. Variabilité spatio-temporelle des pluies de 1932 à 2014 dans le bassin versant du fleuve Kayanga/Gêba (République de Guinée, Sénégal, Guinée- Bissau). Physio-Géo. 2018;**12**:61-78. Available from: http://journals.openedition.org/physio-geo/5798

[9] Brunet-Moret Y. Étude Hydrologique en Casamance. Édit. ORSTOM, rapport final, Dakar. 1970. p. 52

[10] STUDI International: Projet de gestion intégrée des ressources en eau dans le bassin versant du fleuve Kayanga/Geba. Elaboration de l'étude GIRE du bassin versant du fleuve Kayanga/Geba et études de faisabilité et d'Avant-Projet Détaillé (APD) des ouvrages et aménagements hydro-agricoles en Guinée-Bissau. OMVG, Version Provisoire. 2011. p. 392

[11] UNESCO: Glossaire International d'Hydrologie. 2012. Available from: http://unesdoc.unesco.org/images/0022/002218/221862M.pdf

[12] UNESCO. Méthodes de calcul de la sédimentation dans les lacs et les réservoirs. In: Contribution au Programme Hydrologique Internationale (PHI) –II Project A.2.6.1 Panel. 1986. p. 238

[13] Soumaré PO. Modélisation du fonctionnement hydrologique de l'hydrosystème Anambé-Kayanga et gestion intégrée des ressources en eau. Rapport d'étape de thèse de doctorat. 2001

[14] Dacosta H, Coly A, Soumaré PO. Adéquation de l'offre et de la demande en eau: Dynamique hydrologique de l'Anambé (Sénégal). In: Gestion intégrée des ressources naturelles en zones inondables tropicales. Paris (FRA); Bamako: IRD; CNRST. (Colloques et Séminaires). IRD Éditions; 2002. pp. 395-410

[15] Mballo I, Sy O, Thior M. Dynamique des ressources en eau dans le bassin de l'Anambé (Haute Casamance-Sénégal). Larhyss Journal. 2019;**40**:7-25

[16] OMVG. Note de synthèse sur les programmes du Schéma Directeur des fleuves Kayanga/Geba et Koliba/Corubal. 2005. p. 34

[17] FAE. Projet de Gestion Intégrée des Ressources en Eau dans le bassin versant du fleuve Kayanga/Geba. Rapport d'évaluation. Sénégal: OMVG; 2008. p. 32

[18] Zisopoulou K, Panagoulia D. An in depth analysis of physical blue and green water scarcity in agriculture in terms of causes and events and perceived amenability to economic interpretation. Water (Switzerland). 2021;**13**(12):51

[19] Zisopoulou K, Zisopoulos D, Panagoulia D. Water economics: An in-depth analysis of the connection of blue water with some primary level aspects of economic theory I. Water (Switzerland). 2022;**14**(103):1-29. DOI: 10.3390/w14010103

[20] Tortajada C. Dams: An essential component of development. Journal of Hydrologic Engineering. 2014;**20**(1):1-10. DOI: 10.1061/(ASCE) HE.1943-5584.0000919

[21] World Commission on Dams (WCD): Dams and Development: A New Framework for Decision-Making. The report of the World Commission on Dams. 2000. Available from: https:// www.internationalrivers.org/campaigns/ the-world-commission-on-dams

[22] Nilsson C, Reidy CA, Dynesius M, Revenga C. Fragmentation and flow regulation of the world's large river systems. Science (80). 2005;**308**:405-408

[23] Kummu M, Ward PJ, De Moel H, Varis O. Is physical water scarcity a new phenomenon? Global assessment of water shortage over the last two millennia. Environmental Research Letters. 2010;**5**(3)

[24] Rosen M.: Future Population Growth. Our World in Data. 2022. Available from: https://ourworldindata.org/future-population-growth. [Accessed: May 24, 2022]

[25] Chen J, Shi H, Sivakumar B, Peart MR. Population, water, food, energy and dams. Renewable and Sustainable Energy Reviews. 2016;**56**:18-28

[26] Shi H, Chen J, Liu S, Sivakumar B. The role of large dams in promoting economic development under the pressure of population growth. Sustainability. 2019;**11**:2695. DOI: 10.3390/su11102965

[27] World Bank. Challenges and Recommendations for Water Security in Senegal at National Level and in the Dakar-Mbour-Thies Triangle. 2022

[28] FAO. Profil de Pays – Sénégal. Rome, Italie: Organisation des Nations Unies pour l'alimentation et l'agriculture; 2005

[29] Panagoulia D, Vlahogianni EI. Nonlinear dynamics and recurrence analysis of extreme precipitation for observed and general circulation model generated climates. Hydrological Processes. 2014;**28**(4):2281-2292. DOI: 10.1002/hyp.9802

[30] Sarr MA, Zoromé M, Seidou O, Robin C, Gachon P. Recent trends in selected extreme precipitation indices in Senegal-A changepoint approach. Journal of Hydrology. 2013;**505**:326-334

[31] UNEP-DHI. Roadmaps for Water Management in West Africa. Denmark: Hørsholm; 2013

[32] USAID. Climate Risk Profile. USAID; 2018

[33] Scudder T. The Future of Large Dams Dealing with Social, Environmental, Institutional and Political Costs. Milton Park. U.K.: Routledge; 2005. DOI: 10.4324/9781849773904

[34] Scudder T. Resettlement outcomes of large dams. In: Tortajada C, Altinbilek D, Biswas AK, editors. Impacts of Large Dams: A Global Assessment. Heidelberg, Germany: Springer; 2012. DOI: 10.1007/978-3-642-23571-9

[35] Bhalla SS, Mookerjee A. Big dam development: Facts, figures and pending issues. International Journal of Water Resources Development. 2001;**17**(1):37-41. DOI: 10.1080/713672558

[36] Gup T. Dammed from here to eternity: Dams and biological integrity. Trout. 1994;**35**:14-20

[37] Mohamed Y, Alla K, Liu L. Impacts of dams on the environment: A review. International Journal of Agriculture Environment and Biotechnology. 2021;**6**(1):64-74. DOI: 10.22161/ijeab

[38] Spoor WA, Spoor WA. Distribution of fingerling brook trout, Salvelinus fontinalis (Mitchill), in dissolved oxygen concentration gradients. Journal of Fish Biology. 1990;**36**(3):363-373

[39] Naja GM, Volesky B. Toxicity and sources of Pb, Cd, Hg, Cr, As, and radionuclides in the environment. In: Wang LK, Wang M-HS, Hung Y-T, Shammas NK, Chen P, Jiaping P, editors. Heavy Metals in the Environment, 8, 16-18, Boca Raton. Florida, U.S.A.: CRC Press; 2009. p. 49. DOI: 10.1201/9781315117423

[40] Zimmer M. Detritus. In: Jørgensen SE, Fath BD, editors. Encyclopedia of Ecology. 1st ed. Amsterdam, Netherlands: Elsevier; 2008. pp. 903-911

[41] Westrup T. Options for Achieving and Maintaining Low Salinity in Agricultural Dams Report 343. Western Australia, Perth: Department of Primary Industries and Regional Development; 2009

[42] Hingston FJ, Gailitis V. The geographic variation of salt precipitated over Western Australia. Australian Journal of Soil Research. 1976;**14**:319-335

[43] Tullos D, Tilt B, Liermann CR. Introduction to the special issue: Understanding and linking the biophysical, socioeconomic and geopolitical effects of dams. Journal of Environmental Management. 2009;**90**:203-207. DOI: 10.1016/j.jenvman.2008.08.018

[44] Kotchen MJ, Moore MR, Lupi F, Rutherford ES. Environmental constraints on hydropower: An ex post benefit–cost analysis of dam relicensing in Michigan. Land Economics. 2006;**82**(3):384-403

[45] Yang Z, Wang H, Saito Y, Milliman JD, Xu K, Qiao S, et al. Dam impacts on the Changjiang (Yangtze) river sediment discharge to the sea: The past 55 years and after the Three Gorges Dam. Water Resources Research. 2006;**42**(4):W04407. DOI: 10.1029/2005WR003970

[46] Mumba M, Thompson JR. Hydrological and ecological impacts of dams on the Kafue flats floodplain system, Southern Zambia. Physics and Chemistry of the Earth. 2005;**30**:442-447

[47] Kingsford RT. Ecological impacts of dams, water diversions and river management on floodplain wetlands in Australia. Austral Ecology. 2000;**25**(2):109-127

[48] Panagoulia D. Artificial neural networks and high and low flows in various climate regimes. Hydrological Sciences Journal. 2006;**51**(4):563-587. DOI: 10.1623/hysj.51.4.563

[49] Panagoulia D, Tsekouras GJ, Kousiouris G. A multi-stage methodology for selecting input variables in ANN forecasting of river flows. Global NEST Journal. 2017;**19**(1):49-57

[50] Yang ZF, Yan Y, Liu Q. Assessment of the flow regime alterations in the Lower Yellow River, China. Ecological Informatics. 2012;**10**:56-64

[51] Qiting Z, Yaobin C, Jie T. Climate change and its impact on water resources in the Huai River Basin. Bulletin. Chicago Academy of Sciences. 2012;**26**:32-39

[52] Hu W, Wang G, Deng W, Li S. The influence of dams on ecohydrological conditions in the Huaihe River basin, China. Ecological Engineering. 2008;**33**:233-241

[53] Salinas-Rodriguez SA, Barba-Macías E, Mata DI, Nava-López M, Neri-Fores I, Varela RD, et al. What do environmental flows mean for long-term freshwater ecosystems' protection? Assessment of the Mexican water reserves for the environment program. Sustainability. 2021;**13**:28

[54] Ennesser Y. Études Complémentaires et Plan Directeur du Bassin de l'Anambé. Groupement BCEOM-SID International: Impact sur l'environnement; 1994. p. 93

[55] Payan J-L. Prise en Compte des Barrages-Réservoirs Dans un Modèle Global Pluie-débit [Thèse de Doctorat unique]. AgroParisTech, GRN, Cemagref, ENGREF; 2007. p. 256

[56] SOFRECO. Etude du Schéma Directeur d'aménagement et de développement intégré des bassins des fleuves Kayanga/Gêba et Koliba/Corubal. Ressources et besoins en eau et option d'aménagement hydraulique et hydro-agricole. 1993;**3** Annexe B

[57] UICN, IIED. Etats des lieux autour des barrages de Niandouba et du Confluent au Sénégal. In: Rapport définitif. 2010. p. 122

[58] UICN, GWI, IIED. Atelier de restitution de l'étude de référence de la situation de partage des bénéfices autour des barrages du Confluent et de Niandouba au Sénégal. "Projet utilisation multiple et partage des bénéfices autour des réservoirs de barrage en Afrique de l'Ouest". In: Rapport de l'atelier. 2011. p. 22

[59] CSE. Projet d'Appui au Développement Rural du Bassin de l'Anambé: Plan de Gestion Environnementale et Sociale. Ministère de l'Agriculture et de l'Hydraulique, SODAGRI. 2005. p. 69

[60] Diao O. Gestion des produits chimiques agricoles dangereux dans les milieux irrigués: le cas du bassin de l'Anambé. Mémoire de DEA, département de Géographie, FLSH, UCAD; 2008. p. 73

[61] OMVG. Projet d'Energie de l'OMVG. Ligne d'interconnexion. Plan d'Actions de Réinstallation (PAR). Poste de

transformation électrique au Sénégal. 2018. p. 325

[62] OMVG. Organisation de développement du bassin fluvial de Gambie (OMVG). In: Extrait du site Web de l'OMVG. 2020. Available from: https://www.omvg.org/

[63] OMVG. Etude d'impact environnemental et social de l'aménagement hydroélectrique de Sambangalou et de l'interconnexion. In: Rapport Final. 2014. p. 637

[64] CNU. Convention sur le droit relatif aux utilisateurs des cours d'eau internationaux à des fins autres que la navigation. 1997. pp. 286-305

[65] Turton AR. Water wars in Southern Africa: Challenging conventional wisdom. In: Water for Peace in the Middle East and Southern Africa. Genève: Green Cross International; 2000. pp. 112-130

[66] Wolf A. Water, conflict, and cooperation. In: Meinzen-Dick RS, Rosegrant MW, editors. Overcoming Water Scarcity and Quality Constraints. 2020 Focus 9: October. 2001. Available from: http://www.ifpri.org/2020/focus/focus09/focus09_14.htm

[67] Postel SL, Wolf AT. Dehydrating conflict. Foreign Policy. 2001. Available from: . DOI: 10.2307/3183260http://www.globalpolicy.org/security/natres/water/2001/1001fpol.htm

[68] Wolf AT, Yoffe SB, Giordano M. International waters: Identifying basins at risk. Water Policy. 2003;**5**:29-60

[69] Fall AM. La gouvernance de l'eau en Afrique de l'Ouest: Le modèle de l'Organisation pour la Mise en Valeur du fleuve Gambie (OMVG). IUCN. 2004:183-192

[70] Niasse M. Prévenir les conflits et promouvoir la coopération dans la gestion des fleuves transfrontaliers en Afrique de l'Ouest. Vertigo - la revue électronique en sciences de l'environnement. 2004;**5**:1. DOI: 10.4000/vertigo.3979